DISTRIBUTION SWITCHGEAR

TELEPEN
00006305

Consulting Editor: E. A. Reeves

DISTRIBUTION SWITCHGEAR

Construction, Performance, Selection and Installation

R. W. Blower

TD, BSc (Eng), FIEE

COLLINS

8 Grafton Street, London W1

Collins Professional and Technical Books
William Collins Sons & Co. Ltd
8 Grafton Street, London W1X 3LA

First published in Great Britain by
Collins Professional and Technical Books 1986

Distributed in the United States of America
by Sheridan House, Inc.

British Library Cataloguing in Publication Data
Blower R. W.
Distribution switchgear : construction, performance,
selection and installation.
1. Electric switchgear 2. Electric controllers
I. Title
621.31'7 TK2821

ISBN 0-00-383126-4

Typeset by V & M Graphics Ltd, Aylesbury, Bucks
Printed and bound in Great Britain by
Mackays of Chatham, Kent

Contents

Preface

In conceiving this book, I am aware that there are books already available that delve in some detail into the mysteries of switchgear application and performance, but these tend to concentrate on the higher voltage end of the market, where the technical problems become magnified, and demand detailed study of high voltage dielectric characteristics, short-line faults, etc.

Yet the customers for these transmission switchgear equipments are few in number and tend to be specialists themselves as they are, almost without exception, large supply authorities. Distribution switchgear equipment in the voltage range up to 36 kV, on the other hand, is purchased by a wide range of users, particularly in the large process industries such as steel, oil, chemicals, etc. Even quite small industrial sites are fed at voltages above 1 kV and while the user may not be involved in the purchase, he is going to be involved in application, perhaps installation, and certainly maintenance, to some degree.

So it seems that there is a use for a book that concentrates on the distribution voltage range up to 36 kV, and deals with those matters that an interested engineer, who would become involved in the plant that is used in that area, might like to know more about.

This book is an attempt to develop that theme and is written to acquaint users with the subjects that have to be considered in the design and application of such equipment. Some knowledge of electrical engineering is assumed, but mathematical treatment is kept to a minimum, and where it is introduced, I have usually placed it in an appendix so that it is available to those interested without interrupting the narrative.

With this as the objective, and with 40 years in the industry, mostly associated with design and development, it has been difficult to decide what to leave out!

However, the treatment begins with definitions and the factors that affect the design of the various types of switchgear, through the methods available to the manufacturer of satisfying the needs of the user, to a consideration of the selection from what is available to meet a particular application.

Chapter 1 begins by defining terms used throughout the book, which in some cases have specialised meanings in the switchgear field. In particular, it may be worth mentioning that I have used the term 'medium voltage' to mean voltages in the range from 1 kV up to 36 kV. This is contrary to its usual meaning in the UK, but is in common usage in other parts of the world, e.g. in the USA and on the Continent. In a similar way, 'low voltage' is used to mean voltages up to 1 kV.

Chapter 2 explains the various things that can happen when a switching device operates, leading to Chapter 3, which discusses the parameters of the system which need to be evaluated before specifying a switchgear requirement. This is deliberately kept in its simplest form to illustrate principles rather than being a detailed guide covering all possible conditions. The engineer who wants such detail will find it in books devoted to this subject. Chapter 4 discusses the various ways in which the switchgear designer matches the equipment to the task.

Chapters 5–9 introduce peripheral subjects that influence the design and construction of switchgear and illustrate how many disciplines play a part. Chapters 10 and 11 assemble all this information into the form of switchgear used in outdoor and indoor, primary and secondary substations. Chapter 12 is allied to these and covers low voltage switchgear.

Chapters 13–15 introduce other aspects of application, from examining how the protective aspects are controlled to how the actual operation of the switchgear is dealt with.

In Chapter 16, the factors to be considered when choosing switchgear for a particular requirement are discussed and this is followed, in Chapter 17, by a study of the processes of installation, commissioning and testing for the equipment eventually selected.

Finally, the activities associated with proving that the equipment supplied will do the job it is designed for, and the range of international, national and user standards that control many aspects of design, application and operation, are covered in Chapters 18 and 19.

Throughout the book I have tried to explain differences between national practices where these arise, and I hope that has given an international flavour to the work. I have been fortunate in visiting many parts of the world to see at first hand how engineers in other countries address themselves to the common task of supplying electricity safely to their consumers.

I must acknowledge the help and encouragement I have received in producing this book from many colleagues in the industry both at home and abroad. To those who have supplied illustrations to supplement the text go my warmest thanks. I also owe a debt of gratitude to the editor, Eric Reeves, who has had to put up with my erratic output and has introduced some consistency of presentation.

Last, but far from least, I have to thank my wife, Joan, and our children for cheerfully accepting the almost total absence of a husband and father from the family scene throughout most of 1985!

RWB, Grasscroft, 1985

CHAPTER 1

Switchgear - What It Is and What It Does

In November 1932, Mr J.H. Holmes had occasion to write to the late Mr H.W. Clothier concerning the origin of quick-break switches, and the following is quoted from his letter:

> 'The ceremony of the opening of the Wallsend Cafe took place about Christmas 1883, and a great attraction was the Electric Light, which was installed by my firm. In the course of my conversation with the Architect on that occasion he remarked that he did not know that electric light could be turned down like gas: to which I replied that you could not do so. He said that he would show me, and proceeded to turn one of the switches off very slowly, drawing an arc inside the wooden switch box, and thus lowering the voltage on the lamps. I hurriedly stopped his experiment, but this experience urged me to devise some device to prevent this sort of thing happening. This resulted in my Patent for the Quick Break Switch dated February 14th, 1884, No. 3256, the specification of which enunciated for the first time, the principle upon which such switches are constructed, and sixteen British firms took out licences to work under the Patent.'

Figure 1.1 shows a typical quick-break switch of that time.

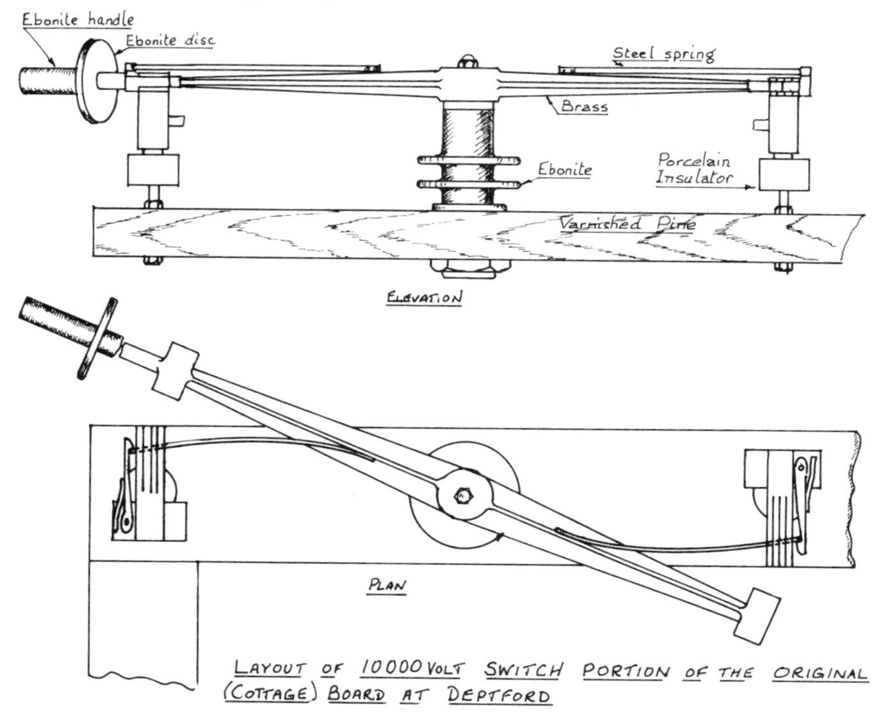

Fig. 1.1 An early quick-break switch.

Just over 100 years ago the principles which today are incorporated without question in the design of distribution switchgear were just beginning to be developed. In the intervening period, just about every possibility for opening and closing electric circuits has been tried, and the prospects now of patenting a principle so fundamental that 'sixteen firms took out licences' are virtually nil!

This book is devoted to describing the characteristics of distribution switchgear, and the upper service voltage limit which has been chosen to define 'distribution' in this context is 36 kV. The vast majority of distribution switchgear is employed in alternating current (a.c.) systems, but there is an important minority of applications still where direct current (d.c.) is used.

DEFINITIONS

Reference will frequently be made to *low voltage*, *medium voltage*, etc., and as confusion exists concerning the exact meaning of these terms, they are defined here as they will be used in this book. Whilst the International Electrotechnical Commission (IEC) defines low voltage as extending to 1000 V, and all voltages higher than that as 'high voltage', there is considerable usage of the term 'medium voltage' in Europe and America, referring to those voltages used for distribution and extending from 1000 V up to about 72.5 kV or so. An attempt has been made by CIRED (the international conference on distribution systems) to define the range of application of medium voltage, but the IEC, as the world electrical standards body, has not yet ratified anything. However, because of the general usage of the term in distribution circles, it will be used here as follows:

Low voltage: up to and including 1000 V.
Medium voltage: above 1000 V up to and including 36 kV (38 kV in America).
High voltage: above 36 kV.

Switchgear covers quite a variety of devices, all intended to carry out the functions of controlling and protecting electrical networks so that the electricity supply can be safely utilised.

The British Standard definition of the general term 'switchgear' is contained in BS 4727: Part 2: Group 6 as:

> *Switchgear*: A general term covering switching devices and their combinations with associated control, measuring, protective and regulating equipment, and also assemblies of such devices and equipment with associated interconnections, accessories, enclosures and supporting structures, intended for use in connection with generation, transmission, distribution and conversion of electrical power.

The switchgear associated with the transmission of electricity over long distances at high voltages has design characteristics arising from the high fault ratings and the high voltages which are particularly associated with switchgear rated at about 145 kV and above. The numbers of these high voltage switching devices are small compared with the numbers of switching devices used to distribute electricity to the industrial or domestic user. The design of distribution switchgear is consequently strongly influenced by the needs of economy and the limitation of the need for frequent attention.

Depending on factors associated with the form of the connected network, the need to pay regard to the amenities of the environment, and the facilities that need to be provided by the particular switchgear installation, switchgear may be installed either indoors or outdoors. Obviously, the need to withstand the rigours of the outdoor environment in different parts of the world will play a large part in determining the design and materials used on outdoor switchgear (see Fig. 1.2).

Fig. 1.2 An outdoor substation. (*Courtesy Eastern Electricity Board.*)

Before proceeding further with the description of the functions performed by switchgear, it will be useful to introduce a few more definitions.

Disconnector (isolator): A disconnector (see Fig. 10.5) is a device used to create a safe gap in an electrical system, and its important characteristics are to withstand the system voltage and any overvoltages which may be generated in the system. When closed, it must carry for a predetermined time any fault current within its rating that may develop in that system. It is intended to be operated off load, i.e. other switching devices must be used to break any current flowing in the circuit before the disconnector is operated, unless the system is such that there is negligible voltage across the contact gap. Typically this means that the disconnector is in a parallel circuit. To ensure that any voltage surges which may be applied from the system do not break down the gap, the withstand voltage of the gap is specified 15% greater than the withstand voltage of the clearance to earth. Disconnectors are covered by BS 5463. (The English word to describe a 'disconnector' used to be 'isolator', and this word is still in frequent use. The present usage, which is derived from the IEC Standards, is more closely related to overseas use, e.g. the French 'deconnecteur'. Isolator is perhaps also too close to the French 'isolateur' which means 'insulator'.)

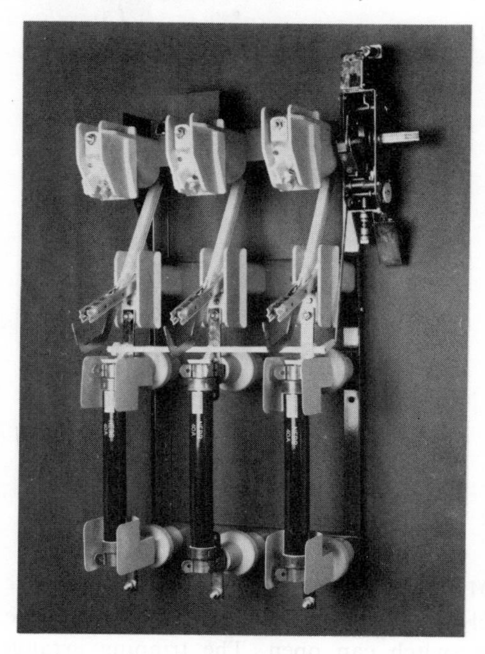

Fig. 1.3 A continental type air switch. (*Courtesy Norwegian Brown Boveri Ltd.*)

Switch: A switch (Fig. 1.3) is required to be able to open an electrical circuit carrying any current up to its rated continuous current, or up to some defined level of overload. Because it is used to perform switching operations on loaded circuits, there is the possibility that it might be closed on to a faulty circuit, e.g. one which has an earth or a phase to phase fault or similar. Therefore it has to be able to close safely on to a peak current corresponding to its rated fault level as well as be able to carry that fault current for a predetermined time. If the switch can also perform the function of a disconnector, i.e. it has the required insulation levels to earth and across the gap, it is called a *switch-disconnector*. Switches and switch-disconnectors are dealt with in BS 5463.

If fuses are added to a switch so that it can then interrupt short-circuit currents, it becomes a switch and fuse combination, and two forms of this device are recognised:

Fuse-switch: A fuse-switch is a switching device in which the fuse-link, or the fuse-carrier with the fuse-link, form the contacts of the combination.

Switch-fuse: A switch-fuse (Fig. 1.4) is a switching device in which the fuses are connected in series with a switch in a combined assembly.

In commonly used UK versions of these devices at medium voltages, the fuse is only intended to deal with short circuits and the switch has a breaking capacity in respect of overloads that overlaps the fuse performance. The timing of the trip feature of the switch is such that the switch opens before the fuse element can

Fig. 1.4 A UK type metal-clad switch-fuse. (*Courtesy Yorkshire Switchgear Ltd.*)

melt, up to a level defined as the 'take-over current'. Above this level the fuse blows before the switch can open. The tripping arrangements of such a combination have to be so designed in relation to the time/current characteristics of the largest current rating of fuse that can be fitted that the area of 'take-over', where the fuse takes over the interrupting duty from the switch, has an adequate margin in both directions. At the time of writing there is no British Standard for these devices, although an IEC recommendation is in preparation, which will no doubt become a BS when it has been published.

Earthing switch: An earthing switch (Fig. 10.9) is a special form of switch, one side of which is permanently connected to earth, and which therefore forms a convenient means of earthing the system conductors when necessary, to allow work to be safely carried out on the electrical system. Earthing switches are dealt with in BS 5253.

Circuit-breaker: A circuit-breaker (Figs 1.5 and 1.6) has the most onerous duty to perform of any of the switching devices, in that it has to be capable of interrupting any current that may pass through it, up to its rated short-circuit breaking current. It must also be able to close on to the peak fault current and carry that fault current for a predetermined time. Medium voltage circuit-breakers are dealt with in BS 5311, and low voltage circuit-breakers are covered by BS 4752.

Table 1.1 summarises the differences in terms of the expected duty to be performed by the foregoing devices, to which reference will continually be made in this book.

Switching devices are rated in terms of a number of electrical parameters, and

Fig. 1.5 A 36 kV outdoor SF_6 circuit-breaker. (*Courtesy South Wales Switchgear.*)

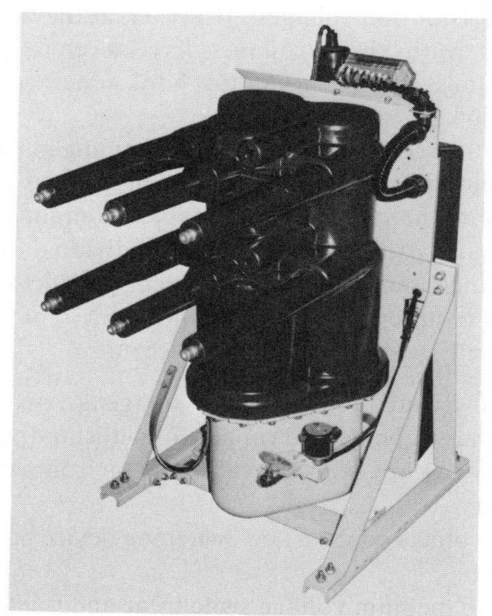

Fig. 1.6 A 24 kV SF_6 indoor circuit-breaker. (*Courtesy Yorkshire Switchgear Ltd.*)

the international, national and customer standards to which reference will be made in a later chapter have set out standard values for many of these.

The kinds of parameter that are of importance to a potential purchaser of distribution switchgear are summarised below. Much more detail will be given in later chapters concerning the effect of these parameters on the design and characteristics of different forms of distribution switchgear.

Table 1.1 Required duty of different switching devices

Device	*Carry fault current*	*Make on fault*	*Break load current*	*Break fault current*
Isolator (disconnect)	yes	no	no	no
Switch	yes	yes	yes	no
Circuit-breaker	yes	yes	yes	yes
Fuse-switch / Switch-fuse	yes*	yes	yes	yes

* Until the fuse clears.

Rated voltage: The rated voltage is stated as the highest system voltage for which the equipment is designed to be used. The values that apply to distribution equipment are from 415 V to 36 kV, although in many industrialised countries with large concentrations of power in main centres of population, voltages up to 145 kV are in use as 'distribution voltages'. However, as the characteristics of the switchgear associated with these voltages have a close affinity with the characteristics of transmission switchgear, this book concentrates on switchgear up to and including 36 kV.

Tables 1.2 and 1.3 list a selection of the most common combinations of rated voltages, rated currents and rated breaking currents used in distribution switchgear, taken from the corresponding tables contained in the relevant International, British and American Standards. Individual manufacturers will obviously determine their own standard ranges of products to suit their own markets and customer requirements, but these tables are representative of rating combinations commonly used throughout the world.

Rated insulation level: Related to the system voltage are two values of voltage which the switchgear has to be proved capable of withstanding as an indication of its long term ability to cope with the electrical environment in which it will operate for many years.

The first is a 50 Hz voltage which a new switching device has to withstand for 1 min as a routine test.

The second relates to medium voltage switchgear and is intended to simulate the effect of lightning and is based on the application of a surge voltage, usually rising to the rated peak value in a time of 1.2 μs or so and then decaying to 50% of its peak value in 50 or 60 μs. This surge voltage is applied as a type test and is often referred to as the Basic Insulation Level (B.I.L.).

Rated normal current: The normal current rating is based on the current which the equipment can carry continuously without any of the parts reaching a

Table 1.2 Commonly used combinations of rated voltage, normal current and short-circuit current for medium voltage distribution circuit-breakers

Rated voltage kV	*Short-circuit current* kA	*Normal current* (A)					
		400	630	800	1250	1600	2000
3.6	8.0	*					
	16.0			*	*	*	*
	25.0			*	*	*	*
	40.0			*	*	*	*
7.2	8.0	*		*	*	*	*
	12.5	*		*	*	*	*
	16.0		*	*	*	*	*
	20.0		*	*	*	*	*
	25.0			*	*	*	*
	40.0			*	*	*	*
12.0	12.5		*		*	*	*
	16.0		*		*	*	*
	20.0		*		*	*	*
	25.0			*	*	*	*
	31.5			*	*	*	*
	40.0				*	*	*
17.5 and 24.0	8.0		*		*	*	
	12.5		*		*	*	
	16.0		*		*	*	
	25.0		*		*		*
36.0	16.0			*	*	*	*
	25.0			*	*	*	*
	31.5			*	*	*	*

Table 1.3(a) Rated voltages and corresponding insulation levels for distribution switchgear – European range, 50 Hz

Rated voltage kV rms	*Impulse voltage withstand, +ve and −ve wave* kV peak	*One minute power frequency voltage withstand* kV rms
3.6	45	21
7.2	60	27
12.0	75	35
17.5	95	45
24.0	125	55
36.0	170	75

Table 1.3(b) Rated voltages and corresponding insulation levels for distribution switchgear – American range, 60 Hz

Rated voltage kV rms	*Impulse voltage withstand +ve and −ve wave* kV peak	*One minute power frequency voltage withstand* kV rms
4.76	60	19
8.25	75	26
15.0	95	36
15.5	110	50
25.8	150	60
38.0	200	80

temperature which might degrade any component. It follows from this that the rated normal current is influenced by the likely ambient temperature in which the equipment will operate. For outdoor switchgear, the effect of solar radiation can add to the ambient temperature.

For the purpose of creating standard conditions for current rating, standard ambient temperatures have been established which are a maximum 24 h temperature average of 35° C with occasional peaks of 40° C. While it has no bearing on the current rating of the switchgear, it is also important to specify the minimum temperatures at which the equipment is designed to operate. As climatic conditions can vary over a wide range, there are several values in the IEC or BS standards relating to the specific equipments, from which the designer can choose. These are −5 and −10° C for indoor gear and −10, −25 and −50° C for outdoor gear.

Standard current values are nowadays based, at least in Europe, on the R10 series, using multiples by 10 of the numbers: 1, 1.25, 1.6, 2, 2.5, 3.15, 4, 5, 6.3, 8 (see Table 1.2).

Rated short-circuit breaking current: The rated short-circuit breaking current relates to devices such as fuses and circuit-breakers which are intended to break fault current. In the past, such devices were rated in terms of their power interrupting capability, which was arrived at as the product of the rated fault currents the device was capable of breaking and the rated voltage, expressed in MVA. Nowadays it is recognised that the current which a circuit-breaker can interrupt is not related to the voltage and the two rating values of breaking current and system voltage are given separately. However, this issue is worth mentioning since the old practice is by no means dead, particularly in American markets, and system planning engineers tend to find it convenient to calculate fault levels in MVA as shown in Chapter 3. The rated short-circuit breaking current is the maximum symmetrical rms current that the circuit-breaker can interrupt at its rated voltage. It is usually quoted in kA.

Rated breaking currents are also based on the R10 series, as is seen from Table 1.2.

Rated making current and short time current: The rated making current and short time current are related. The peak current which can arise when a short circuit occurs has to be withstood by the switching device. If that device is not called upon by the protective system to clear the fault, it will also have to withstand the thermal effects of carrying the short-circuit current for a short time, until some other device interrupts it.

The ratio of reactance to resistance of the system in which the short circuit occurs, together with the relationship between the time when the fault is made and voltage zero, control the magnitude of the making current peak (see Chapter 2).

International standards make assumptions concerning the ratio of inductance to resistance of the system under short-circuit conditions which will cover the vast majority of electrical networks.

The most usual figure for the X/R ratio on which the standard figures are based is 14 in the IEC and BS standards and 15 in the American standards. This gives a value of about 2.5 for the ratio of the peak current to the rms value of the symmetrical fault rating.

The short time rating is practically always the same as the rated breaking current for circuit-breaking devices and an identical scale of values is used for the non-fault breaking equipment. Standard values for the duration of the short time current are 1 and 3 s, the value chosen being dependent on the characteristics of the protection system being used. In assessing the need for one value or the other it is usual to assume that the 'first line of defence' in the protection system might fail and the time for which the fault current persists then depends on the back-up protection. As a general rule, the more extensive the network, the more likely it is that the higher figure will be required. Also, in networks where installation began many years ago, with earlier forms of relays or other protective devices, the need for wider tolerances at the different levels of discrimination again leads to choosing the higher value.

The foregoing covers the essential rating parameters of switchgear, and now consideration can be given to the functions performed by these switching devices in the distribution network.

SWITCHGEAR FUNCTIONS

The principal functions of switchgear are to facilitate the control of the power supply, the protection of the network and to provide convenient access to the conductors of the system so that operations such as earthing and testing of those conductors and maintenance of the switchgear and connected plant can be carried out. In addition, switchgear provides a convenient point at which to measure current, voltage and power being supplied. To provide the signals that are required to drive the measuring and protective devices, current and voltage transformers are usually incorporated in the switchgear. These serve two important purposes, first to reduce the high currents and voltages at the main conductors to more suitable levels for the equipment they are driving, and second to provide an interface which allows standard instruments, meters and relays to be employed. In low voltage switchgear some switches have direct acting releases and can dispense with current transformers. Also, at these voltages, voltage transformers are rarely necessary.

In the majority of switchgear applications the device is not called upon to operate very frequently, even in its role as a control device, and there are many circuit-breakers in service which have never been called upon to interrupt fault current. So the most important characteristic of any switchgear is the ability to stand up to environmental conditions for long periods of time, yet operate as required in accordance with its design parameters whenever the protective system gives a signal, or the operator initiates an open or close movement.

With a circuit-breaker in particular, it is often a surprise to a newcomer to realise the sort of operating times that are involved. If the protection is instantaneous, a relay will give a trip signal usually within 10 ms of the initiation of a fault. Some 40–50 ms later the circuit-breaker contacts part and arcing commences. With the new technologies currently in use in circuit-breakers, this arc will be extinguished at the first current zero in each phase, giving a maximum arc duration of 13 ms on a 50 Hz system, or 11 ms on a 60 Hz system.

The total time which will thus elapse between the inception of a fault and its extinction will be in the order of 65 ms, i.e. just over 3 cycles on a 50 Hz system, or just under 4 cycles on a 60 Hz system. There will be cases when such speed is not necessary or desirable and the protective system will control the operation accordingly. This subject is dealt with in detail in Chapter 13.

When considering the question of access to the electrical system through the switchgear, mention must be made of safety requirements. To ensure that the operator is protected as far as possible from the risk of touching live conductors, codes of practice and safety rules have been drawn up, particularly by utility engineers, and operators are rigorously trained to follow them. This subject is covered in detail in Chapter 15. The design of the switchgear also plays a part in safety, as interlocks of various kinds are usually incorporated to prevent dangerous conditions from arising.

These questions of reliability and safety will continually recur in all the chapters of this book as they are overriding factors in the design and performance of all switching devices. As the ultimate reason for installing switchgear, particularly circuit-breakers, is as the last line of defence against serious system failure, it must operate without any hesitation when a trip or close operation is initiated.

The vast majority of switchgear is installed on three-phase alternating current systems. Different methods are used in various parts of the world for the earthing of the neutrals of such systems, and the effect that these may have on switching phenomena are introduced in later chapters. System frequencies of 50 or 60 Hz are in use and the difference has only a marginal effect on switchgear performance.

Finally, there are two other definitions of terms which are used in the rest of the book, which relate to the importance of the role played by the switchgear in the overall distribution system.

Primary switchgear: This describes the switchgear closest to the point where the distribution system receives its main power supply. It is generally made up of circuit-breakers and has the highest ratings in terms of fault level and continuous current of any switchgear on the system. It is most likely nowadays to be at 36 kV, 24 kV or 12 kV. Its importance in terms of the size of the part of the network that would lose power in the event of a fault justifies the most sophisticated protection, essential back-up protection and perhaps a double busbar arrangement giving the maximum flexibility of switching. Figure 1.7 shows a typical primary switchboard.

Fig. 1.7 An indoor primary substation. (*Courtesy GEC Distribution Switchgear Ltd.*)

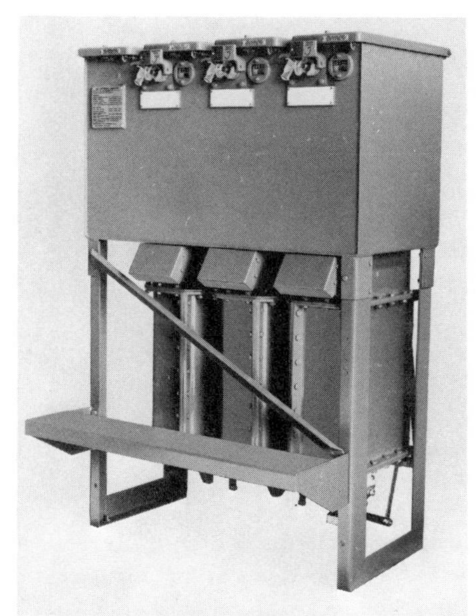

Fig. 1.8 An outdoor oil-filled ring main equipment. (*Courtesy NEI-Reyrolle Distribution Ltd.*)

Secondary switchgear: This describes the switchgear closest to the eventual user and it derives its supply from one or more primary switchgear installations. The switching devices used are often of the simplest, made up of switches and fuse-switch combinations. Manual operation is the rule and as secondary switchgear equipments are the most numerous on the network, particularly on a supply utility, economy is very important. Figure 1.8 shows the medium voltage switchgear part of a secondary substation.

Direct current switchgear is rather specialised in its application, e.g. some railway supply applications, and presents different problems in current interruption because there are, of course, no current zeros, which are used by all a.c. circuit-breakers to achieve interruption.

CHAPTER 2

Switching Phenomena

When a circuit is made or broken, a change takes place in the electrical parameters of that circuit. All practical circuits contain energy storing elements, such as capacitance or inductance, so the change is accompanied by transient phenomena of one sort or another as these energies are redistributed, before the system settles down into its new steady state.

The making or breaking of direct current circuits is a special case, which is only met in a very limited number of applications, such as power supplies for electric traction or for electrolytic processes. When the switch closes, the current rises exponentially to its steady state value at a rate which depends on the system inductance. In order to switch off a direct current circuit, when the system voltage is high enough to maintain an arc between the contacts, a switching device capable of influencing the arc path to produce an arc voltage in excess of the system voltage is required in order to interrupt the arc. This is usually done by providing an arc chute which causes the arc to be lengthened and cooled. This process is dealt with more fully in Chapter 4. When direct current supplies are required for scientific applications, where the voltages are often higher, the system of providing an auxiliary alternating current supply which can be injected into the direct current system to produce an artificial current zero is often used.

Most networks of interest to the switchgear designer and end user are alternating current systems, and the frequencies in general use for the distribution of electrical power are either 50 Hz or 60 Hz. The former is in use in most of Europe, Asia, Africa, and Australasia, whilst the Americas, Japan and some Pacific countries use the 60 Hz system.

The important point about all circuit-breaking devices designed for interrupting alternating current is that they make use of the fact that the current passes through zero 100 or 120 times every second. In the discussions which follow a 50 Hz system will be assumed, the difference between the two frequencies being so small as to have an insignificant influence on the phenomena which accompany the switching function. In fact all testing of switchgear to prove its switching capability, when done at the indigenous frequency of the country of manufacture, is fully accepted as proving its capability for the alternative frequency.

Chapter 4 deals with the processes which take place in the arc which accompanies any practical switching operation, but it will be noted here that the arc is an essential part of the switching process as it enables the current to continue flowing after the contacts have parted until a current zero is reached. In an ideal circuit-breaking device the current carrying path should be a perfect conductor until the current reaches zero and then instantaneously become a perfect insulator. As this cannot happen in a practical world, the way in which the switching device operates has an influence on the transient phenomena which accompany the switching process.

CIRCUIT PARAMETERS

Figure 2.1 shows a simple single-phase alternating current circuit, which can represent any typical distribution system. The supply is from a source of reactance X_S and resistance R_S, and in most cases the resistance is less than a tenth of the value of the reactance X_S. As a result, the power factor in the event of a short circuit across the load is of the order of 0.1.

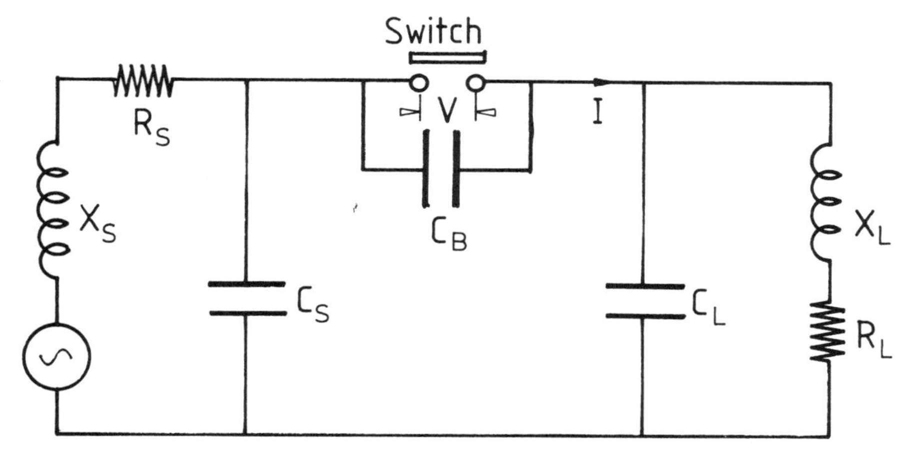

Fig. 2.1 Illustrating switching phenomena – typical circuit conditions.

If f is the power system frequency (Hz) and L is the source inductance then $\omega=2\pi f$ (=314.16 in a 50 Hz system or 377 in a 60 Hz system), and $X_S=\omega L$. The load is shown also as a combination of resistance R_L and reactance X_L, but the user of the system tries to keep the power factor of this part of the network as close to 1.0 as possible to minimise his losses and avoid tariff penalties. Finally, leakage capacitance C_L, C_S and C_B is shown in each major part of the circuit, made up of the capacitance between the conductors and between the conductors and the ground.

If the voltage of the supply is represented by $e=E \sin \omega t$, where E is the peak value of the sinusoidal voltage wave, t represents the time and ω has the value already defined, then the current in the circuit when the load is switched on will be

$$i=E/Z\ (\sin(\omega t+\phi-\theta)-e^{-\omega t \cot\phi} \sin(\phi-\theta))$$

where $\cos\phi$ is the power factor and θ represents the phase angle of the voltage when the switch contacts touch. This is illustrated in Fig. 2.2(a) for a single-phase circuit under normal load conditions, with the circuit being made at voltage zero

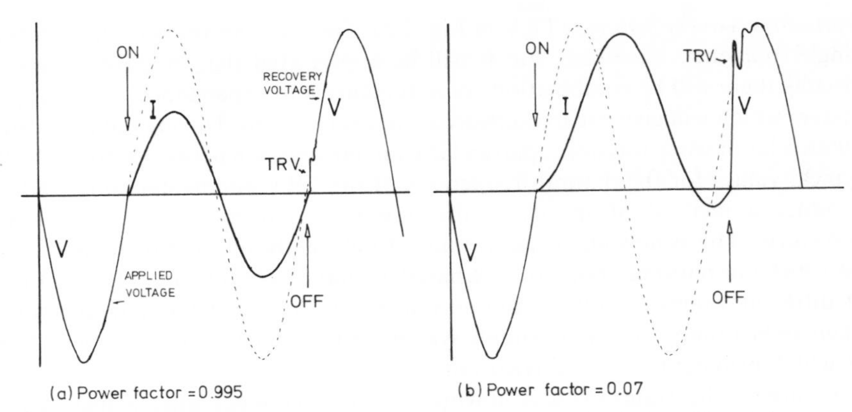

(a) Power factor = 0.995 (b) Power factor = 0.07

Fig. 2.2 Switching-on and switching-off an a.c. circuit (e.g. Fig. 2.1).

so there is no transient current. Even if the circuit had been made at voltage peak, the transient current would have been insignificant because of the very rapid decrement resulting from the high value of cot ϕ.

The corresponding current that will flow if the load is short-circuited is much larger in amplitude and has a larger and more slowly decaying transient component, since as already mentioned cos ϕ, and therefore also cot ϕ, is much smaller in value. This is illustrated in Fig. 2.2(b).

Clearly also, there is quite a difference in the phase displacement between the current and voltage in the two cases of normal load and short circuit. This has an important effect when the current is being interrupted, as the instantaneous value of the voltage from the source is quite near its peak value when the short-circuit current passes through zero. Examination of the relationship between the current and voltage (dotted) also shows differences between the situation when there is a large transient (offset) content in the current waveform and when this has died away. As is to be expected, the voltage cannot recover instantaneously to the required level if the current is interrupted at its passage through zero, and a transient voltage is added to the 50 Hz value as illustrated in Fig. 2.2, which shows the difference between the very small phase displacement with the load circuit and the large phase displacement under short-circuit conditions, but modified because of the distortion introduced by the asymmetrical current waveform.

The frequency (f_N) of the sinusoidal transient voltage is a function of the supply inductance and the leakage capacitance and is given approximately by the formula

$$f_N = \frac{1}{2\pi\sqrt{(LC)}}$$

TRANSIENT RECOVERY VOLTAGE

The damping of the waveform is due to the resistance in the system, and the course of the voltage which appears after current interruption is called the *transient recovery voltage* (TRV in Fig. 2.2). The illustrations here show only a single frequency waveform, but it will be appreciated that in many practical circuits there will be combinations of inductance and capacitance in the supply system which will give rise to multi-frequency waveforms. During the 1950s and 1960s a lot of work was done, particularly on transmission networks, to establish typical values for the natural frequencies of the TRV and its damping, and to produce a method of specifying the important parameters of the voltage waveform. This is nowadays clearly understood and the international standards, on which our national standards are based, contain tables of values for the TRV at different values of voltage and current which represent the highest figures likely to be found in ordinary supply systems, and which are therefore the values at which proving tests are carried out.

Figure 2.3 illustrates the way in which a single frequency wave is specified in BS 5311 and Table 2.1 is an extract of the standard values for t_3 and U_c for voltages up to 36 kV. In the standard a 'delay line' is also specified for the standard TRV, but as that has more to do with establishing the correct conditions on the high power test plant, it is dealt with in more detail in Chapter 18.

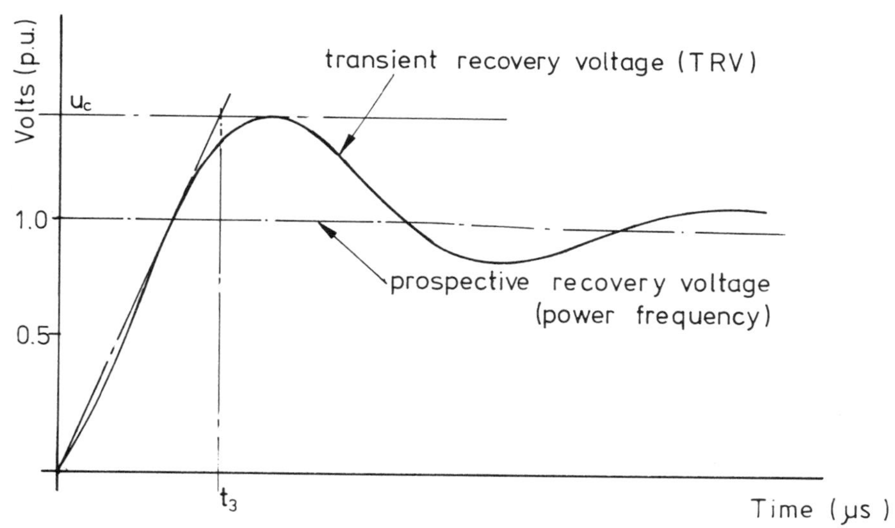

Fig. 2.3 Defining 'Transient Recovery Voltage'.

Returning to the single-phase circuit shown in Fig. 2.1, two other factors which have an effect on the transient recovery voltage may be studied and are illustrated in Fig. 2.4. Once the switching device contacts have parted, an arc is formed and there is a voltage drop across that arc as indicated in the illustration. This arc voltage is in phase with the current and leads to the recovery voltage transient beginning, not at voltage zero, but at some value as shown in the diagram at the point (2) where the current comes naturally to zero. This leads to an increase in the peak value of the TRV from point (5) to point (4).

The second condition which can arise is where the circuit interrupting action is so efficient that the current is forced prematurely to zero, as at point (1) in Fig.

Table 2.1 Standard values for TRV parameters at distribution voltages (BS 5311)

System voltage kV	*Peak line voltage (U_c)* kV	*Time to peak (t_3)* μs	*RRRV** U_c/t_3 kV/μs
At 100% of fault rating:			
3.6	6.2	40	0.154
7.2	12.4	52	0.238
12	20.6	60	0.345
17.5	30	72	0.415
24	41	88	0.47
36	62	108	0.57
At 60% of fault rating:			
3.6	6.6	17.2	0.385
7.2	13.2	22.2	0.59
12	22	25.5	0.86
17.5	32	31	1.04
24	44	37.5	1.16
36	66	46.5	1.42
At 30% of fault rating and below:			
3.6	6.6	8.6	0.77
7.2	13.2	11.2	1.18
12	22	12.8	1.72
17.5	32	15.4	2.08
24	44	18.8	2.34
36	66	23.2	2.85

*RRRV = Rate of Rise of Restriking Voltage.
= Slope of tangent to TRV curve (Fig. 2.3).

2.4. This may be viewed in practical terms as the current being diverted into the capacitance which is in parallel with the circuit-breaking gap. It follows then that what happens depends to some extent on the values of inductance and capacitance of the system.

CURRENT CHOPPING

One way of looking at the consequence of this premature interruption is to consider the energy balance of the system. If current flow ceases at some small current i_c, energy is left in the inductance which will be transferred into the capacitance and appear as a voltage across the capacitance, and hence across the circuit-breaker contact gap.

This voltage is derived from the energy balance equation as follows:

$$0.5LI_c^2 = 0.5CV_c^2, \text{ therefore } V_c = i_c\sqrt{(L/C)}$$

If the system surge impedance $Z_S=\sqrt{(L/C)}$, $V_C=i_C Z_S$.

This phenomenon is known as *current chopping* and is referred to again in discussing the relative merits of different kinds of circuit-breakers in Chapter 4. The effect of this premature extinction on the transient recovery voltage is shown in Fig. 2.4, where the 'suppression voltage' is shown with the peak value of V_C and this leads to an augmentation of the peak of the TRV at point (3) in that diagram.

It should be pointed out that the equation used to derive the transient voltage peak above is a simplification of the actual circumstances as it ignores the losses in the system. The sort of system and load condition that gives rise to this current chopping phenomenon is where a transformer or motor is switched off-load and, because of the iron circuits, is highly inductive. However, the iron circuits themselves absorb energy in the form of hysteresis losses and that reduces the potential peak voltage.

Also, the waveform of the magnetising currents contains a considerable proportion of harmonics and therefore there are long periods during each half-cycle when the current is quite low. This increases the probability of the extinction taking place at these low levels, again leading to lower values of transient voltage peak.

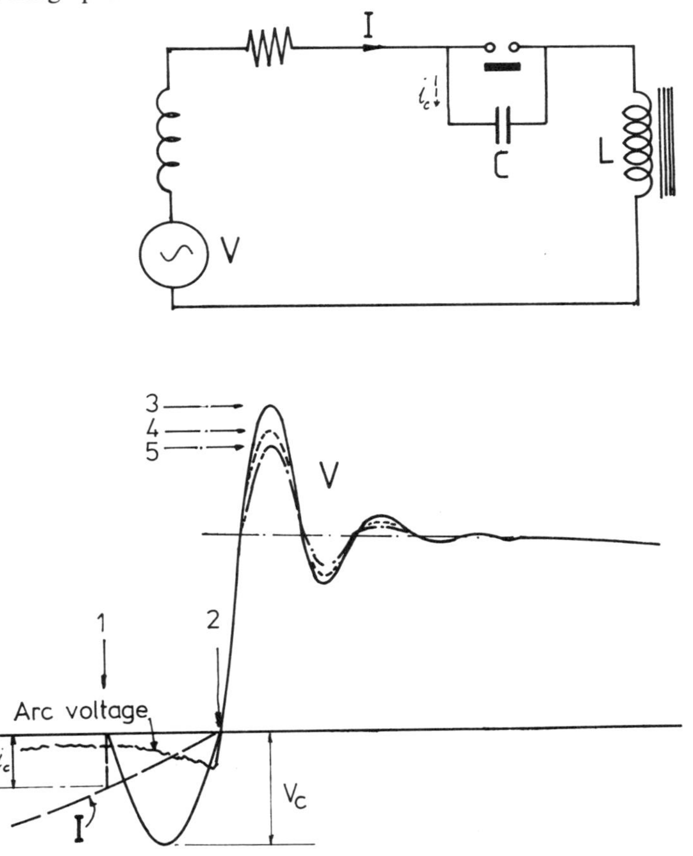

Fig. 2.4 Effect of arc voltage and current chopping on the recovery voltage.

The worst condition usually occurs if the circuit is switched immediately after it is made and while the inrush magnetising current is still flowing. This is more sinusoidal in waveform and 7–10 times greater in magnitude than the steady state magnetising current. The resistance in the system also has an effect in reducing the transfer of energy into the leakage capacitance, and thus reducing the peak voltage.

VOLTAGE ESCALATION

Figure 2.5 is a single line diagram of a circuit where a motor is fed through a circuit-breaker which has a particularly effective interruption system. The no-load current of the motor is only a few amps, so that if the circuit-breaker is opened it is probable that the current will be extinguished prematurely with some overvoltage generation, and while the contact gap is small. The effect could be that the overvoltage thus generated causes the small contact gap to break down and current flow to resume round the circuit illustrated by the chain dotted line. As this only includes a small leakage inductance, the frequency of that current will be quite high, possibly in the 10–100 kHz region.

Many circuit-breakers will be unable to interrupt at this frequency, and the current will flow until the natural 50 Hz zero. However, some circuit-breakers do have the ability to interrupt this current at a high frequency current zero, and this creates another rising voltage. This may again break down across the now slightly larger contact gap, at a correspondingly higher voltage, and the process is repeated until the gap can withstand the transient recovery voltage, as illustrated in Fig. 2.5. This phenomenon is referred to as *multiple restriking*, or *voltage escalation*. Depending on the circuit constants, it can give rise to quite high transient voltages, but as the inductance is relatively small, the surge impedance of the part of the circuit involved is low and this has a limiting effect.

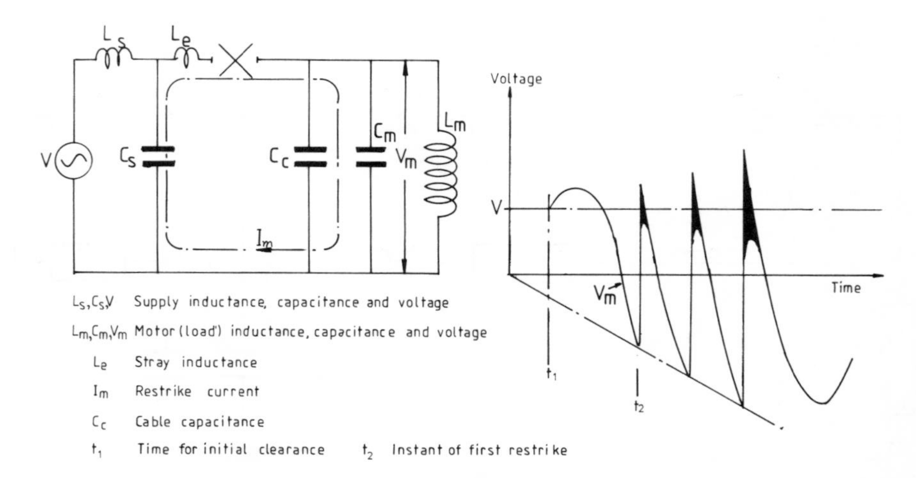

Fig. 2.5 Repeated restriking from current chopping.

VIRTUAL CURRENT CHOPPING

In circuits with significant amounts of interphase capacitance, which may arise if additional capacitors are added to the system for such purposes as power factor correction or to reduce the system surge impedance, a phenomenon known as *virtual current chopping* can take place.

Figure 2.6 is a three-phase diagram of a motor switching circuit and the capacitors C_1, C_2 and C_3 are connected as described. If now the circuit-breaker is opened, it will interrupt at a current zero as shown in Fig. 2.7 at point t_1. If this is very early in the opening stroke so that the contact gap is small, a restrike can occur at point t_2, and a high frequency current will begin to flow. Because the impedance of the capacitors at the high frequency will be low, this current can flow as shown; through phase 1, the neutral impedance Z_E and the capacitance C_1. Balancing currents then flow round phases 2 and 3. The effect of these, when added to the 50 Hz currents still flowing in those other phases, can force a premature current zero, as illustrated at point t_3 in Fig. 2.7.

This virtual current chop could be several times greater than the circuit-breaker is capable of producing naturally. However, the peak voltage is limited because the surge impedance is low, as it includes the relatively high capacitance of the capacitor C_3. Also the value of the current in a motor circuit when the machine is unloaded is quite low, so that, even if chopped from its peak value, the result need not be dangerous.

CAPACITANCE SWITCHING

Another load condition which gives rise to rather different conditions is the switching of capacitive loads. Figure 2.8 shows a single-phase circuit where the

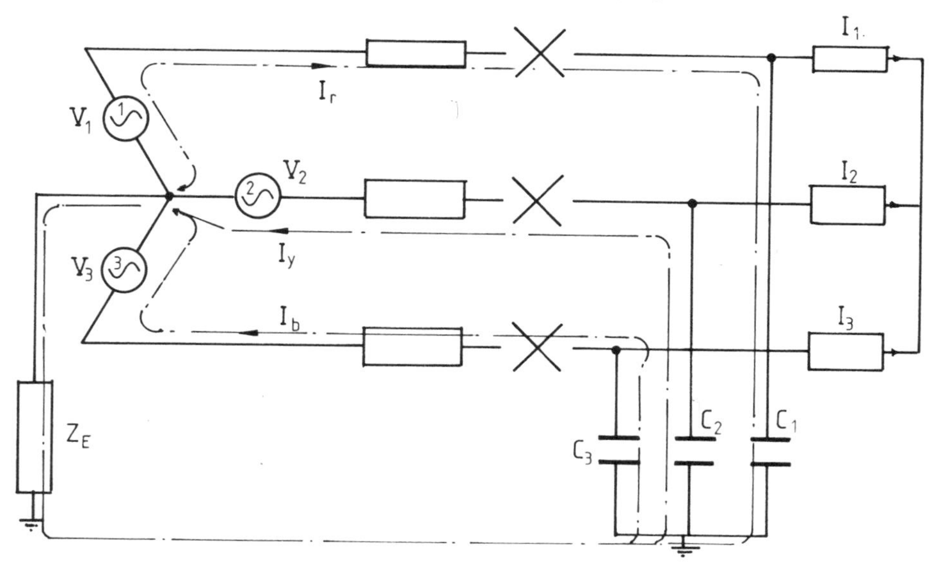

Fig. 2.6 Conditions for 'virtual' current chopping.

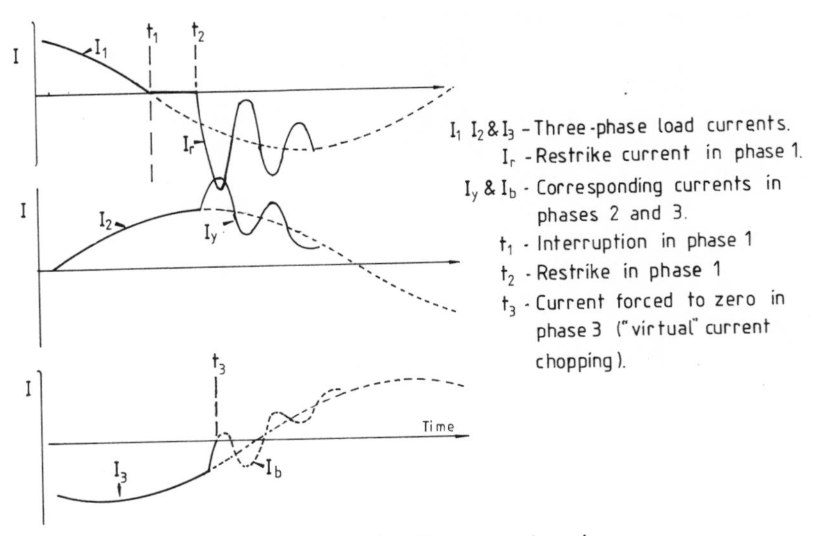

Fig. 2.7 'Virtual' current chopping.

load is a capacitor. This is in a discharged condition before the circuit-breaker is closed, because industrial capacitors are usually fitted with discharge resistances. The capacitor acts as a short circuit across the supply when switched on, the flow of current being limited only by the supply resistance and reactance which could be quite low. This situation can be aggravated when another capacitor is connected on the supply side of the circuit-breaker and is therefore already charged; it will then discharge into the load capacitor when the switch is closed. This can often be the case on industrial systems which use banks of capacitors for power-factor correction and switch them to suit the load conditions during the daily load cycles. In some applications resistors or reactors need to be fitted to limit current surges to acceptable values.

Now consider the interruption of the capacitance current. It will be seen that there are two important differences from the inductive circuit. First, the capacitor is left with a trapped charge when current flow ceases, so the voltage change across the circuit-breaker gap is not a high frequency, high amplitude transient recovery voltage, but a cosine wave with a very slow initial rate of rise of voltage. In practice there is a small transient voltage as shown at (1) in Fig. 2.8, the amplitude of which is negligible at high fault levels, when the source inductance is low. As a result the interrupting conditions are relatively easy and can lead to a particularly early interruption when the contact gap could be quite small.

As the voltage changes polarity this contact gap is stressed to a voltage almost twice the system voltage peak 10 ms after clearance. Unless the circuit-breaker has a rapidly increasing dielectric recovery in the contact gap, there is a danger of it breaking down. If this occurs at or near the voltage peak, then a high frequency current will flow, at a frequency determined by the value of the capacitor and the supply inductance. This frequency is likely to range from a few hundred hertz to the low thousands. If the circuit-breaker is efficient enough to interrupt this

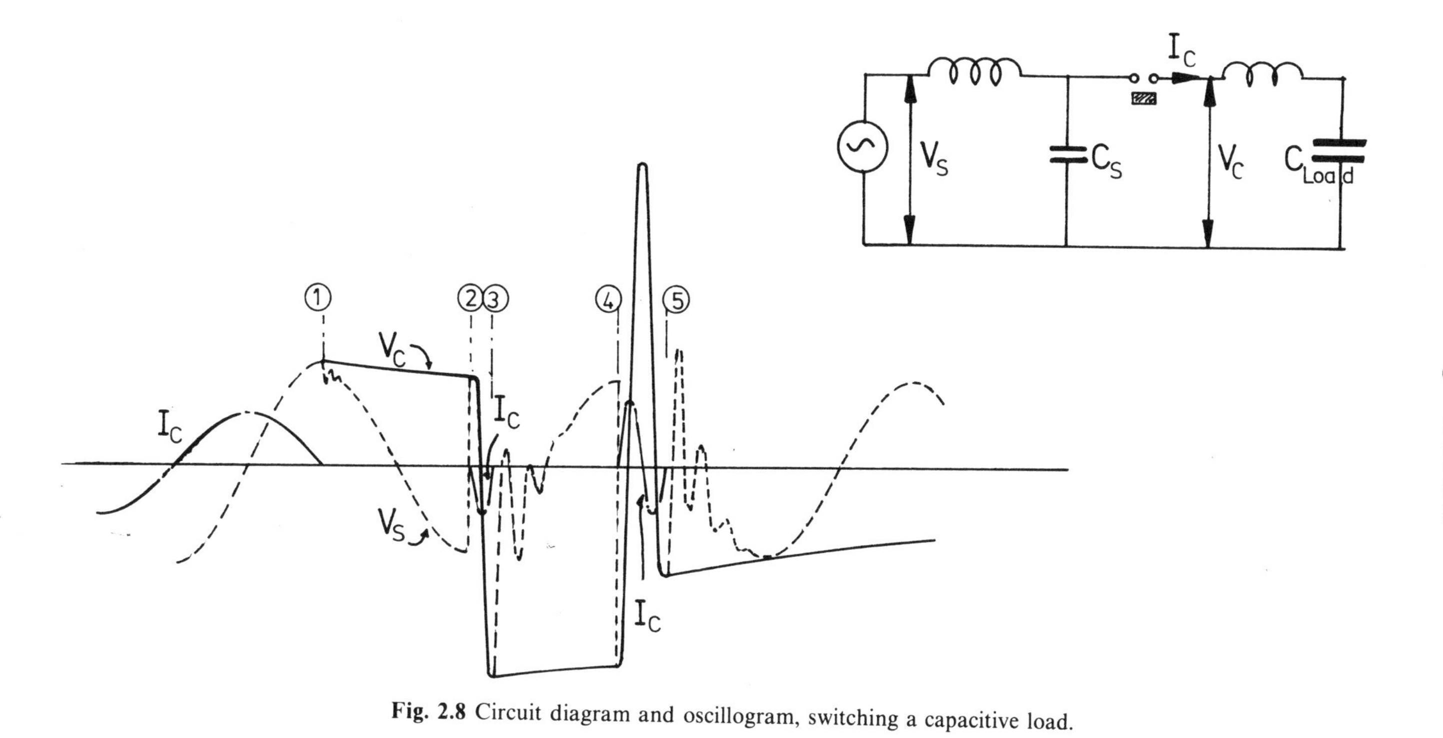

Fig. 2.8 Circuit diagram and oscillogram, switching a capacitive load.

current at a high frequency zero, then the conditions are established for the classic voltage build-up, which is usually illustrated in textbooks for a lossless case. If the gap always breaks down at the peak voltage, and the high frequency current is always interrupted at the first current zero, then it is easy to see that the voltage on the capacitor will successively be 1, 3, 5, etc., times the peak system voltage.

In practice there will be discharge of the capacitor, and random choice of current zero by the circuit-breaker, with the result that the course of the voltages and currents will be more like those illustrated in Fig. 2.8. Here are shown, successively, a breakdown of the contact gap at point (2) followed by an interruption at the first current zero at point (3). This is followed by a transient recovery voltage oscillation and the capacitor is left with a trapped charge some 2.5 times the peak voltage. At point (4) the contact gap is stressed to over three times peak voltage at a time still only 20 ms after the initial clearance. This results in a further restrike at point (4), causing the voltage on the capacitor to swing to a value in excess of three times peak. However, the possible, even probable, statistically random behaviour of the circuit-breaker may now lead to a second current zero clearance at point (5), resulting in a trapped charge of much lower magnitude and consequently a successful clearance with no further restrikes.

However, because of the random nature of this behaviour, if the circuit-breaker gap strength characteristic is such that restrike can occur, steps would need to be taken to protect the system from the consequences of the statistically unlikely, but still possible, voltage build-up. Nowadays modern technology can provide circuit-breakers that are completely restrike-free. The phenomena described here could occur on some oil circuit-breakers or air circuit-breakers, particularly at the higher distribution voltages.

THREE-PHASE FAULT SWITCHING

So far the description of switching phenomena has been restricted to single-phase circuits, where the interaction of the currents and voltages can be followed more easily than in the three-phase circuits that are in general use in distribution systems.

Figure 2.9 illustrates a typical three-phase system with the neutral point earthed through an impedance Z_{O_E}. This can have the value zero for a solidly earthed neutral system, be almost entirely resistive for a resistance earthed system, or be almost entirely inductive for a system earthed through a 'Petersen' coil with the objective of allowing the system to remain energised for a few hours when there is an earth fault on one phase.

This diagram also illustrates the five ways in which a fault may arise on a three-phase system, and the effect on the switching device will vary both with the type of fault and the earthing arrangements.

Probably the most common fault is the single phase to earth type as indicated by the arrow (1). Here the fault current is evaluated by imposing the phase voltage V_{B_O} across the supply impedance in series with the earth impedance Z_{O_E}. Z_{O_E} is often designed to significantly limit such earth fault currents, particularly in systems having a high fault level, such as auxiliary switchgear in power stations.

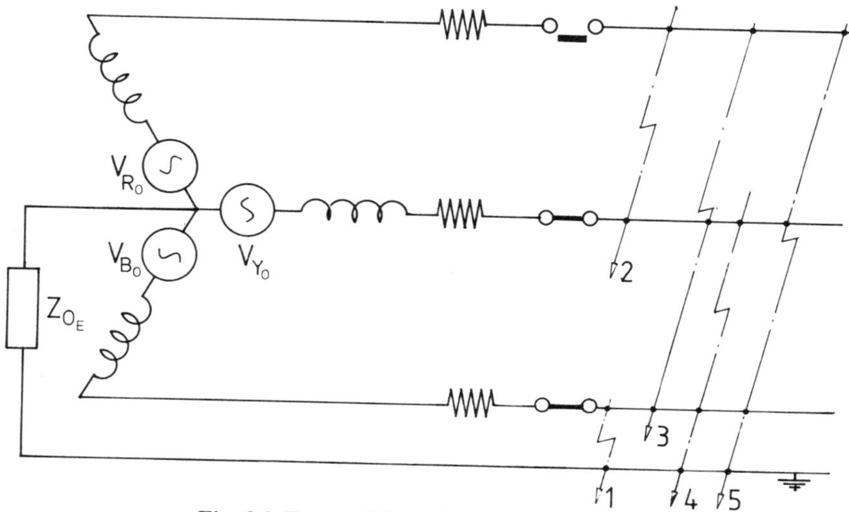

Fig. 2.9 Types of faults in three-phase systems.

Then there are faults involving two phases, with or without earth, as shown by arrows (2) and (4). In the two-phase fault, the line voltage is impressed across twice the phase impedance as can be seen from the diagram, resulting in a fault current equal to 86.6% of the three-phase rated breaking current. The currents in the two phases are equal in value but opposite in phase and when the arc is extinguished the recovery voltage is equal to the line voltage, divided across two gaps in series.

If the capacitance distribution in the system is symmetrical, then each gap will have a recovery voltage power frequency component of amplitude equal to 86.6% of the phase voltage. It can be seen therefore that this is an easier condition than a phase-to-earth fault without limiting earth impedance, both in terms of maximum fault current and recovery voltage.

However, if the fault involves earth as well, as illustrated at point (4) in the diagram, then it becomes two independent single-phase faults from the point of view of the recovery voltage, but is more severe than the single-phase condition if a limiting impedance Z_{O_E} is fitted, as a phase-to-phase fault current is also present.

Finally, we have the possibility of a complete three-phase fault with or without earth, as shown at points (3) and (5) in the diagram. Looking first at the case where earth is involved, the fault current can be the full value for the three-phase system, but the recovery voltage for each phase is limited to the phase voltage, and the three phases will each extinguish separately at the usual current zeros which occur at intervals of 60 electrical degrees.

In the case where earth is not involved, then a special condition exists when the first phase clears, leaving the condition illustrated in Fig. 2.9, where *R* phase is shown open but *Y* and *B* phases are still conducting. Under this condition the fault side of the contact gap in R phase will take up a voltage midway between the phase voltages V_{B_O} and V_{Y_O}, while the voltage on the supply side of the gap is the phase voltage V_{R_O}. The vector sum of these voltages is equal to 1.5 times the phase voltage, which gives the highest voltage stress on the switching device of any of the possible three-phase fault conditions. This is why the high power tests on circuit-breakers are mostly made in a three-phase insulated neutral test circuit.

D.C. COMPONENT

The existence of three phases with voltages at 120° phase displacement means that whenever the fault is created, there will be an asymmetrical component in at least two phases. Figure 2.10 shows one possibility when a three-phase short circuit takes place under the usual highly inductive condition. In this instance the voltage in Y phase is at zero, leading to a maximum d.c. component in that phase current. A corresponding d.c. component of half the amplitude and opposite polarity exists in the other two phases.

Obviously, any combination of degrees of asymmetry can occur depending on the switching moment, the only criterion being that the sum of the d.c. components in all three phases at any point in time must be zero – assuming that there is no fourth wire for neutral current flow.

An important aspect of the d.c. decrement, from the point of view of the circuit-breaker performance, is the amount of offset in the a.c. wave during the arcing period, as this controls the amount of arc energy that will be liberated. As can be noted from the curves in Fig. 2.10, the current wave is not only of greater amplitude when it contains a significant d.c. component, it is also of longer duration between current zeros. The degree of asymmetry seen by the circuit-breaker when the contacts open depends on the time which elapses between the instant when the fault is established and the opening of the contacts and also on the rate of decrement of the d.c. component.

Looking at the equation on page 15, it will be seen that the coefficient of t in the exponential term is $\omega \cot \phi$, where ϕ is the power factor angle and L, R and Z represent respectively the inductance, resistance and impedance of the fault limiting elements of the circuit. It is usually an acceptable simplification to ignore the capacitance of the system.

The following relationships between these parameters are well known:

$$Z=\sqrt{((\omega L)^2+R^2)}, \cos \phi=R/Z, \text{ from which } \cot \phi=R/\omega L$$

In the national and international standards governing the issues which concern users of switchgear, the factor which is used to define the rate of decay of the d.c. component is the ratio of reactance to resistance ($\omega L/R$). As has already been mentioned in Chapter 1, this value is set at 14 as being representative of the majority of normal conditions. Figure 2.10 has been drawn using this value and it will be seen that there is quite a degree of offset remaining in Y phase (which is fully offset initially due to the choice of switching moment) after some 45 ms. A fast modern distribution circuit-breaker just about reaches contact separation at this point, and has to interrupt an a.c. fault current containing about 35% d.c. component. The d.c. component is usually stated as a percentage of the peak value of the a.c. current. Therefore a d.c. component of 100% is a fully offset

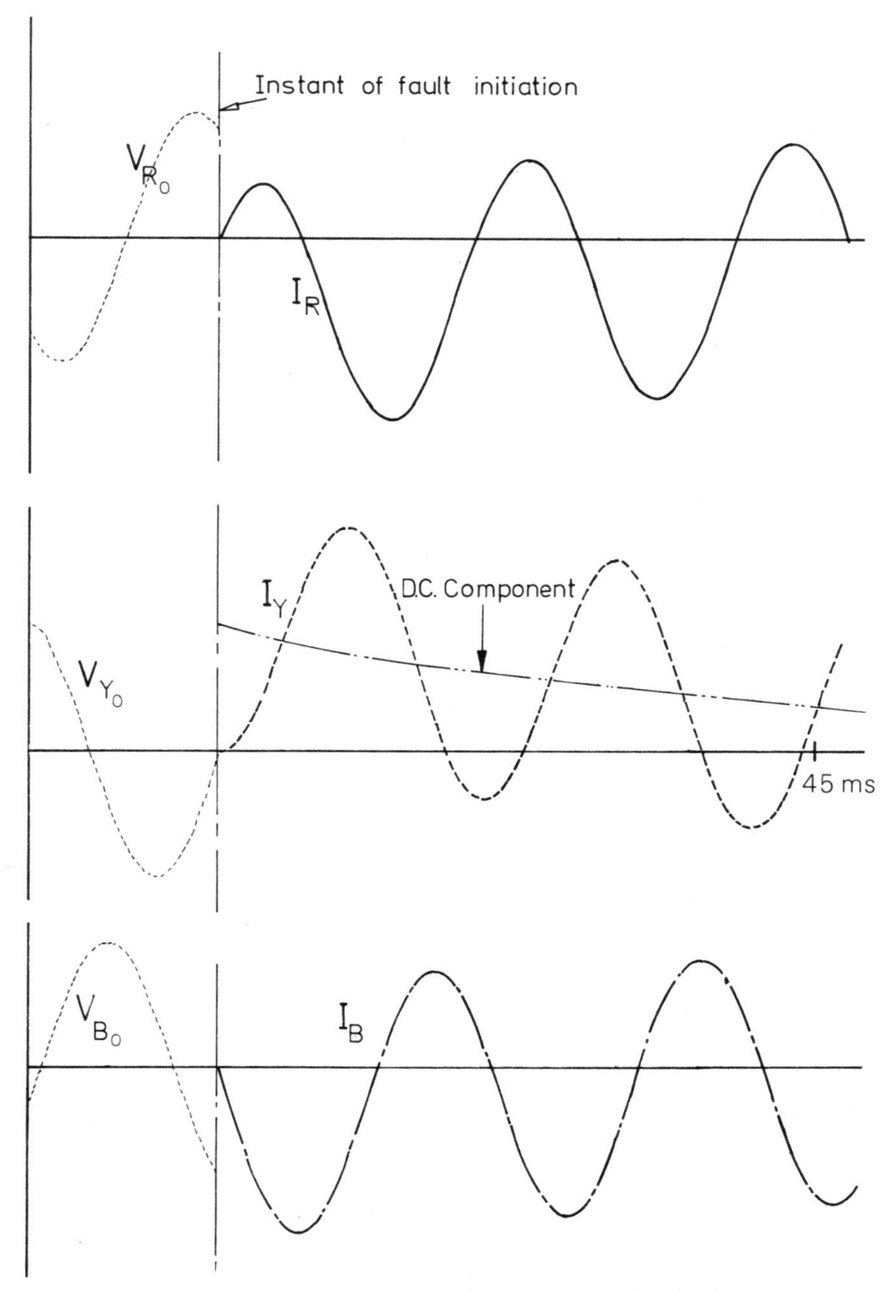

Fig. 2.10 Fault currents in a three-phase reactive circuit.

wave, as at time zero in Y phase in Fig. 2.10. In the USA the total rms a.c. current is specified, which is the rms value of the offset wave. These differences between specifications are considered in more detail in Chapter 19.

There are circumstances in which the circuit has a particularly high inductance and/or a very low resistance. Typical of such cases are the heavy duty auxiliary switchboards in a large power station. These are close to the source of generation and therefore the resistance of the heavy current conductors is low, and the relatively large transformers supplying the auxiliary power have high inductances. In this type of installation the X/R ratio might significantly exceed the 'standard' value of 14. In practice there are other factors at work which mitigate the worst effects of this, notably the need to have most of the power sources in service in parallel to provide the full rated fault level. The highest ratios of X/R will thus occur at fault levels well below the maximum possible.

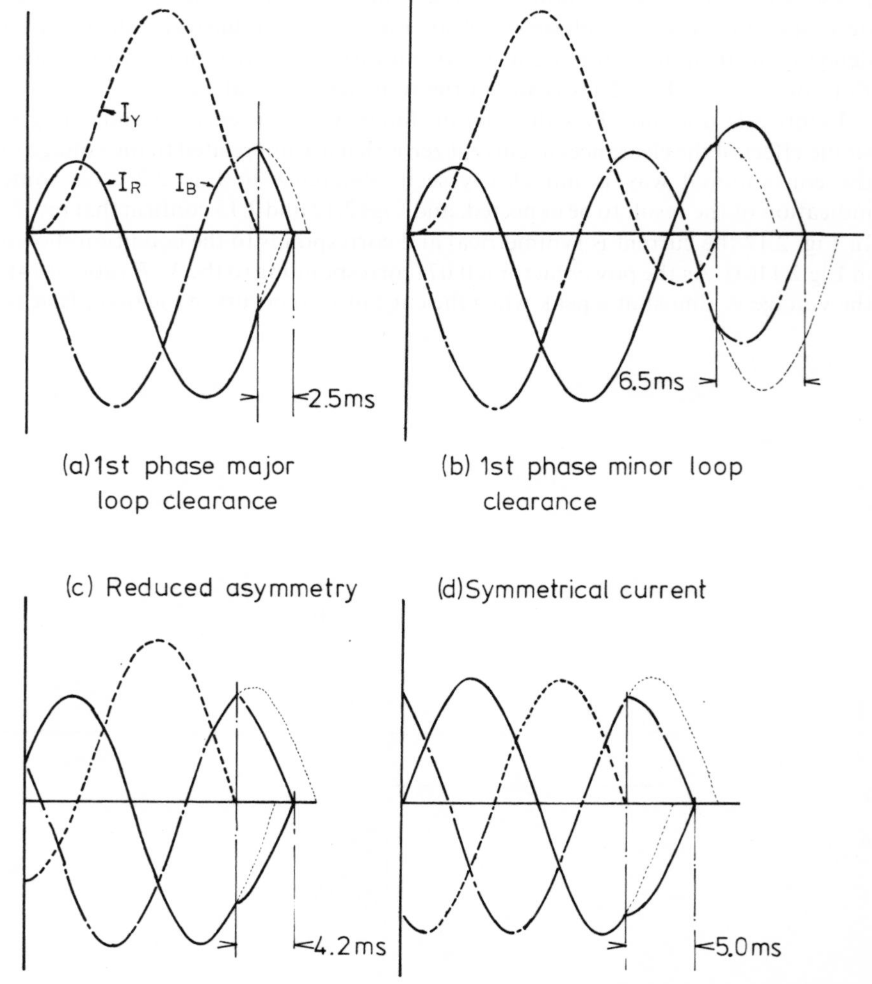

Fig. 2.11 Current distortion in second phases to clear, showing the effect of asymmetry.

EFFECT OF ASYMMETRY

It is interesting to look briefly at the effect that the asymmetry has on the interrupting parameters in each phase of a circuit-breaker. Figure 2.11 shows four possible conditions that might arise when a circuit-breaker opens on a fault. The effects are deliberately exaggerated by assuming a very early time for the contact parting in order to demonstrate the differences. As will be seen the factors that vary are the size of the current loops, both in the first phase to clear and the second phases to clear, the duration of those current loops and the shape of the TRV wave (Figs 2.12 and 2.13). It is obvious that there is quite a large difference depending on whether the first phase clears after a major loop or a minor loop.

Figures 2.11(a) and 2.11(b) show interruptions in the most asymmetrical phase soon after fault initiation. The important difference is the energy developed in the second phases to clear. Figure 2.11(c) illustrates the result of delaying the tripping of the circuit-breaker until the current is more symmetrical. For comparison, Fig. 2.11(d) shows the fully symmetrical case.

Figure 2.11 does not show the relationship between the currents and voltages, so the effect of the clearances at current zeros that are not related to the voltage in the conventional way is not clearly seen. Reference to Fig. 2.2 gives some indication of the result to be expected, and Figs 2.12 and 2.13 confirm that result. In Fig. 2.12 the current is symmetrical and corresponds to the condition shown in Fig 2.11(d). As the power factor is 0.07 (corresponding to the X/R ratio of 14), the voltage is almost at a peak when the current zero occurs in the first phase to

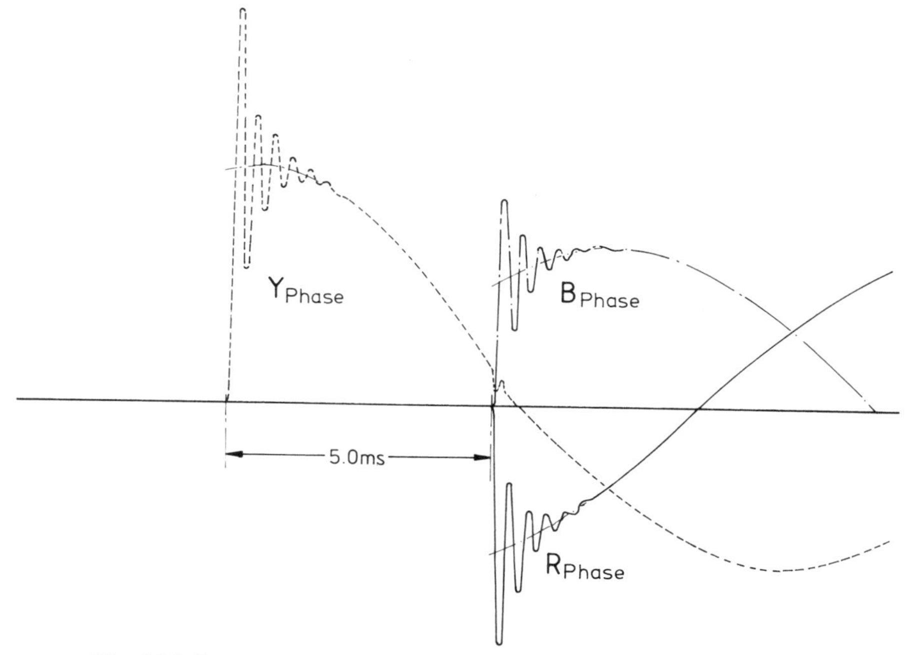

Fig. 2.12 Recovery voltage transients with symmetrical fault current.

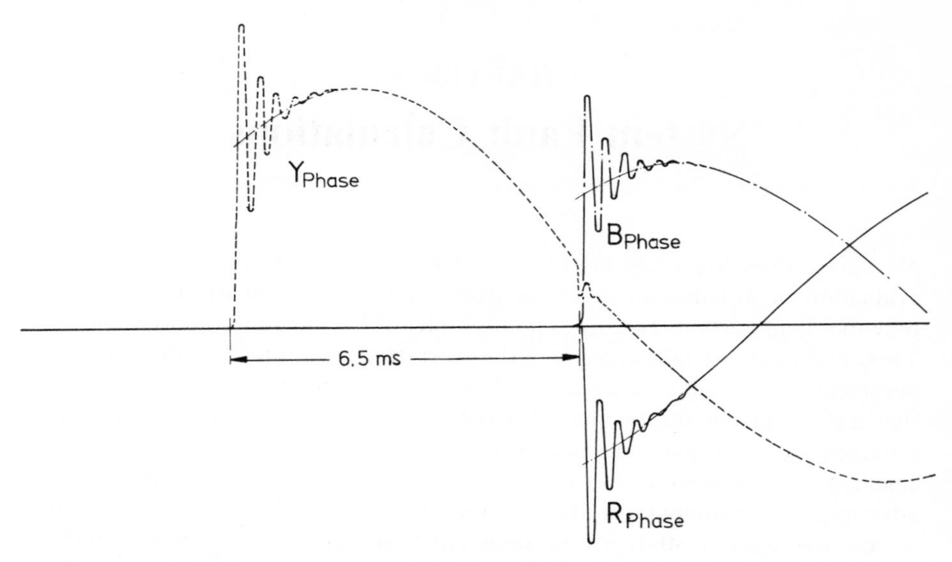

Fig. 2.13 Recovery voltage transients with asymmetrical fault currents.

clear. Since Fig. 2.11 represents a three-phase fault condition with only one neutral earth, the second and third phases have to clear at the same time, each having to change phase angle in order to achieve this. The resultant time change, with one zero being delayed and the other being advanced, means that the recovery voltage transient in the second and third phases to clear cannot occur near the voltage peak, as is seen in Fig. 2.12.

With the further time changes introduced by the current asymmetry, none of the current zeros corresponds to a voltage peak in Fig. 2.13, which is drawn to represent the clearance conditions in Fig. 2.11(b).

CHAPTER 3
System Fault Calculations

Nowadays there are a number of computer programs available to help in the evaluation of distribution networks from the point of view of analysing the potential fault currents at various points under different connection conditions. These programs will take account of the exact steady state and even the transient parameters of the connected plant and can therefore give an accurate picture of the likely fault conditions, provided these parameters are known or can be obtained from the plant manufacturer. With many industrial networks, the calculations can become too involved for longhand solutions, and another advantage of the computer study, once the constants have been entered, is that it is easy and quick to study several design alternatives. However, even with the ready availability of such machine methods, it is important to understand the principles involved in fault calculations, and this chapter sets out to investigate the factors that control the system behaviour under various switching conditions, including those of fault. Simple examples are used in order to emphasise these principles, but the methods described are adequate for a first assessment of the fault levels in a system, and can therefore serve as a check on the more accurate computer type of analysis.

Before starting the fault analysis of any distribution system, it is necessary to prepare a single line diagram, showing all the switchboards and the associated plant (transformers, generators, reactors, capacitors and motors). Figure 3.1 is such a diagram, which is part of a distribution system typical of a modest industrial site. As can be seen, it is assumed to take its supply from the local utility at 33 kV, and the declared fault level at that point is 750 MVA, which corresponds to a rated breaking current of 13.1 kA(rms). This declared value is almost certain to be a rounded up figure from other fault calculations performed by the supply authority and if used as given will create a safety margin above the true level. As the choice of switchgear at the 33 kV level is not in question in this example, the given value will be used. However, if the end result of the calculation gives a value close to, or slightly above, a standard rating of switchgear, then more exact values should be obtained. In this case it can be seen that a slight dilemma would arise with the 33 kV switchgear since the nearest standard rated short-circuit current is 12.5 kA. Reference to the actual calculations in this case would almost certainly show the 600 A margin that would allow this standard rating to be used.

The reactances of the principal items of plant are usually stated as a percentage, at the rated power level of the plant. The MVA or kVA rating of each item is also required. Where significant lengths of cable or overhead line are used for interconnection of parts of the system, the resistance and reactance per unit length is needed. For use at the planning stage of a system, this information is

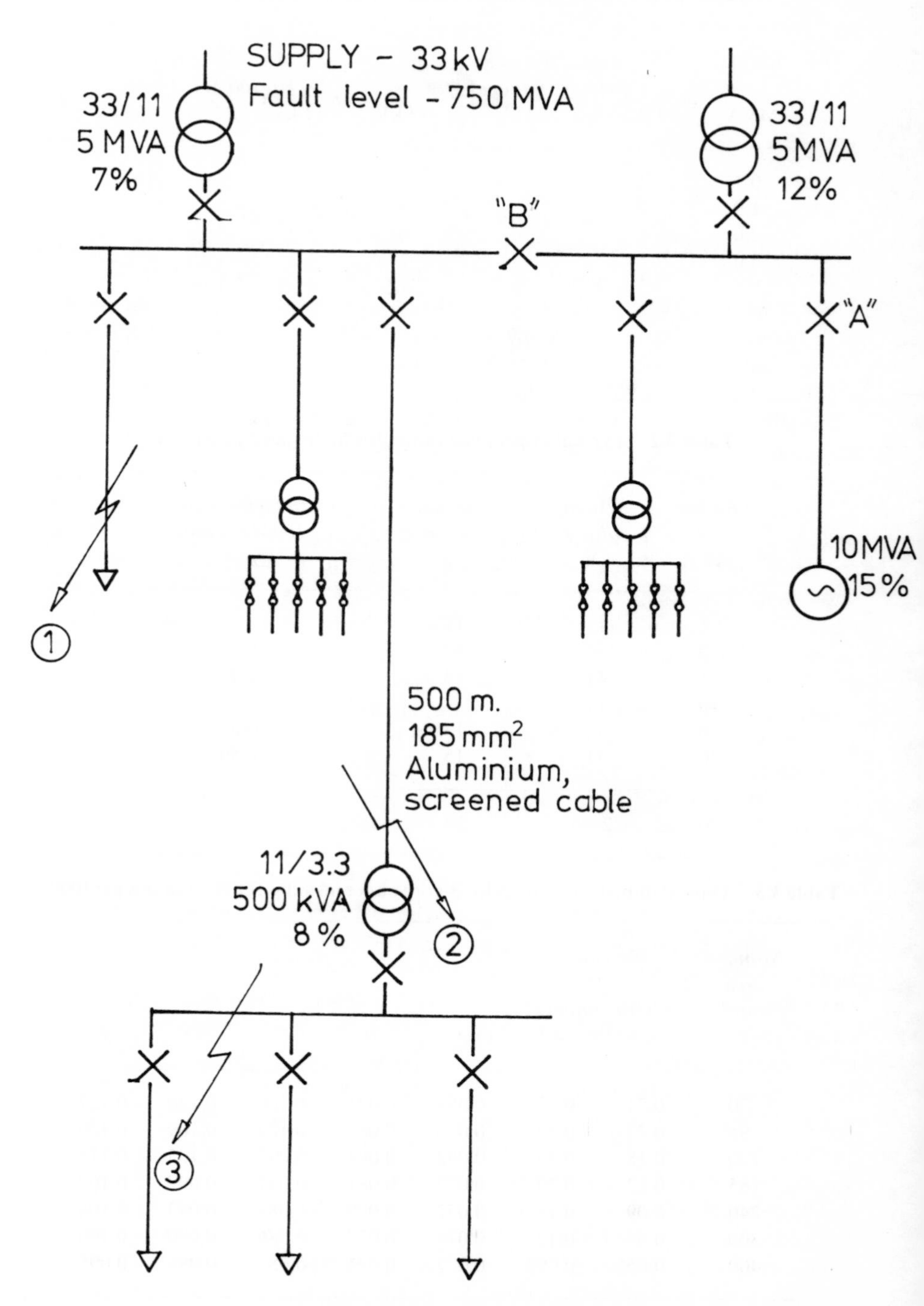

Fig. 3.1 A typical distribution system.

Table 3.1 Typical impedance values for distribution transformers

Rating MVA	*Actual reactance (%) at primary voltages of:* 11kV	22 kV	33 kV	*Reactance (p.u.) to 100 MVA base* 11 kV	22 kV	33 kV
0.1	5	5	–	50	50	–
1.0	5	5	5	5	5	5
2.0	6	6	6	3	3	3
5.0	6	7	7	1.2	1.4	1.4
7.5	7	8	8.5	0.93	1.06	1.13
10	9	9	9	0.9	0.9	0.9
20	–	10	10	–	0.5	0.5
30	–	–	10	–	–	0.33

Table 3.2 Typical impedance values for high-speed generators

Rating MVA	*System voltage* kV	*Actual reactance* %	*Reactance on 100MVA base* p.u.
1	11	12	12
2	11	12	6
5	11	12	2.4
10	11		1.2
20	11	13.5	0.625
30	11	15	0.50
30	22/33	20	0.67

Table 3.3 Typical impedance values for 3-core paper insulated cables (values per 1000 m)

Nominal area mm²	*Resistance* Ω *(All voltages)* Cu	Al	*Reactance* Ω A	B	C	D	E
70	0.31	0.52	0.089	0.091	0.106	0.120	0.127
95	0.23	0.39	0.086	0.087	0.107	0.109	0.120
120	0.18	0.30	0.082	0.084	0.097	0.105	0.115
185	0.12	0.20	0.077	0.081	0.091	0.097	0.107
240	0.09	0.15	0.077	0.079	0.088	0.093	0.104
300	0.07	0.12	0.076	0.077	0.086	0.090	0.100
400	0.055	0.095	0.075	0.075	0.083	0.088	0.095

A = 11 kV belted cables
B = 11 kV screened cables
C = 22 kV screened cables
D = 33 kV screened cables
E = 33kV SL cables

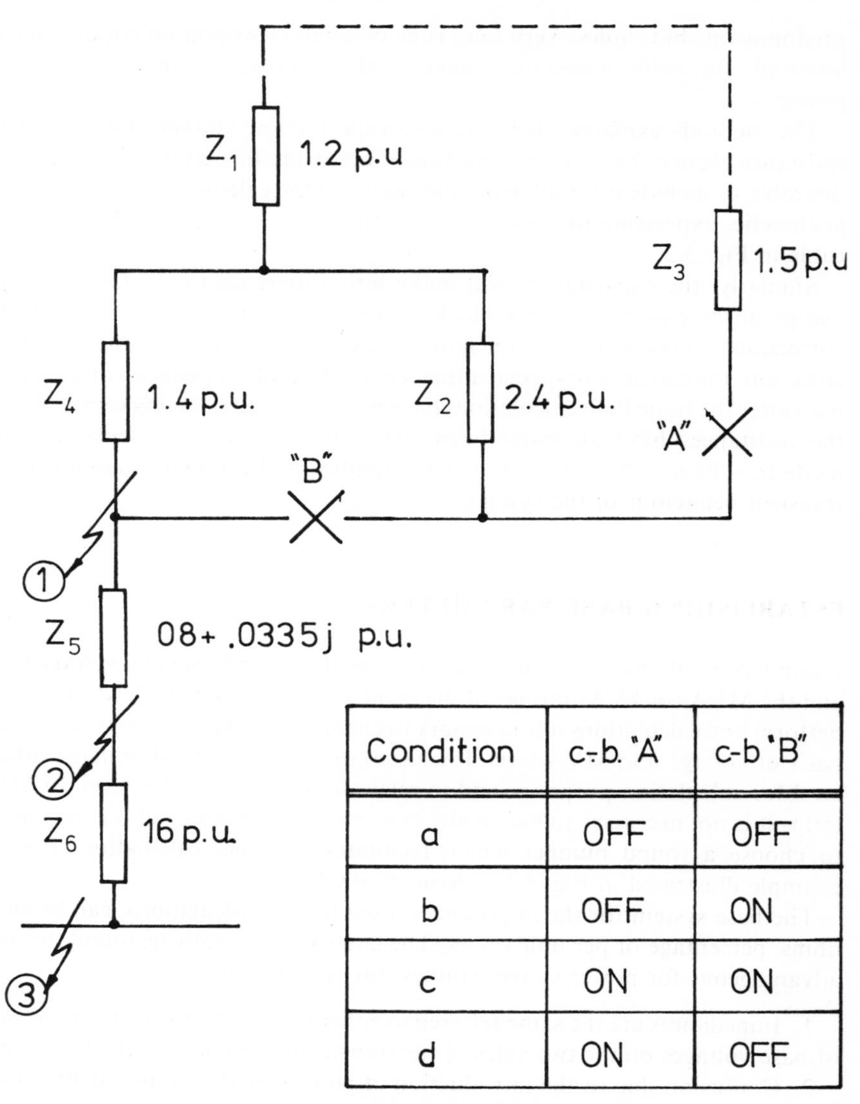

Condition	c-b. "A"	c-b "B"
a	OFF	OFF
b	OFF	ON
c	ON	ON
d	ON	OFF

Fig. 3.2 Impedance diagram corresponding to Fig. 3.1.

usually available from the manufacturer's published literature, and typical figures are listed in Tables 3.1–3.3. Obviously, in the case of an existing system, the actual figures stated by the manufacturer would be used.

In the evaluation of the values of fault current to be expected at various points on a medium voltage system, the dominant parameter is the reactance of the components, except in the case of long runs of cable or overhead line. Therefore, to simplify the calculations, it is usual to assume that the percentage impedances quoted for transformers, generators, etc., are purely inductive. Cables are represented by resistance and reactance, since it will be seen that the former is predominant, but, unless very long runs of small cross-section conductors are involved, the cable impedance makes little difference to the resulting fault power.

The methods explained here are also valid for low voltage systems but the resistance element has a more important effect on limiting fault currents, and it is desirable to include it for all plant and carry out the calculations using complex arithmetic, expressing the impedances in the form $A+jB$, as is shown for the cable in Fig. 3.2.

Similarly, the capacitances will make little difference to the fault currents, except in the case where large blocks of capacitors are used for power factor correction and could increase the initial inrush current. Such conditions as these come into the category of special influences on the fault assessment, and are dealt with after the basic fault calculations have been made. In the same vein, although the resistances have an insignificant effect on the current calculations in a medium voltage system, they do have a significant effect on the damping of the transient behaviour of the system.

ESTABLISHING BASE PARAMETERS

Examination of any system will indicate that there can be several voltage levels and the MVA (or kVA) ratings of the plant items vary quite widely. In order to perform any calculations it is necessary to change these parameters to a common basis of voltage and MVA level. It is usual to choose the predominant voltage level for calculation purposes, and to refer all the values to this level. In MVA terms it is not necessary to match any existing item of plant, and it is customary to choose a round number which facilitates the final calculation. For the example illustrated in Fig. 3.1, a basis of 100 MVA is chosen.

The base system for the impedances, used for the calculations, can be either ohms, percentage or per unit values. The latter will generally be found the most advantageous for power system studies, for two reasons:

1. Impedances are the same referred to either side of a transformer, if the ratio of base voltages on the two sides of the transformer is equal to the turns ratio.
2. Confusion due to the introduction of powers of 100 in the calculations is avoided.

With the base quantities being given in terms of power and voltage the base impedance becomes:

$$Z_B = \frac{(\text{kV})^2 \text{ ohms}}{\text{MVA}} \quad (3.1)$$

The per unit impedance is given by:

$$Z(\text{p.u.}) = Z(\text{ohms}) \times \frac{\text{base MVA}}{(\text{base kV})^2} \quad (3.2)$$

$$(Z\% = Z(\text{p.u.}) \times 100) \quad (3.3)$$

The following equations are used to adjust the impedance values to the base values of voltage and impedance:

$$Z(\text{p.u.})\ (\text{new base MVA}) = Z(\text{p.u.})\ (\text{given base}) \times \frac{\text{new base MVA}}{\text{given base MVA}}$$

(3.4)

$$Z(\text{p.u.})\ (\text{new base kV}) = Z(\text{p.u.})\ (\text{given base}) \times \frac{(\text{new base kV})^2}{(\text{given base kV})^2}$$

In the example of Fig. 3.1 we use Equation 3.1 to evaluate the input impedance based on the fault level, and Equation 3.2 to relate it to the correct bases.

$$Z(\text{ohms}) = 33^2/750 = 1.45$$

$$Z(\text{p.u.})(\text{on a base of 11 kV and 100 MVA}) = (1.45 \times 100)/11^2 = 1.2\ \text{p.u.}$$

Similar calculations for all the plant items on the diagram of Fig. 3.1 will produce the impedance diagram of Fig. 3.2.

The fault levels at the three points where arrows have been drawn can now be calculated, using simple series and parallel impedance formulae to work out the total impedance Z_T between the input and the fault point, with and without the bus-section circuit-breaker closed, and with and without the generator infeed.

As an example, consider the fault condition at the arrow (1) where the input impedance is in series with the input transformer to the 11 kV busbars.

The impedance is 1.2 + 1.4 p.u. = 2.6 p.u. Therefore:

$$\text{The fault MVA} = \frac{\text{Base MVA}}{Z_T}\ \text{MVA} \qquad (3.5)$$

$$= 100/2.6 = 38.5\ \text{MVA (approx)}$$

To take another example, consider point (2) in Fig. 3.2 under condition (d) in the table.

The impedance down to point (1) is made up of Z_1 in parallel with Z_2 and Z_3 in series, all in series with Z_4, which all works out to 2.318 p.u. Add in the cable and the result is 0.08+2.3515j, the magnitude of which is $Z_T = \sqrt{(0.08^2 + 2.3515^2)} = 2.3529$. The fault level is therefore 100/2.3529=42.5 MVA. This illustrates how little effect the resistance even of 500 m of cable has on the fault level; in fact as the fault level at point (1) under the same connection conditions is 43.1 MVA (see Table 3.4), it also indicates how little effect the cable has at all.

Table 3.4 summarises the results of the calculations for the fault levels at the three positions marked on Fig. 3.1 under the four conditions of connection listed on the impedance diagram. After the first series of results it is clear that the maximum fault level occurs with the bus section closed and the generator running. Subsequent calculations were therefore restricted to this condition.

In the example the worst condition gives rise to a fault level of 115 MVA at 11 kV, which corresponds to a short-circuit current of 6.0 kA. This very modest requirement leaves any 12 kV circuit-breaker with a handsome margin.(See Table 1.2.)

Table 3.4 Calculated fault values for the points indicated in Fig. 3.1

Fault point	*Item*	*Condition* (a)	(b)	(c)	(d)
1	Z_T(p.u.)	2.6	2.08	0.87	2.32
	P(MVA)	38.5	48.0	114.6	43.1
2	Z_T(p.u.)	—	—	0.91	—
	P(MVA)	—	—	110	—
3	Z_T(p.u.)	—	—	16.9	—
	P(MVA)	—	—	5.9	—

STAR-DELTA TRANSFORMATION

In reducing the impedances of such a system diagram as Fig. 3.2, there are two useful transformations that can be made to simplify the arithmetic. These are known as the *star-delta transformations*, and are processed in accordance with the formulae below, with reference to Fig. 3.3(a) and (b), which shows three system points 1, 2 and 3 linked respectively in a delta or star system of impedances. Either system can replace the other if the impedances are adjusted according to the formulae.

These formulae are in general terms where the three impedances are different, as they usually would be when the formulae are used to ease calculation in a single-line network diagram. As can be seen, if they are used to change the impedances of a balanced three-phase system, the ratio between the star and delta impedances is '3'.

Delta to star conversion

$$Z_1 = \frac{Z_{12} \times Z_{31}}{Z_{12} + Z_{23} + Z_{31}}$$

$$Z_2 = \frac{Z_{23} \times Z_{12}}{Z_{12} + Z_{23} + Z_{31}}$$

$$Z_3 = \frac{Z_{31} \times Z_{23}}{Z_{12} + Z_{23} + Z_{31}}$$

If all the delta impedances are equal to Z_D then $Z_S = Z_D/3$.

Star to delta conversion

$$Z_{12} = Z_1 + Z_2 + \frac{Z_1 \times Z_2}{Z_3}$$

$$Z_{23} = Z_2 + Z_3 + \frac{Z_2 \times Z_3}{Z_1}$$

$$Z_{31} = Z_3 + Z_1 + \frac{Z_3 \times Z_1}{Z_2}$$

If all the star impedances (Z_S) are equal, then the delta equivalent, (Z_D), $= Z_S \times 3$.

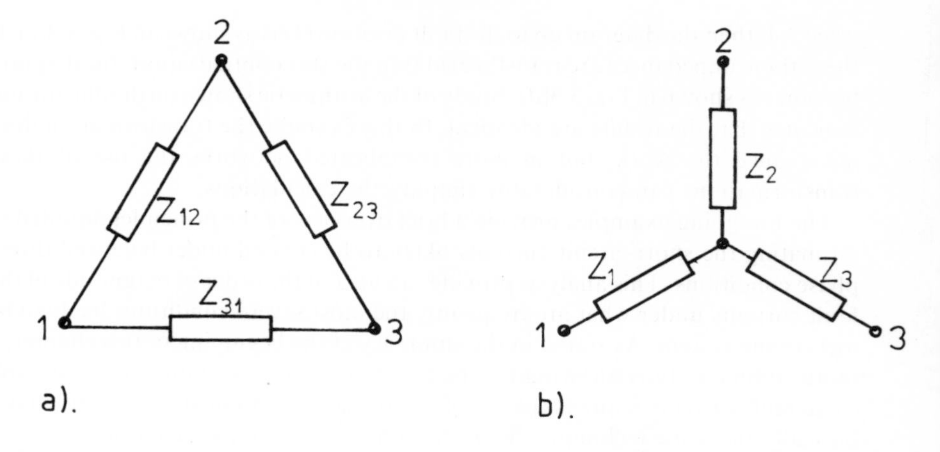

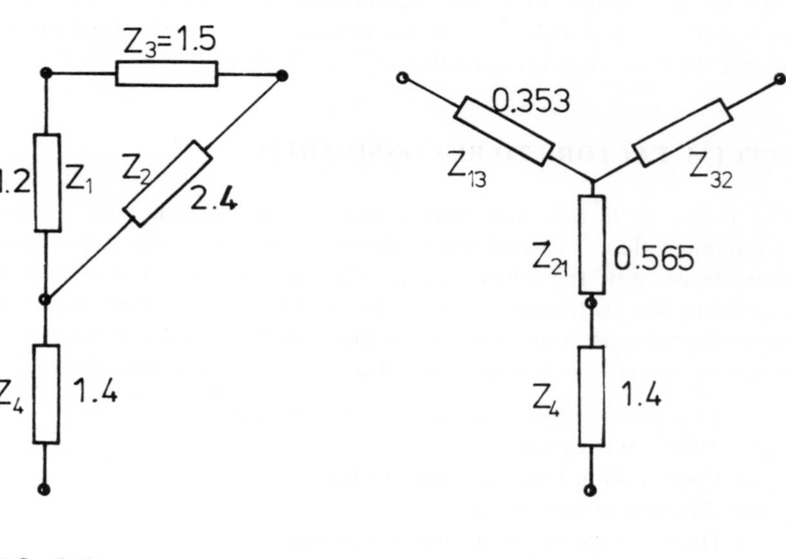

$$Z = \frac{1.2\times3.9}{5.1} + 1.4$$
$$= 2.318$$

c).

$$Z = 0.353 + 0.565 + 1.4$$
$$= 2.318 .$$

d).

Fig. 3.3 Star-delta transformation.

As an example, consider the impedance diagram of Fig. 3.2, where the three impedances Z_1, Z_2 and Z_3 are in delta connection under the condition (c) in table 3.4, then the diagram up to the fault position (1) is as shown in Fig. 3.3(c). If these three impedances are transformed into the star configuration, the diagram becomes as shown in Fig. 3.3(d). Study of the arithmetic shown on the illustration indicates that the results are identical. In this example the transformation does not reduce the work, but in more complicated networks the use of these transformations can considerably simplify the calculations.

The foregoing examples provide a brief overview of the principles adopted in estimating the short-circuit currents likely to be created under balanced three-phase conditions. This analysis provides an idea of the order of magnitude of the fault currents under what are frequently the most severe conditions likely to be met on the system. As stated in the summary at the beginning of this chapter, a more complete study allowing for the effect of unbalanced fault conditions and earth fault currents requires the use of symmetrical components, but still follows basically the same techniques as in the simplified example given here.

For many distribution systems such a steady-state analysis will suffice as the remaining parameters involved in specifying the switchgear may safely be left to the relevant national or international standard. These parameters include TRVs, asymmetrical currents, etc., for which values are given in the standards covering all but the most onerous conditions.

SPECIAL FACTORS TO BE CONSIDERED

The following factors may have a more severe influence on the switching requirement than is catered for by the standard conditions, depending on the characteristics of the network, and may have to be taken into account in specifying the switchgear, or even the plant. These usually arise when the network contains an unusually large proportion of inductive plant, e.g. motor loads, or a particularly large concentration of power in a small area:

(a). Low power-factor and high d.c. component.
(b). High TRV values.
(c). Contribution from rotating machines.
(d). Absence of current zeros.
(e). High voltage surges during switch closing.

Fortunately, as will be explained in each case, the worst combinations of conditions are unlikely to occur, e.g. very low power factors require high inductance and low resistance, a condition not compatible with high fault levels.

LOW POWER FACTOR AND D.C. COMPONENT

In conditions where there is a large concentration of power, which usually means that the resistance is low (short conductors of large cross section) compared with the inductance of the plant, the ratio of reactance to resistance will be higher than the normally assumed value of 14 or so. Figure 3.4 shows the difference in decrement for the fully offset current wave in a circuit where the X/R ratio is 14 compared with the case where this ratio is 64. There are conditions in power stations under which such values can arise. If the opening time of the circuit-breaker is 50 ms, a not unusual value, then the percentage d.c. component seen by a 'standard' circuit-breaker is 32%, as measured at A in Fig. 3.4. The actual value that occurs in the network is 75% measured at B in Fig. 3.4.

In the great majority of cases, even in large power stations, the fault level which is available under the conditions of connection which give rise to these high X/R ratios is well below the maximum fault rating required with all plant connected in the most onerous way. Once the reactance levels have been established for the network diagram, resistance values can be estimated and a rough approximation to the X/R ratio can be established. From this it can be

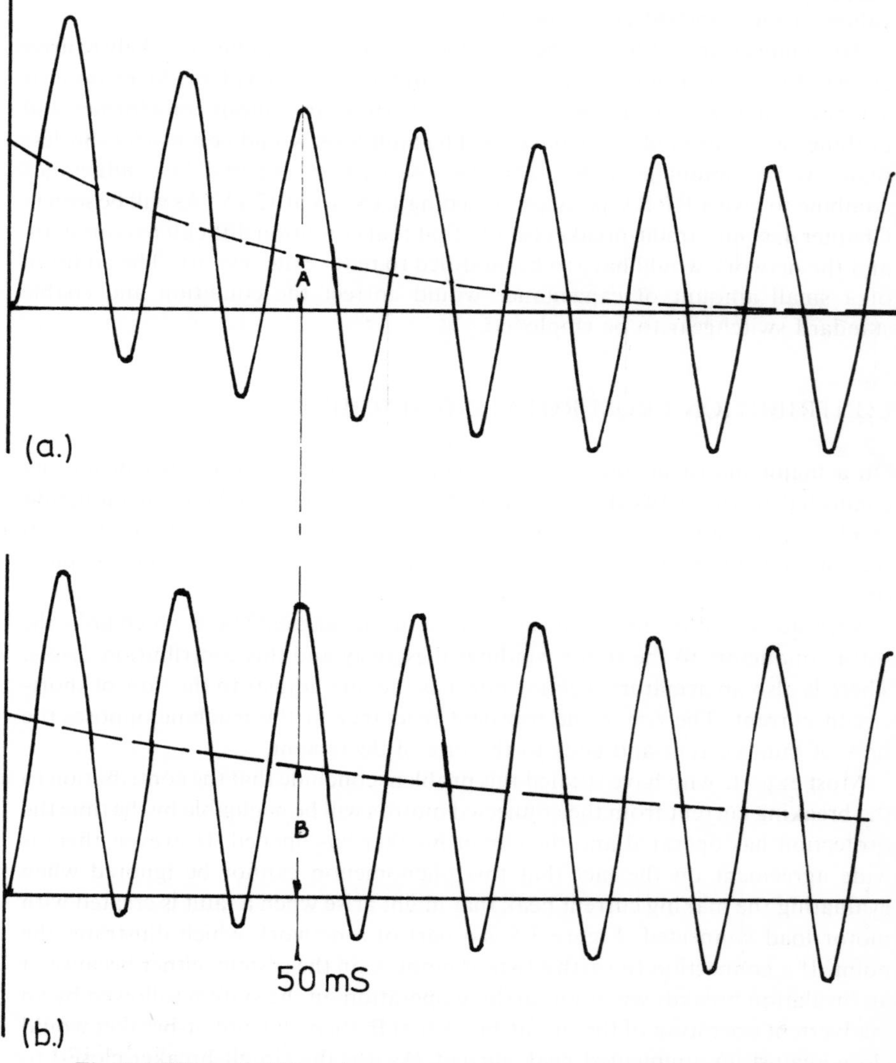

Fig. 3.4 Effect of X/R ratio on the d.c. decrement of the fault current.

seen whether the required asymmetrical current rating falls under the envelope of the standard rated circuit-breaker, chosen according to the maximum fault level that the system can deliver.

HIGH TRV VALUES

Under the same conditions that can give rise to very low fault power-factors, the leakage capacitance of the system may also be low compared to a more extended arrangement of busbars and connecting cables. This may possibly give rise to a higher than usual restriking voltage frequency and again be outside the tested values of the standard circuit-breaker.

The connection conditions that give the most onerous values are likely to arise close to the point of supply, with the minimum of connected plant. An example is a main supply board in a power station, with a large input transformer and nothing else connected to the busbars. The fault level would be considerably less than the maximum available, but the low capacitance and low inductance combine to give a RRRV perhaps exceeding 1 kV/μs at 12 kV. As will be seen in Chapter 4, some circuit-breakers could find that condition difficult to cope with, and the network would have to be modified to reduce the severity. The addition of a small amount of capacitance would correct the condition and enable standard switchgear to be employed.

CONTRIBUTION FROM ROTATING MACHINES

On a major industrial site it is very likely that there are a number of motors connected to the network. There may be synchronous machines or induction machines, or both, and at the instant that the fault is created, the energy stored in the rotating masses of these machines feeds power into the system and enhances the fault level.

Very quickly after the initiation of the fault the magnetic field which links the rotor conductors to the stator windings dies away and this contribution ceases. There is also an armature reaction effect as the flux adjusts to the flow of short-circuit current. The rise in the transient reactance of the machine opposes the flow of fault current and adds to the rate of decrement.

Most experts who have studied this problem conclude that the contribution to the breaking current from the connected motors will be negligible by the time the protection has operated and the circuit-breaker has opened. However, there is wide agreement on the fact that this phenomenon cannot be ignored when evaluating the making current peak that might arise when a fault is created with motor load connected. Figure 3.5 is a part of a network which illustrates the point. If a connection to earth exists at point A on the system, either because of an insulation breakdown or an earthing operation on the system followed by an inadvertent operation of the circuit-breaker at B, then that circuit-breaker would close against an augmented peak current. As it is the circuit-breaker closest to the fault point, it ought also to be the device that is tripped first to clear the fault.

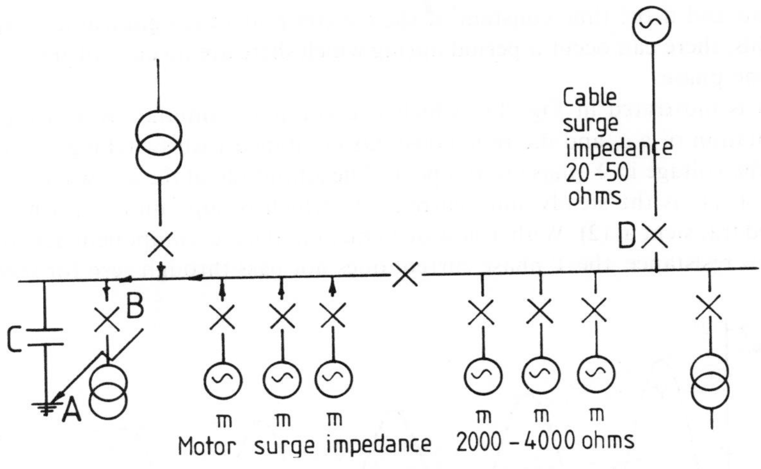

Fig. 3.5 A typical industrial motor network.

The augmentation in rating required to meet this condition is usually specified as an increase in the ratio between the peak making current and the rms symmetrical value of the rated breaking current. The accepted value for this is 2.5 in European standard specifications and these documents contain a rider to the effect that this value (and indeed the values of TRV and X/R ratio) might have to be increased in unusual circumstances. For sites where there is a high proportion of motor load, e.g. rolling mills, oil pumping installations, etc., this factor could be 3.0 or even 3.5. With modern circuit-breakers the achievement of this enhanced making capacity is not usually a problem, but it will be realised that correct representation of the true conditions on a high power testing station is not possible, so the circuit-breaker would not be opened on such a test.

ABSENCE OF CURRENT ZEROS

Another condition that can be demonstrated in systems which include local generation is the possibility that the special reactance conditions in a generator when it is short circuited lead to the current wave in at least one phase being so offset that it does not pass through zero for several cycles.

This is not the place to examine in detail the operating flux and excitation conditions in a short-circuited generator, but a simplified view of the situation is as follows. If a generator is excited with its stator short circuited, armature reaction has time to establish itself and the resulting current at full excitation is comparatively low, say four or five times full load current. However, if a generator is running at speed and already excited, and then is short circuited, it takes time for the flux created by the short-circuit current to penetrate the rotor and increase the machine reactance. The result is that the generator presents a changing reactance to the system, and there is also a d.c. component depending on the relationship between the rotor magnetic axis and the stator windings when the short circuit occurs. The transient reactance has a very short time constant and if the time constant of the d.c component is significantly greater than this, there can occur a period during which there are no current zeros in at least one phase.

This is illustrated in Fig. 3.6, which is a computer simulation of a typical combination of a.c. and d.c time constants combined with switching at a time when the voltage in *Y* phase is at a peak. The amplitude of the a.c. wave shown dotted at (1) is the steady state current, on which is superimposed a heavily damped transient at (2). With a slow decrement in the d.c. component due to the very low resistance, the *Y* phase current does not pass through zero for several cycles.

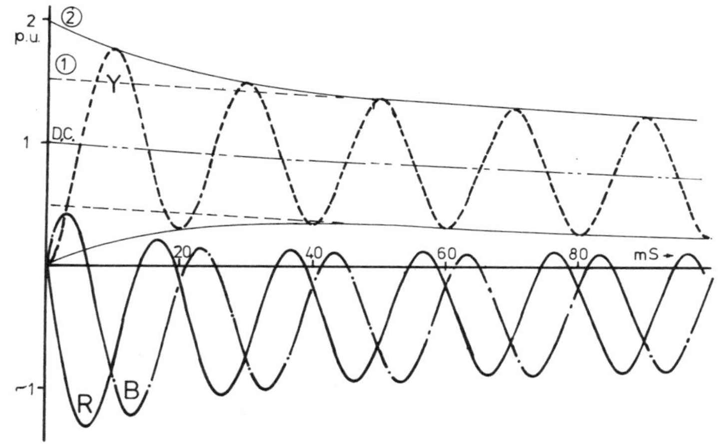

Fig. 3.6 Generator short-circuit current.

Since any circuit-breaker that might be called upon to clear this fault current depends on the current passing through zero, the possibility of this situation occurring is a source of concern. In practice this condition is very rare since it depends on the coincidence of a number of factors for serious loss of zeros to occur. The first condition is for a short circuit to occur with virtually no load connected to the system and for the generator in question not to be in parallel with another source of supply. The excitation conditions required for the worst situation are also unusual, as a very low and possibly leading power factor is needed for this. With the increasing use of gas turbine generators for peak lopping and for power supply on off-shore oil installations, there is a possibility of such a fault developing while the machine is being run up to speed and excited, before it is synchronised to the busbars.

It is interesting, then, to examine what effect the switching operation itself may have on the electrical parameters. It modifies the system parameters in two ways, both of which will alleviate the condition. First, however pronounced the condition may be, there is always one phase in which the current passes through zero in a near normal manner and current interruption will take place. This causes a redistribution of the d.c. components in the remaining phases, that which was most asymmetrical becoming less so and the other more so. If there is only one earth on the system, the remaining two-phase currents are equal in value and opposite in sign.

In addition, there is a voltage drop across the arc in all phases, the value depending on the type of circuit-breaker in use. This arc voltage, which can be regarded as equivalent to an arc resistance as it is in phase with the current, is smallest in a vacuum circuit-breaker and largest in an air circuit-breaker.

This resistance has a negligible effect on the value of the current as it is in quadrature with the main current limiting reactance, but it can have a significant effect on the damping of the d.c. component, which has the effect of bringing the offset current wave quickly down towards the zero axis and introduces current zeros in both remaining phases.

This can be approximated in a computer study by introducing a steady additional resistance into the system as soon as the circuit-breaker contacts open. In practice, as the arc voltage is constant, indeed it usually rises as the current approaches zero, the arc resistance reaches a quite high value close to current zero; therefore the simulation illustrated here is pessimistic, and the resumption of current zeros occurs even quicker than shown in Figs. 3.7–3.9.

Figure 3.7 represents the case of a vacuum circuit-breaker in which the arc does not become constricted at high current densities. (See Chapter 4.) The arc voltage is assumed constant at 50 V, and it will be seen, by comparison with Fig. 3.6, that the redistribution of the d.c. components and increased damping does lead to current zeros, but not for 45 ms after the first phase clearance. However, if the fault current is not too great, and the conditions for this lack of current zeros to arise are not usually associated with maximum fault conditions, the vacuum circuit-breaker has the best chance of all types of circuit-breaker to cope with this condition. Still it would make sense in this case deliberately to introduce a time delay in the tripping circuit.

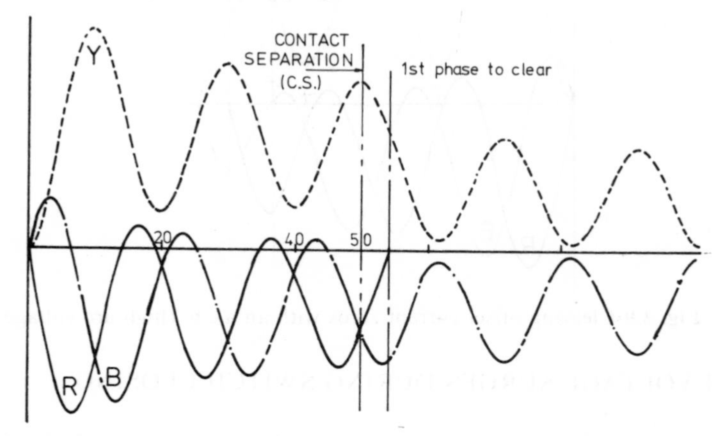

Fig. 3.7 Clearing offset current zeros with a v.c.b. (low arc voltage).

Figure 3.8 is a computer simulation of an oil or SF_6 circuit-breaker interrupting the same circuit of Fig. 3.6 and it can be seen that the greater arc resistance leads to interruption at the next current zero after the first in this case. Finally, Fig. 3.9 illustrates the case for the air circuit-breaker with a very high arc voltage, here taken as 3 kV, which has a much greater impact on the current waveform.

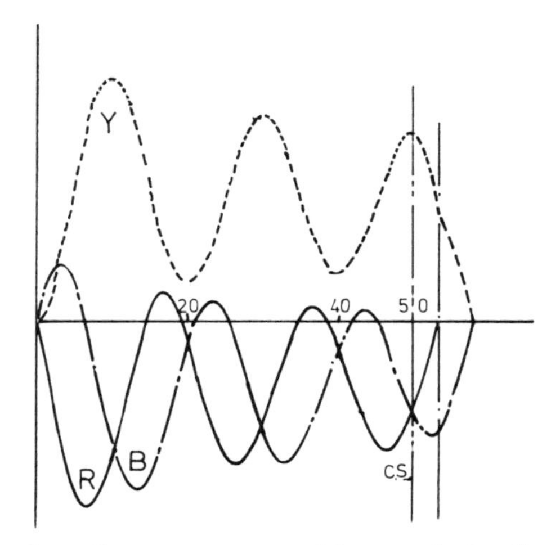

Fig. 3.8 Clearing offset current zeros with an o.c.b. (medium arc voltage).

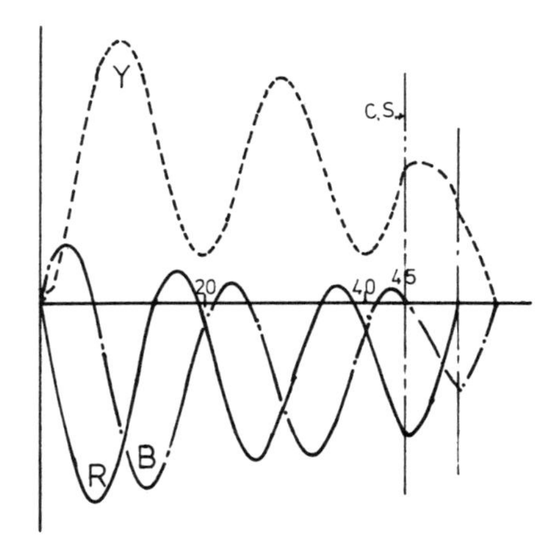

Fig. 3.9 Clearing offset current zeros with an a.c.b. (high arc voltage).

HIGH VOLTAGE SURGES DURING SWITCH CLOSING

In recent years there has been an increase in the number of insulation failures in industrial motors supplied from medium voltage sources (3.3 kV–11 kV). These failures have occurred during switching operations, and have led many engineers to attribute the reason for the failures to overvoltages created by some of the newer switching techniques, particularly vacuum. However, there are technical papers in existence that illustrate the fact that the phenomenon which gives rise to these voltage surges has been known for the last 30 years, but it has never been a frequent problem until now. The reasons for the greater incidence of such insulation failures are:

(a). A larger number of electric motors in the size range of 75 kW upwards are being supplied at medium voltage.
(b). Competitive pressures and technical advances in insulating materials have led to much more compact designs for a given output and in consequence have created higher normal voltage stresses in the windings.
(c). Direct-on-line starting has become more prevalent as systems have higher fault levels and can therefore more easily supply the heavier starting currents without significant voltage dips.
(d). The increasing use of cables insulated with polymeric materials with lower losses have ensured that voltage transients developed at the switching device are transmitted to the machine windings with little change.
(e). As motors are always connected to the system by cables, the need to design in an adequate impulse level to withstand voltage surges of atmospheric origin has not been brought to users' attention in the same way that applies to transformers, which are often used in electrically exposed situations.

The electrical conditions occurring when a motor load is directly connected to the supply can be explained with reference to Figs. 3.5 and 3.10. Figure 3.5 shows a typical system containing a number of cable connected motors, denoted by 'm'. Figure 3.10 illustrates the voltage conditions during the switching process. The description which follows relates to closing the circuit-breaker D in Fig. 3.5, with most of the other motors already connected. This circuit-breaker is also shown in Fig 3.10.

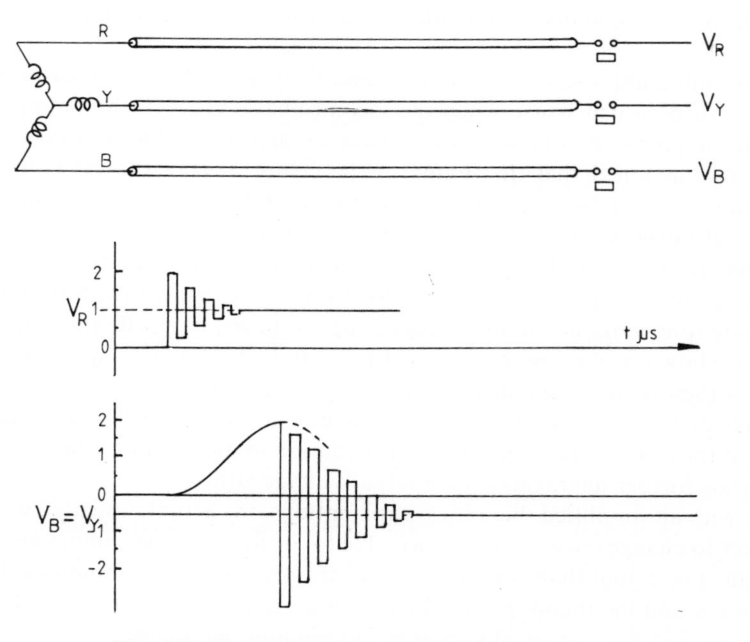

Fig. 3.10 Illustrating motor closing transients.

As soon as the contacts of the circuit-breaker approach so closely that the gap can no longer withstand the peak phase voltage, that contact gap will break down and the voltage at point D almost instantaneously reaches a value somewhere between the instantaneous value of the system phase voltage and the initial voltage at the motor, assumed to be zero. By the usual travelling wave theory, the ratio of the voltages at the junction point between the system and the branch at point D will be the ratio of the surge impedances at that point. As the surge impedance of the motors is probably some thousands of ohms, and the duration of the events being studied is in the microsecond region, the motors can be regarded as open circuits at the remote ends of their cables. The voltage at the input to the cable at D in Fig. 3.5 is $V \times Z_C/(Z_C + Z_C/n)$ where Z_C is the cable surge impedance (of the order of 50Ω) and n is the number of parallel cables connected to the busbars. If there is a capacitor connected to those busbars, which is quite likely for power-factor correction, then that can be regarded as a virtually infinite number of cables for the very short time scale being considered.

It follows that the voltage at the input to the motor cable suddenly rises to a value very close to that of the phase voltage at the instant of switching. That value will also be close to the peak value, as the rate of change of gap strength due to the closing of the contact gap is slower than the rate of change of the power frequency voltage. The result is likely to be that a surge voltage with a rise time of a microsecond or less will travel down the cable to the motor terminals.

On arrival, it will be reflected and the magnitude of the reflected wave again depends on the surge impedances. If Z_M is the motor surge impedance, the voltage at the motor terminals rises to $V_M = V \times 2Z_M/(Z_M + Z_C)$. As Z_M is very much larger than Z_C, this approximates to $2 \times V$. This voltage and following reflections of attenuated magnitudes are shown in the upper voltage trace of Fig. 3.10.

This initial impulse voltage is of acceptable magnitude in most cases, but the effect can be much worse when the second phases close, if the timing of that closure happens to coincide with voltage oscillations in the windings. In Fig. 3.10, it is assumed that the timing of the changes being studied is so fast in relation to the power frequency that the system voltage is constant for all practical purposes. If the first pole closure occurs at R-phase peak, then the voltage at the Y- and B-phase terminals of the switch will be -0.5 p.u. The windings of the machine, assumed not to have an earthed neutral point, will oscillate following the impulse voltage applied to R-phase with the amplitude 2 p.u. as shown in the lower curve of Fig. 3.10. If the contact gaps in Y- and B-phases then break down at the peak value of that oscillation, a step function voltage of 2×2.5 p.u. is applied to the machine windings. As this again has a rise time of the order of 1μs, the voltage distribution in the winding is far from linear and thus further aggravates the internal voltage stresses.

As with all simplified theoretical explanations, the actual measured waves are subject to changes from those shown, due to the losses in the components of the system. The actual shape of the pre-strike transient voltage resembles a damped sine wave and the theoretical peaks are not normally reached because of system damping and the statistical variation in switching moments.

Now that the reason for the high voltage stresses in motor windings resulting from direct connection to the supply is understood, improvements to the insulation of those windings have been made and voltage impulse tests now form part of the specification for medium voltage motors for direct-on-line starting. It is still important for the user to be aware of all aspects that can have a bearing on the fault conditions in a distribution system.

CHAPTER 4
Circuit-breaking

In all forms of circuit-breaking devices, an arc is formed when the contacts part. The presence of this arc is, at the same time, an advantage and a disadvantage. Since it is necessary to create a contact gap large enough to withstand the restriking voltage before an attempt is made to extinguish the arc, and the extinction should occur at a current zero in an a.c. circuit in order to avoid creating overvoltage problems, some means of prolonging current flow after the contacts have parted is essential. On the other hand, a high current electric arc creates a large quantity of ionised gas which has to be dispersed in an extremely short time.

It is not the intention here to delve deeply into the physics of arcs, but a general overview of their characteristics, and a brief summary of the main theories relating to the process of arc extinction are helpful in understanding the reasons behind the construction of the various forms of circuit-breaking device.

PHYSICS OF GASES

In all circuit-breaking devices the arc burns in a gas. This gas will be mainly composed of nitrogen in air circuit-breakers, hydrogen from vaporised and dissociated hydrocarbon oil in oil circuit-breakers, vaporised metal in vacuum circuit-breakers and SF_6 in SF_6 circuit-breakers. Gases have the property that all the molecules and other atomic particles move freely under the influence of forces derived from the environment. The ability of a gas to conduct is a function of the mobility of electric charge carriers amongst its particles. The velocity with which these particles move is a function of the temperature, and quite high temperatures are encountered in an arc discharge. The concept of temperature in a gas discharge is more a measure of particle velocity than of heat in the usual sense, and it is stated in Kelvin, which is the Celsius scale starting at absolute zero (−273° C).

The molecules of the gas are composed of atoms, which comprise a nucleus made up of a specific number of *neutrons* (having no electrical charge) and *protons* (having a positive charge). Rotating in fixed orbits round the nucleus are a number of very small, light particles, the *electrons*, each carrying a negative charge. In its stable state, the number of electrons in an atom balances the number of protons, so that there is no nett charge. However, the atom has two unstable states, one in which one or more of its electrons is not in its normal orbital shell and it is said to be in an 'excited' state, and the other when one or more electrons have been removed (or occasionally added) when the atom is 'ionised'. Depending on whether the number of electrons is reduced or increased, the atom will have a nett positive or negative charge. Gases which can capture electrons, and thus become negatively charged, are termed *electro-negative* gases. The amount of energy which has to be imparted to an atom to move an electron from one orbit to another, or strip it off completely, is a constant for each case, and these energies have all been measured by physicists. For the purpose of understanding the mechanism of formation of an arc discharge it is only necessary to know that these atomic changes are accompanied by absorption or emission of energy.

Obviously, if the molecules of a gas are in continuous motion, at an average velocity which corresponds to the gas temperature, there will be collisions between the particles. The frequency and vigour of these collisions will depend on the velocity of the particles, which in turn is a function of the distance over which the particles have accelerated since their previous collisions. This distance will be increased if the pressure of the gas is reduced, i.e. there are less particles present in a given space. Conversely, the 'mean free path', defined as the average distance travelled by a particle between collisions, will be reduced if the pressure is increased.

The amount of energy released when one of these particle collisions takes place dictates the likely outcome of the event. At the lowest end of the energy spectrum, a simple elastic rebound will result, the total available energy being shared between the particles which are diverted from their original paths. If the energy release is above one of the threshold levels for the movement of an electron from one shell to another, one of the atoms involved could become excited and some energy will temporarily be stored in this way. Usually the unstable state is of very short duration and the electron resumes its normal orbit, the energy being released again as heat or light. At the highest levels of energy, an electron can be removed from its parent atom and travel freely as a negatively charged particle, leaving behind a positively charged ion.

If now a voltage gradient is created in the gas, by applying a voltage between two electrodes immersed in the medium, any charged particles present in that electric field will be attracted in the direction of whichever of those electrodes is of opposite potential. In particular, electrons will be attracted to the anode, or positively charged electrode, and being of small mass will acquire high velocities. The electrically charged atoms, known as *ions*, will also be accelerated but at a much slower rate, because of their greater mass. Given a mean free path of sufficient length and a sufficiently intense voltage gradient, the accelerated electrons can accumulate sufficient energy between collisions to release further electrons from any other particle that they strike.

If the circumstances are such that this process can become cumulative, then a stream of electrons will flow from one electrode to the other and an electrical breakdown will occur. In practice, in the electric arc, the conditions for continuous ionisation are first created by thermal means. This occurs when the normally closed contacts start to open and their resistance increases due to reducing contact area and reducing contact pressure. The temperature then begins to rise until as the very last point of contact is broken the gas in that region is at a high temperature and thermal ionisation occurs. The external electrical system is also applying a voltage across the contact gap and by mutual reaction between the current flow, the thermal ionisation and the voltage gradient, the conditions for an arc become established.

It will be understood from the foregoing discussion that the higher the pressure in the gas the higher will have to be the temperature and/or the voltage gradient to ensure that the electrons can acquire the necessary ionising energy in the shorter mean free path. This explains why an increase in pressure in any gas leads to an increase in its dielectric strength.

At the other extreme, the reduction in the gas pressure first of all leads to a reduction in the voltage gradient needed for ionisation, as the mean free path increases. However, as the space between the electrodes becomes less and less crowded there comes a time when the mean free path exceeds the distance between the electrodes. Now there will be a proportion of electrons which travel to the anode and become neutralised there without striking another particle on the way. Also there will be particles which travel out of the electrode space and mix with cooler particles and again play no part in ionisation. So there comes a point in the relationship between voltage gradient and gas pressure, when the voltage gradient has to rise with decreasing pressure in order to maintain the *plasma*, as the ionised and conducting body of gas between the electrodes is called.

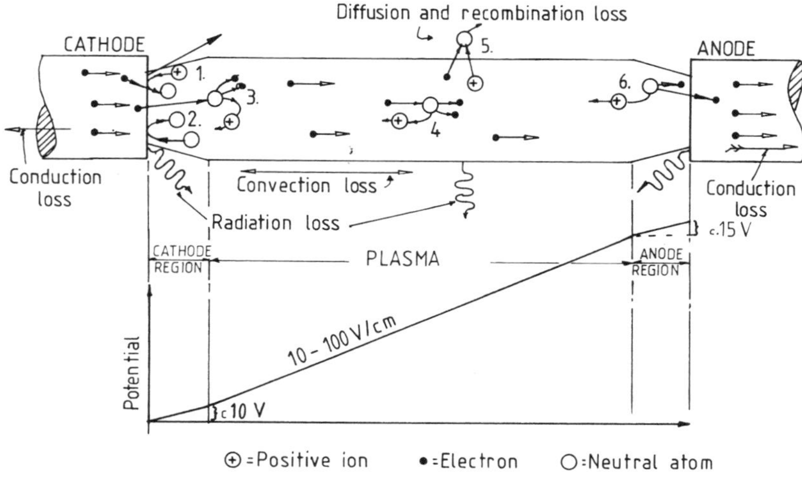

Fig. 4.1 Physical processes in the electric arc.

Figure 4.1 illustrates some of the processes which take place inside the plasma, and also shows, typically, the arc voltage between the anode and cathode. The numbers refer to the following phenomena:

1. A positive ion reaches the cathode, is neutralised by collecting an electron, and returns as an uncharged particle into the plasma. The energy released by the return of the electron heats the cathode.
2. A simple elastic collision of a neutral particle with the cathode.
3. An electron released from the cathode collides with a neutral atom with

sufficient energy to release another electron, resulting in the existence of two free electrons and a positive ion.

4. This shows a similar event to that described in (3), but within the plasma. This is probably the most common event in the hot core of the plasma.

5. This shows an electron and a positive ion being attracted to each other and combining to form a neutral atom outside the plasma. This creates an energy loss from the plasma.

6. Near the anode an ionising collision may not lead to an increase in the electron population as the electron can be absorbed by the anode, or the negative space charge which forms there.

The figure also shows the thermal losses due to radiation, convection and conduction through the contact structure. In practice, conduction plays only a small part in the energy loss from the arc.

The voltage drop along the arc path is in three parts. There is a small voltage drop at each electrode of 10–15 V, that at the anode being higher than that at the cathode, and there is a voltage gradient along the arc which has a much stronger relationship to the pressure and thermal conditions in the plasma than it has to the current flowing.

It will be appreciated that as the current being carried by the ionised plasma is nearly always a.c., it will be continually changing in magnitude and the number of charged particles needed for conduction will need to change accordingly. Thus, conditions in the arc will be transient conditions, as well as a complicated relationship between voltage gradient, temperature, pressure and contact separation.

To give some idea of the current densities involved in an arc, it is noted that for refractory electrodes, such as tungsten, the current density at the cathode will be some $1000\,A/cm^2$, whilst with low boiling point materials such as copper, this figure will be as high as $10^6\ A/cm^2$.

THE SF_6 ARC

Sulphur hexafluoride is a very stable, odourless, non-toxic and non-flammable gas which is one of the heaviest gases known, being approximately five times heavier than air. Its physical, chemical and electrical properties are studied in detail in Chapter 5; here its particular characteristics when used as an arcing medium are discussed.

The ionisation processes in an SF_6 arc plasma are generally the same as in other gases, with one important difference. Study of Fig. 4.2 shows the particle distribution in an SF_6 arc as a function of the plasma temperature, and it will be seen that there is a substantial population of negative fluorine ions. As expected, with increasing temperature the SF_6 molecule is quickly dissociated into its constituent atoms and then, at a temperature of about 3000 K, ionisation begins with the formation of positive sulphur ions.

Contrary to expectation, there is no corresponding appearance of electrons; instead the electrons are captured by the fluorine atoms to form negative ions.

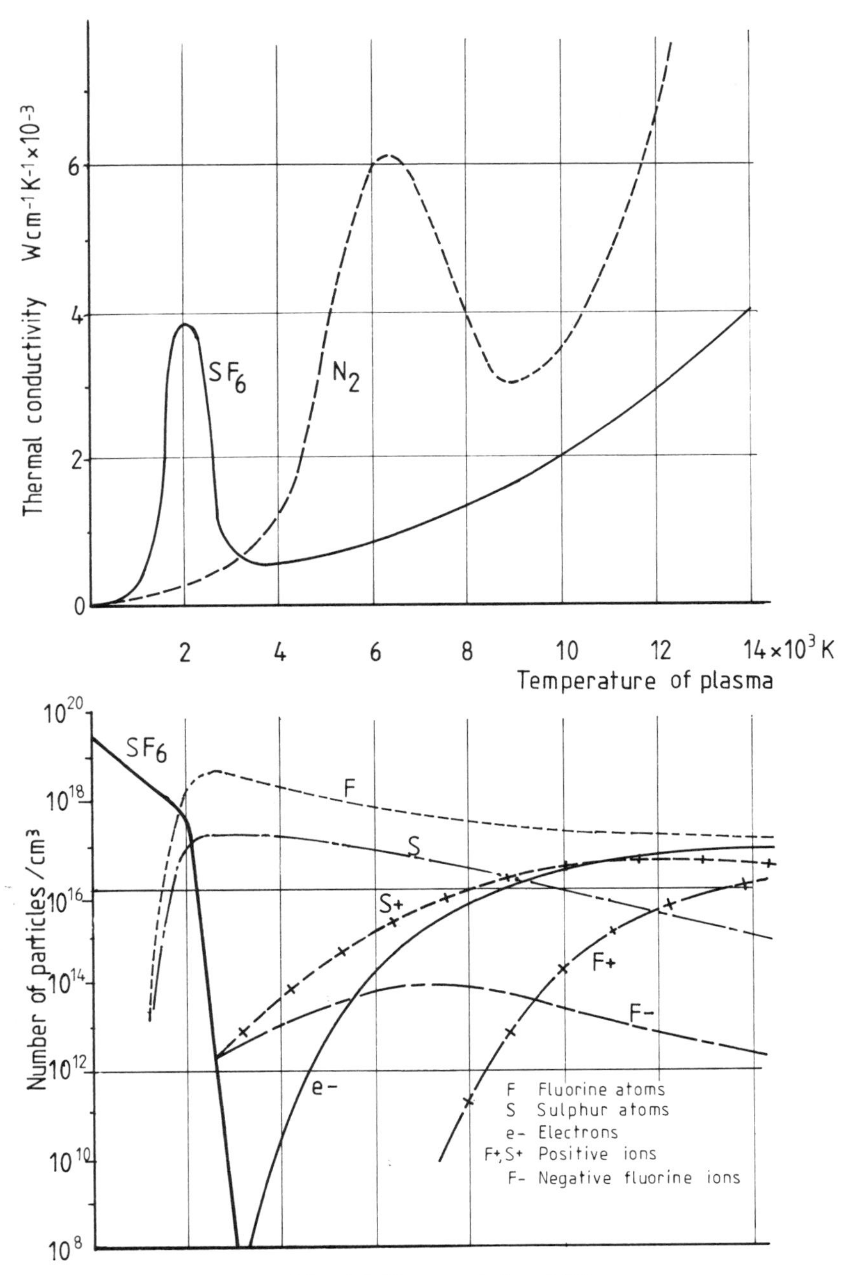

Fig. 4.2 Characteristics of SF_6.

Only when the arc temperature exceeds about 4000 K are there any substantial numbers of free electrons for conductivity to increase significantly. The mass of the negative fluorine ion is 185 times that of an electron and it moves that much more slowly, so that each replacement of an electron by a negative ion is equivalent to a reduction in current flow rate by a factor of 185.

An important characteristic of the arc plasma when considering the arc extinction processes is its thermal conductivity. The upper part of Fig. 4.2 compares the thermal conductivities of SF_6 and nitrogen (N_2). The peaks in the curves correspond to the molecular dissociation temperatures, which can be seen in the lower part of the figure as 2000 K for SF_6, and is 6500 K for N_2. The result of this is that the arc plasma in SF_6 is surrounded by a cylinder of cool gas with a high thermal conductivity and a very low electrical conductivity. In a nitrogen arc the 'mantle', as the outer sheath is known, is at a much higher temperature and correspondingly more conductive.

This is illustrated in Fig. 4.3, which shows temperature profiles for both gases. For SF_6, two profiles are shown, for two levels of current flow. The result of the cooling mantle and the small diameter of the arc core in the SF_6 arc gives it a very short thermal time constant, which is important when considering the extinction process.

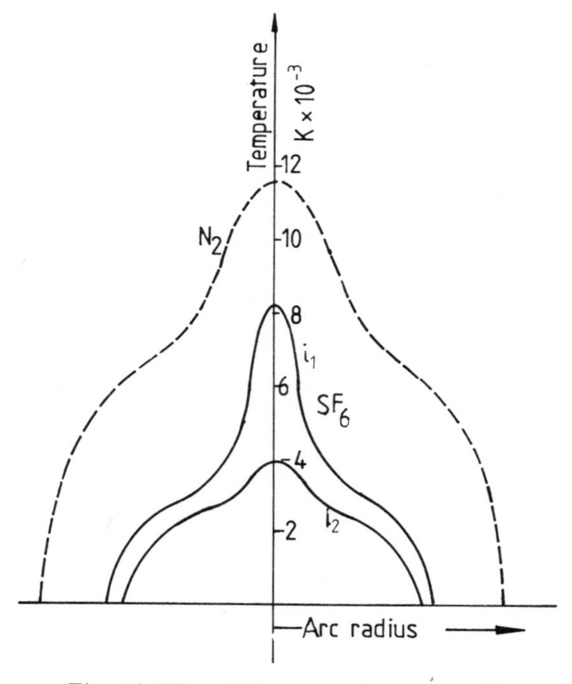

Fig. 4.3 SF_6 and N_2 arc temperature profiles.

THE VACUUM ARC

The earlier discussion concerning voltage breakdown as a function of pressure led to the conclusion that a reducing pressure would first lead to a reduction in voltage strength, but a point would be reached where further pressure reduction would result in an increase in voltage strength.

Table 4.1 compares the mean free path in air as a function of pressure and

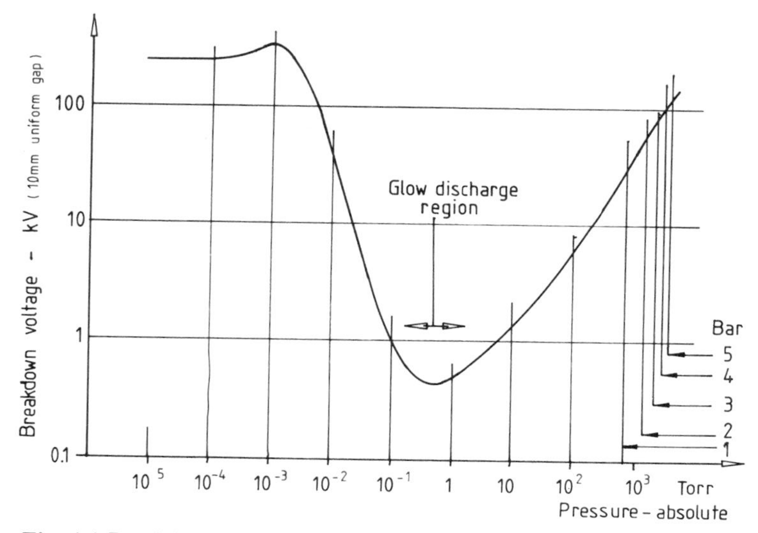

Fig. 4.4 Breakdown strength of air at different pressures (10 mm gap).

illustrates that the mean free path becomes long in comparison with the typical contact gap of 8–12 mm found in vacuum interrupter, at pressures below 10^{-3} Torr (1 Torr is equivalent to a pressure represented by a barometric 'head' of 1 mm of mercury). The effect of this on the breakdown strength of a 10 mm contact gap is shown in Fig. 4.4. To make an effective vacuum interrupter requires a vacuum of the order of 10^{-6} Torr, which represents a condition where, for every billion particles present in the interrupter at atmospheric pressure, only one is left after evacuation!

Table 4.1 Molecular density and mean free path in N_2 at 25° C

Pressure Torr	*Density* molecules/cm³	*Mean free path* cm
760	2.5×10^{19}	6.3×10^{-6}
100	3.3×10^{18}	4.8×10^{-5}
10	3.3×10^{17}	4.8×10^{-4}
1	3.3×10^{16}	4.8×10^{-3}
10^{-1}	3.3×10^{15}	4.8×10^{-2}
10^{-3}	3.3×10^{13}	4.8
10^{-5}	3.3×10^{11}	4.8×10^{2}

Achieving this degree of vacuum economically is just one of the problems that had to be solved before vacuum interrupters could become a commercial proposition.

The vacuum arc started to become important in circuit-breaking in the 1950s, when serious efforts were made to obtain worthwhile breaking capacities from vacuum devices. Since the very existence of an arc depends on the presence of gas molecules, it seems at first sight to be a contradiction to talk of an arc in a vacuum. However, the phenomenon that has already been described for the creation of ionisation in normal gaseous arcs serves equally well to initiate an arc in vacuum. In this case the heat created as the contacts finally part generates metal vapour by 'boiling off' some of the contact material. The presence of an electric field in a very small gap quickly generates free electrons and a localised plasma is created in the ionised metal vapour. Although vacuum interrupter contacts are usually made of alloys (or mixtures) of copper and other metals, the copper has the lowest boiling point and the vapour is thus largely composed of copper atoms.

The fact that the arc medium is dependent on the material of the contacts has many implications for the interrupter performance. The arc generated from a single cathode root, or 'spot', as illustrated in Fig. 4.5, will carry currents of up to about 100 A, at which point it splits into two spots. This figure is true for copper, other materials having different maximum currents. Each individual arc is conical, as the magnetic pinch effect concentrates the plasma into a high pressure region. The positive ions produced close to the cathode by the ionisation are ejected from the magnetically constricted high pressure region towards the anode together with the electron stream which is magnetically attracted to the anode. This results in a neutralisation of the plasma resulting in a very low voltage gradient in the arc.

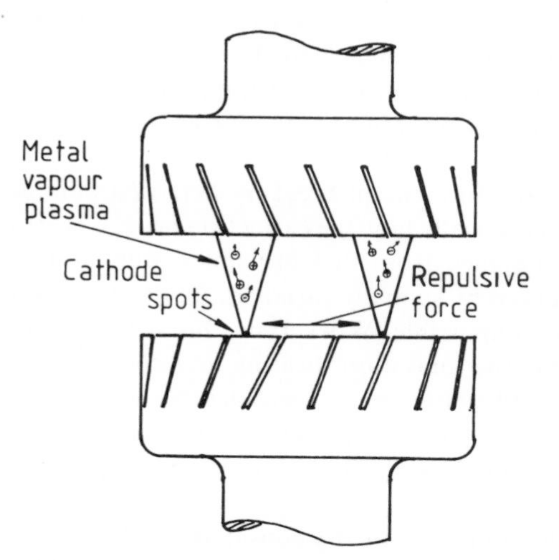

Fig. 4.5 A vacuum arc.

Vacuum arcs also exhibit another unique characteristic in that the separate arc spots repel each other, even though the arcs are carrying currents in the same direction and would normally be expected to attract each other. Whatever the explanation of this behaviour, it is of considerable importance in the process of arc extinction. The individual cathode spots are very small (of the order of

10^{-5} cm^2) and the resulting very high current density (10^5–10^7 A/cm) leads to continued melting and evaporation of the contact material, thus maintaining the supply of vapour for continued arcing. These cathode spots are continually in high speed motion with the result that the depth of erosion from each spot is very shallow, of the order of a few tens of microns. This type of arcing is referred to as the 'diffuse phase'and is recognisable by the very low arc voltage (of the order of 30 V) and a very short thermal time constant of a microsecond or less.

However, when the total current through the interrupter exceeds about 10000 A, in a simple butt contact design, the vapour pressure becomes such that normal Maxwellian laws apply, and the arc becomes magnetically constricted. When this happens, in the absence of any external influences, the arc roots become much larger and more deeply heated, so that the conductivity continues through the current zero, and the extinction is prevented. This type of arcing is referred to as the 'constricted phase', and is recognisable by a much higher arc voltage, in the region of 300 V.

The whole development of satisfactory vacuum interrupters has rested on the creation of contact designs and geometries which prevent the excessive heating of the cathode roots, and thus maintain arc time constants in the microsecond region.

ARC INTERRUPTION THEORY

Chapter 2 shows that in most fault interruption conditions in an a.c. system, a very rapidly rising voltage is impressed between the contacts of a circuit-breaker as current flow ceases at a current zero. For successful arc interruption the circuit-breaker designer has to ensure that the contact gap can withstand that voltage so that the arc does not resume as the current reappears with the opposite polarity.

Over the many years since circuit-breakers were first built, researchers have sought to explain theoretically how the extinction process works so that circuit-breakers could be designed from first principles. But the phenomena are so complex, and influenced by so many variables, frequently of a transient nature, that this has proved impossible. Workers in this field have derived complex equations to describe the ionisation, heating and cooling processes that take place in the arc, and these are a considerable help to the designer in understanding the likely effect of design changes and thus guide him in the largely trial and error process of improving the interrupting performance of a circuit-breaker.

The most generally supported theory considers the thermal balance of the arc column during the period before and after current zero. During this period energy is being removed from the plasma by:

1. Dissociating molecules into atoms.
2. Ionising atoms into ions and electrons.
3. Conduction loss into the contacts.
4. Conduction and convection loss from the arc column.
5. Loss of particles by diffusion.

Energy is being supplied to the arc by:

1. The external circuit by current flow or voltage application.
2. By recombination of charged particles.

These phenomena can be represented by assuming that the energy losses can be equated to a rising arc resistance. Study of the form of the energy losses leads to the conclusion that this rise is probably exponential. The application of the transient recovery voltage then leads to the appearance of a current after the zero point, known as the *post-arc current*, which initially rises and then falls as the resistance rapidly increases. This is illustrated in Fig. 4.6, which shows the effect of impressing TRV waves (V_t) of increasing frequency across a contact gap where the effective post-arc resistance is increasing exponentially, as shown in Fig. 4.6.

The result is plotted as the power developed in the contact zone as a function of time after current zero. If the power removal from this zone by the arc extinguishing means provided by the designer can be represented by a characteristic such as the chain-dotted line shown in Fig. 4.6, then extinction could take place in the first two examples. But the higher power peak in the third case would lead, at point A, to a reversal of the resistance rise, a collapse of the recovery voltage and resumption of fault current flow. This would be a failure to clear at that current zero and would indicate a limiting value of TRV for that design. Depending on the type of circuit-breaker under consideration, the next current zero might exhibit a stronger power removal characteristic (as shown at B in Fig. 4.6) such that clearance would follow, even at the higher frequency TRV.

The alternative theory was advanced at an earlier period than the energy balance theory just described and was derived from a presumed dielectric recovery based on breaking the discharge path of the arc by a 'wedge' of clean dielectric. Depending on the effectiveness of the arc extinguishing device the rate of dielectric recovery would exceed the rate of rise of the transient recovery voltage, and clearance would be effected. Figure 4.7 shows a typical recovery voltage transient, superimposed on which are two possible gap dielectric recovery curves. Following the 'dielectric race' theory, if the gap recovery was as curve (1) clearance would occur, but if it followed curve (2) the gap would restrike and another half period of arcing would ensue.

In practice, calculated rates of increase of dielectric strength fell short of the values demonstrated by the ability of test models to clear, by an order of magnitude or more, which led to greater credence being given to the energy balance theory. In practice, occasions can arise where both theories can play a part in explaining the interruption phenomena, since cases occur where late breakdown of the contact gap takes place, long after deionisation of the gap should have occurred under the energy balance theory. This can only be explained by the fact that the contact gap has not reached the necessary dielectric strength to maintain the clearance.

From these discussions it will be seen that the task facing the switchgear designer is to ensure that the contacts in the switching device establish a gap capable of withstanding the system recovery voltage at an early current zero

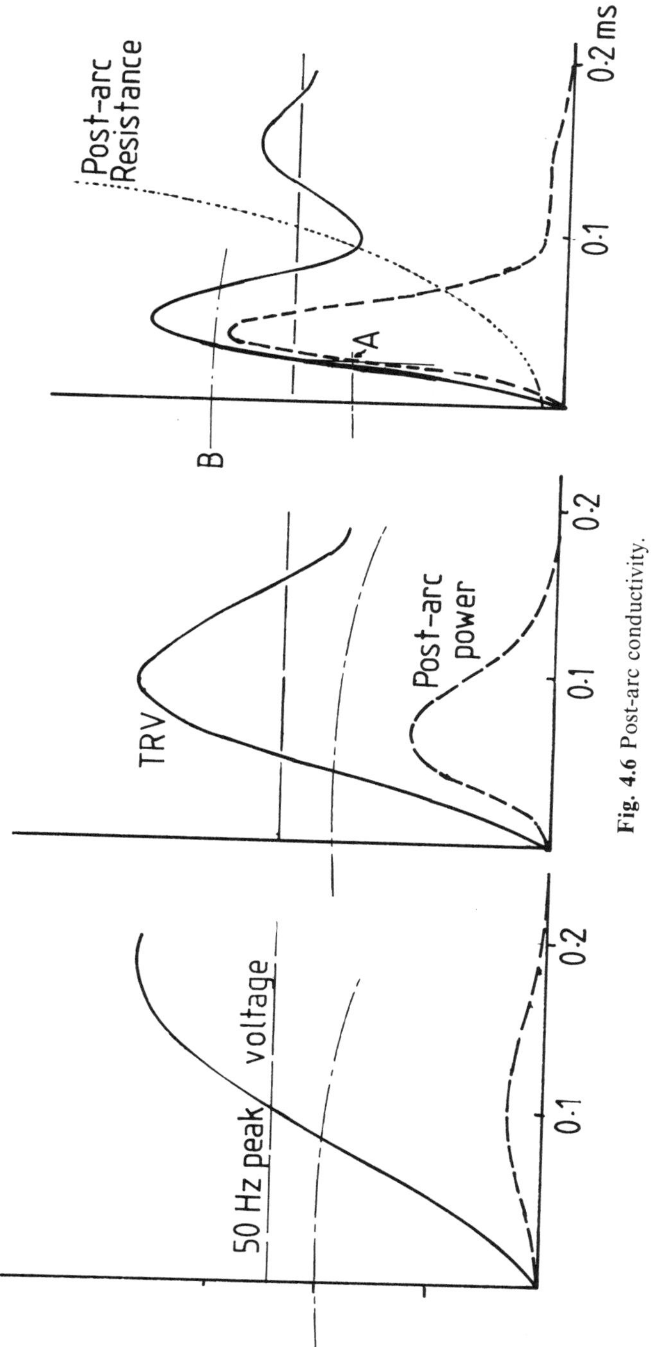

Fig. 4.6 Post-arc conductivity.

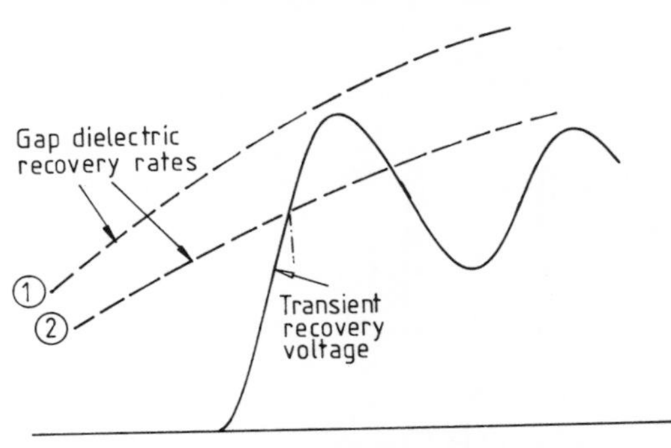

Fig. 4.7 Spark breakdown.

following contact separation, arrange for a very strong cooling action in the contact gap region and/or arrange for the replacement of the semiconducting gas by a clean, insulating medium at current zero. As the time scale is measured in microseconds, this is not an easy task! How it is achieved in typical circuit-breakers, using various arc interrupting media, will now be discussed.

OIL CIRCUIT-BREAKERS

Although the earliest switching devices employed air as the extinguishing medium, it was not long before a better dielectric was needed and insulating oil soon appeared on the switchgear scene. Initially it was used in small quantities just surrounding the contacts. Figure 4.8 shows a 500 A, 2000 V oil switch designed by Ferranti in 1885. As larger powers and voltages were required, the oil was used also as an insulating medium in an earthed metal tank and the bulk oil circuit-breaker was born.

Initially, the contacts were parted in the oil bath and the general movement of the oil, the cooling effect of the hydrogen created from the oil by the arc, and the distance which separated the contacts, were the factors that cooled the arc path and created the dielectric strength needed to interrupt the relatively small fault currents in the early days of power distribution. As fault currents increased, the increase in pressure due to the vaporisation of the oil by the short-circuit current increased the deionising properties of the hydrogen bubble surrounding the arc path, and increased its dielectric strength, thus preserving the breaking capacity of the device. Obviously care had to be taken to ensure the mechanical strength of the container was sufficient to withstand the higher pressures as circuit-breaker ratings increased, and the proportioning of the air space above the oil became a critical design parameter.

Eventually a point was reached when the requirement to restrain the pressure with a fully rated three-phase fault was incompatible with the requirements for reliable and consistent interruption of single-phase fault currents of moderate value, and more sophisticated measures had to be taken. Research work carried out at the Electrical Research Association in the UK led to the development of the side-vented 'explosion pot' or the 'cross jet pot', which forms the basis of just

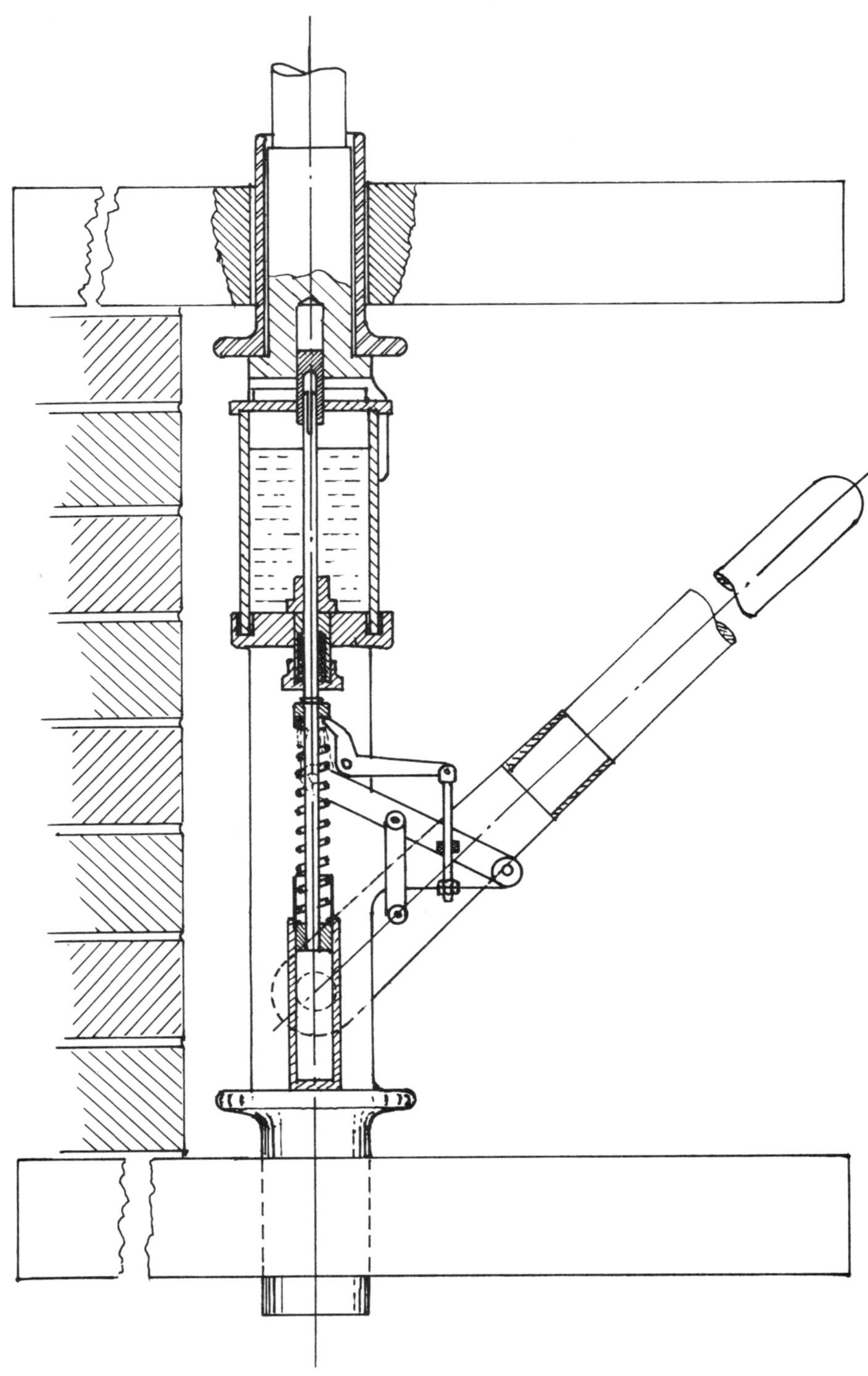

Fig. 4.8 Extract from British Patent 13091/1895: the first oil-break switch: Ferranti 500 A, 2000 V.

about every arc control device used in oil circuit-breakers today.

This development took place towards the end of the 1920s and was followed shortly afterwards by the installation of high power testing stations so that the fault-breaking capabilities of circuit-breaking devices could be scientifically studied. This aspect of switchgear development is dealt with in detail in Chapter 18.

Also at this time the development of oil circuit-breaking techniques took quite different paths in the UK and in Europe. In the UK the metalclad bulk oil design was refined by the use of the side-vented arc control device, whilst in Europe the 'minimum oil' type of circuit-breaker was introduced. This used solid insulation as the oil container and introduced phase segregation, each phase being contained in a separate, cylindrical, insulating housing. Within that housing were contacts and arc control devices very similar to the ERA pattern.

The original idea was to limit the fire and damage caused if a circuit-breaker failed to clear a fault, usually because it was beyond its rating, but anyone who has seen a bulk oil circuit-breaker and a minimum oil circuit-breaker fail on a short-circuit test plant knows that this is erroneous! This type of circuit-breaker is still produced in large numbers, but now using modern high strength insulating materials for the housings. The earliest models had porcelain bodies.

Figure 4.9 is a cross section drawing of a typical side-vented arc control device on which is superimposed the arc length characteristic related to fault current. The chamber at the top, which contains the fixed contact, is usually made of metal and is closed, except for a small hole at the top to make sure that no air is trapped inside when the circuit-breaker is being filled with oil. The principle of operation of this device is that the pressure developed by the vaporisation and dissociation of the oil is retained in the pot by withdrawing the moving contact through a stack of insulating plates having a minimum radial clearance round the contact. Thus there is practically no release of this pressure until the moving contact uncovers one of the side vents, created by cutting a slot in one of the plates.

The compressed hydrogen gas can then escape across the arc path, exerting a powerful cooling action on the ionised column as it does so. When the current zero is reached, the post-arc resistance increases rapidly due to this cooling action and clearance occurs. At currents below the maximum rating of the circuit-breaker, the cooling action is less vigorous, but also the degree of ionisation is less, so clearance is still achieved. It may take a little longer to develop the necessary pressure, however, and clearance may be delayed to a later current zero. Design of an effective, wide range, arc control device requires a careful balance of vent areas, vent spacing and contact speed to ensure consistent performance over the full range of currents.

Figure 4.9 illustrates one of the refinements of the design to improve the interruption of small currents, in that a supplementary oil chamber is shown below the side vents. This is known as a *compensating chamber* and provides a fresh source of oil to be vaporised to feed more clean gas back across the arc path when clearing low currents, such as the load current of the circuit-breaker. The arc length characteristic shows that this feature is brought into use at currents

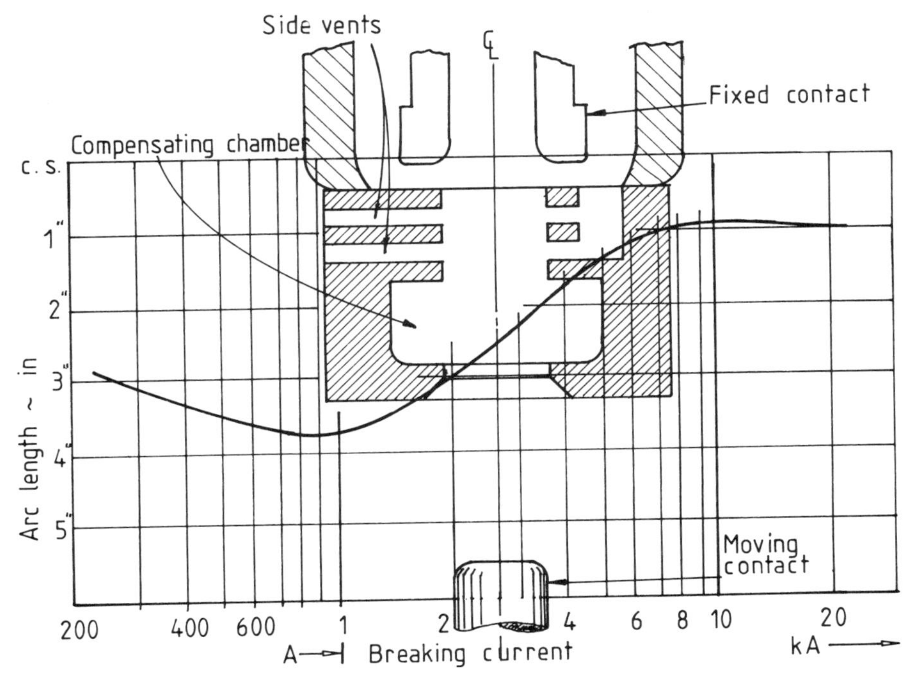

Fig. 4.9 Oil circuit-breaker arc characteristic.

below about 4 kA.

One of the self-compensating features of such a device at the highest fault currents is the fact that the internal pressure, which can reach 30–40 bar, acts on the moving contact to increase its speed and thus open the side vents sooner. As the diagram shows, once the arc has been extinguished, the moving contact leaves a respectable gap in clean oil before coming to rest. One of the desirable features of such a device is that the arc shall always be extinguished before the contact leaves the 'pot'. Other variations include the addition of small vents in the side of the top chamber, situated so that they relieve excess pressure at full short-circuit currents, but do not release too much pressure when interrupting light currents.

As with all types of circuit-breaker which use the power of the arc to provide the means for its own extinction, there exists a 'critical zone' in the breaking current range. In Fig. 4.9 this occurs at currents in the 900 A region. This arises when the range of operation of the arc control device is being extended. As already indicated, steps have to be taken to cater for the very high pressures that occur at 100% fault currents, yet at low currents the design has to ensure the provision of adequate arc extinguishing effort. The art of the design is to make sure that there is not a 'blind spot' along the way between these two extremes. International standards that lay down high power test procedures recognise this problem by ensuring that tests are arranged to discover the point of maximum arc length and prove the performance accordingly.

Bulk oil circuit-breakers usually have a metal top dome, through which six

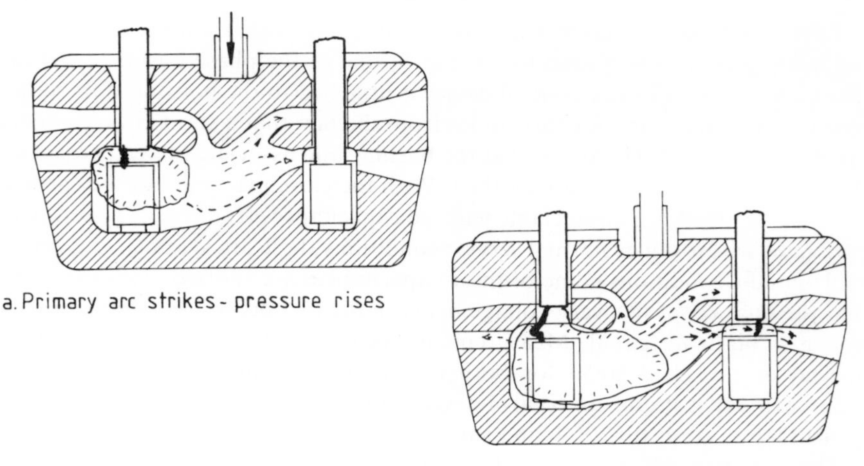

a. Primary arc strikes - pressure rises
b. Pressurised oil flow attacks secondary arc

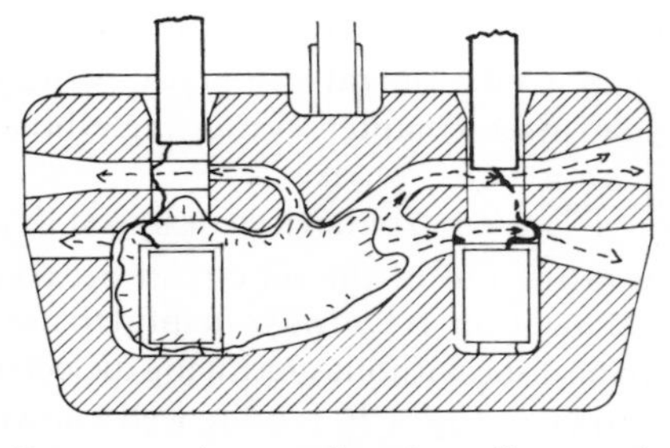

c. Near current zero oil and gas flows cool arcs

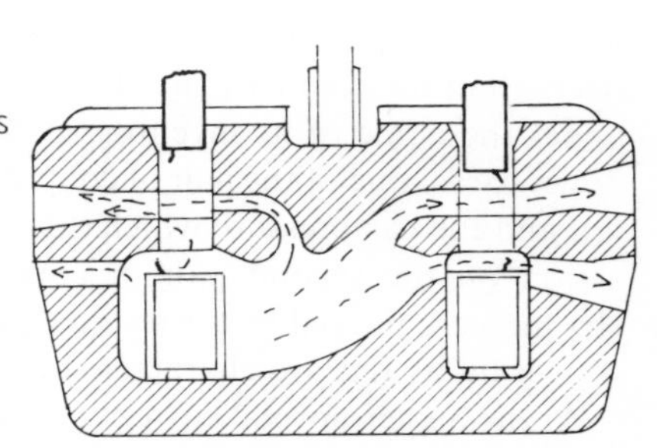

d. Arc trap purged as pressure dissipated

Fig. 4.10 The operation of the Caton arc trap.

bushings pass to connect into the system, and the arc control devices, generally of the form illustrated, are attached to the bottom ends of those bushings. The moving contacts are then arranged on a 'bridge', which in turn is fastened to an insulated drive rod. A removable tank, containing the insulating oil (complying with BS 148) is bolted to the dome. Earlier designs had two sets of contacts and arc control devices in series in each phase, and this type of oil circuit-breaker (o.c.b.) was referred to as 'double break'. However, consideration of the voltage division between these two breaks during the recovery period led to some manufacturers producing single-break designs, particularly at the lower normal current and short-circuit current ratings.

These o.c.b.s are generally the same as the double-break variety, except that the moving contact is hinged to the end of one of the phase bushings and the fixed contact with its arc control device is attached to the bottom of the other bushing. The argument for this comes from the fact that a short circuit usually involves earth, and this means that the bushing and associated fixed contact on the fault side of the o.c.b. are earthed. With the close proximity of the earthed metal tank, there are appreciable leakage capacitances from the conductors to earth,and whilst under normal service conditions these are symmetrical, under short-circuit conditions some of these capacitances are shorted, with the result that the division of recovery voltage between the two breaks of a double-break design is far from equal. Measurements have put this ratio nearer to 80:20 instead of the ideal 50:50 for both devices to be equally effective. So the argument is that if one arc control device can do 80% of the work, it is not too difficult to make it do 100% and save some space and cost as a result.

The only possible weakness of this argument is under capacitance current switching conditions. The leakage capacitance conditions in the o.c.b. are normal when performing this switching duty, so two breaks will give nearly twice the voltage strength as one, for the same gap. Under the rather onerous voltage conditions when switching capacitance loads (see Chapter 2) the single-break breaker is likely to be more prone to re-ignitions or restrikes.

Many and varied have been the ingenious arc control devices designed over the years to try to extract the maximum rating out of the oil circuit-breaker, and one of the most effective deserves a brief mention. It is always helpful to use a pressure generating auxiliary arc, if this can be done without sacrificing dielectric strength in the contact zone. The device shown in Fig. 4.10 is quite well known, under the name of its inventor, as the *Caton arc trap*. When the device moves downwards, one of the pair of contacts, which are in series, parts before the other (a), and the gas pressure which that arc generates creates a strong extinguishing action on the second pair of contacts (b). After arc extinction (c), the gas channels are purged by the residual pressure (d). The effectiveness of the device is shown by its flat arc characteristic.

At the higher normal current or short-circuit current ratings, the complication of the hinge joint in the contact system removes the economies of the single-break design, and heavy duty o.c.b.s are usually double-break. Figures 4.11–4.13 show typical modern bulk oil circuit-breakers.

THE MINIMUM OIL CIRCUIT-BREAKER

As mentioned earlier, since about 1930 the minimum oil circuit-breaker has become the common design of oil circuit-breaker in Europe and is still made in substantial quantities for use in distribution systems. The arc extinction principles are the same as those used in the bulk oil circuit-breaker, baffle plates being placed into an insulated structure which surrounds the contacts and controls the flow of gas in relation to the arc column. As each phase is in a separate, mainly cylindrical housing, the physical layout is designed to suit the

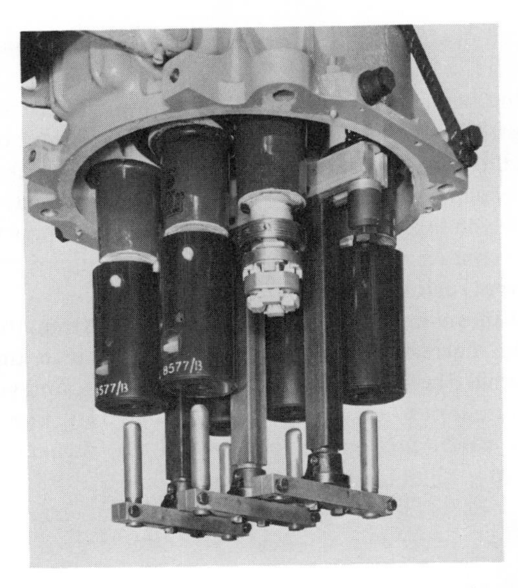

Fig. 4.11 Typical bulk oil circuit-breaker. (*Courtesy GEC Distribution Switchgear Ltd.*)

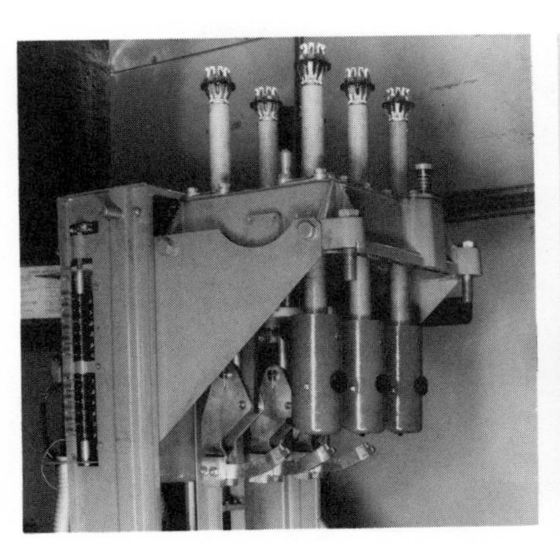

Fig. 4.12 Typical single-break oil circuit-breaker (*Courtesy NEI-Reyrolle Distribution Ltd.*)

Fig. 4.13 Bulk oil circuit-breaker with arc trap. (*Courtesy Yorkshire Switchgear Ltd.*)

geometry of the construction and therefore differs in detail from the bulk oil approach.

One essential difference is that the resultant circuit-breaker has its terminals more easily accessible from the side than from the top, and this leads to considerable differences in the associated cubicle design when used in metal-enclosed switchgear. It also means that all minimum oil circuit-breaker designs are single break. Several ingenious designs of transfer contact for the moving contact have been developed over the years, some of which are mentioned in Chapter 8.

The drive to the moving contact is usually a simple crank mechanism with a drive shaft being taken through the housing, sometimes above oil level, but more often below, requiring oil seals. By dividing the tubular body into two compartments, usually in the area of the transfer contact, it can be made convenient for the change in oil volume of the lower compartment due to the movement of the contact rod into it to result in the pumping of the displaced oil into the arc control region.

This will assist in the rate of growth of the dielectric strength of the gap when interrupting small currents and improve performance in that area. This is particularly valuable when switching capacitive currents, and goes a long way to offset the disadvantage of single-break o.c.b.s for that duty. Figures 4.14 and 4.15 show typical modern designs of minimum oil circuit-breaker.

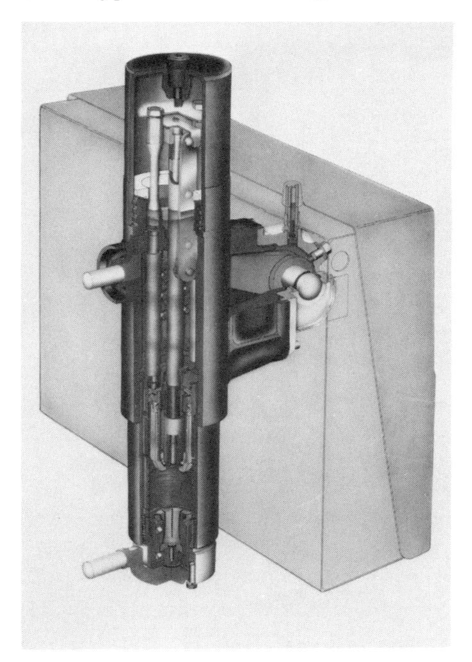

Fig. 4.14 Typical minimum oil circuit-breaker. (*Courtesy Calor-Emag Elektrizitäts-Aktiengesellschaft.*)

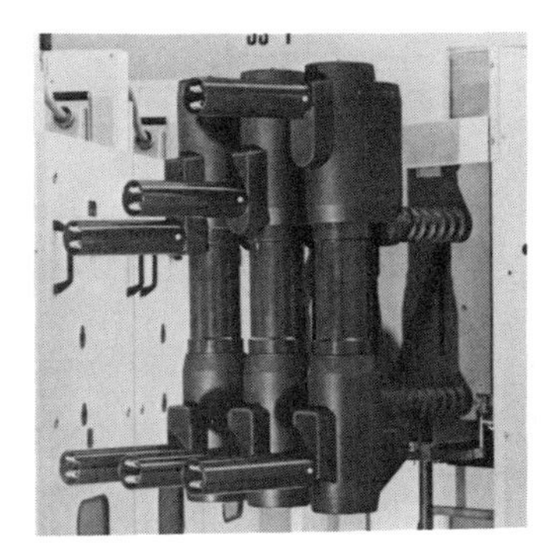

Fig. 4.15 Traditional minimum oil circuit-breaker. (*Courtesy AEG-Telefunken.*)

AIR CIRCUIT-BREAKERS

In Europe the oil circuit-breaker developments described above led to the almost total replacement of the earlier designs of air circuit-breaker for service in medium voltage distribution systems by oil circuit-breakers in many countries. However, in some areas, notably France and Italy, the air circuit-breaker was developed and used for systems up to 15 kV, but in general its use is restricted to low voltage applications or high security installations where the risk of an oil fire or oil contamination of the environment was too high to be tolerated. Certainly that has been the case in the UK.

Countries following American practice, however, have used air circuit-breakers almost exclusively for systems up to 15 kV until the coming of the new technologies of vacuum and SF_6.

The air circuit-breaker follows rather different principles of arc interruption from any other type of circuit-breaker. While the objective is the same, i.e. to prevent the resumption of arcing after current zero by creating a situation in which the contact gap will withstand the system recovery voltage, the air circuit-breaker does this by creating an arc voltage in excess of the supply voltage.

This can be done in three ways:

1. Intense cooling of the arc plasma, so that the voltage gradient is very high.
2. Lengthening the arc path to increase the arc voltage (see Fig. 4.16(c)).
3. Splitting up the arc into a number of series arcs (see Fig. 4.16(a)).

The first objective can be achieved by forcing the arc into contact with as large an area as possible of cool insulating material. All air circuit-breakers are fitted with chambers surrounding the contact and arc zone, usually called the 'arc chute', because the arc is driven into and through it. If the inside is suitably shaped, and the arc can be made to conform to the shaping, cooling from the arc chute walls can be achieved. This type of arc chute needs to be made from some kind of refractory material, asbestos compounds once being a favourite choice. With the understanding, today, of the health hazards associated with the use of asbestos, other materials are brought into use, such as high temperature plastics reinforced with glass fibre, and ceramics.

The second objective can be achieved concurrently with the first, if the walls of the arc chute are shaped so that the arc is not only forced into close proximity with them, but also driven into a serpentine channel. The lengthening of the arc and the simultaneous increase in the voltage drop per unit length soon lead to a high arc voltage, and a high arc resistance. This has the important consequence of substantially changing the system power-factor so that the instantaneous value of the supply voltage as the current approaches a zero is much below its peak value.

Therefore, as the value of the voltage required to maintain the arc is being increased, a simultaneous reduction in the value of the voltage in the system which is available to maintain the arc is taking place. Finally, the high value of resistance represented by the arc so influences the damping of the TRV oscillation that the amplitude factor is reduced almost to unity.

The preceding paragraphs explain what is wanted, and how to set about achieving it, but a heavy current arc is an intractable thing and one of its main tendencies is to keep itself as short as possible as that helps to maintain its ionisation level, and hence its current-carrying capability. So, whilst it is not difficult to design convoluted walls to arc chutes, it is not easy to persuade the arc to enter them! The usual arrangement of an air circuit-breaker is to arrange the

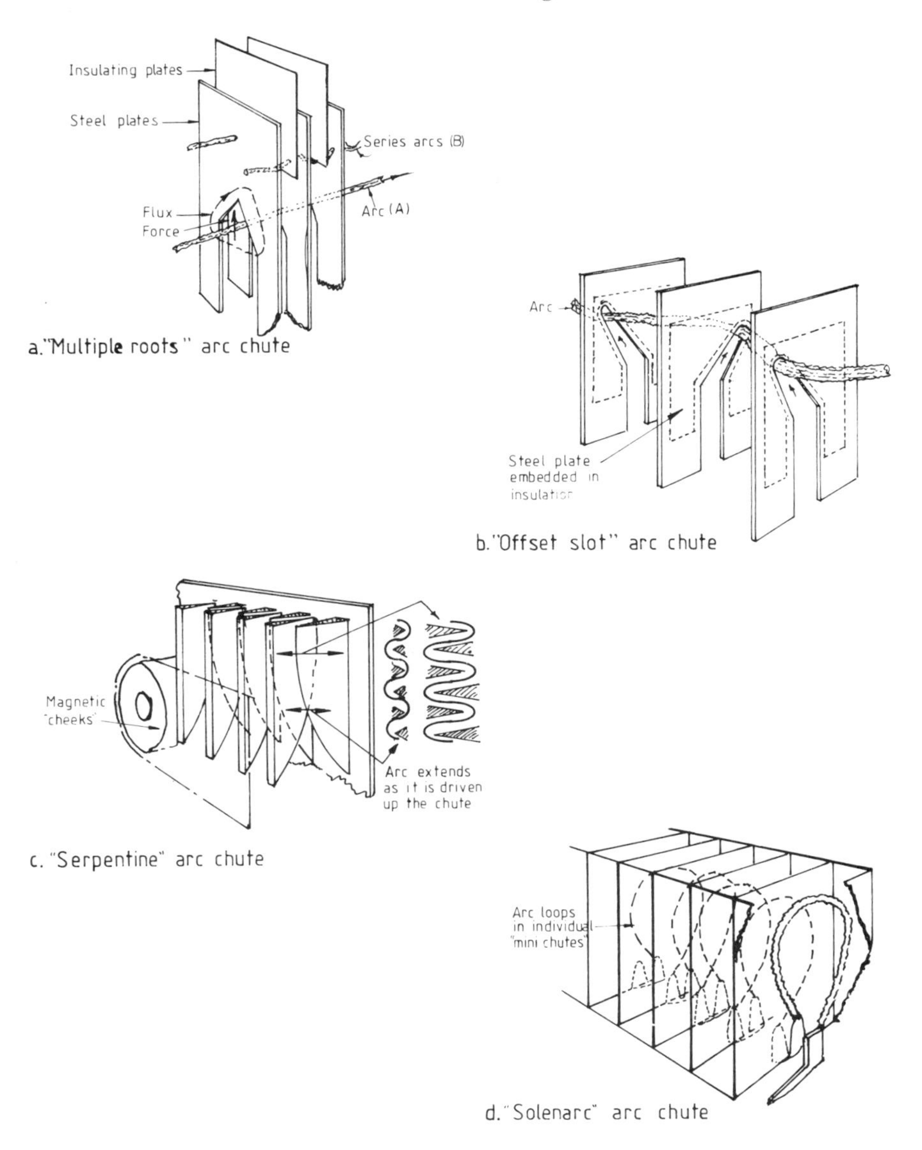

Fig. 4.16 Air circuit-breaker arc chutes – cold cathode and serpentine arc.

arc chute above the contacts, and arrange for the contacts and the connections leading to them to form a tight loop so that a magnetic field is produced within the contact gap acting upon the arc in such a manner as to drive it up into the chute.

The heat of the arc will also soon create a current of convected air which will also play a small part in this activity. But, at the higher distribution voltages, the effort available from these sources will not be sufficient to force the arc to extend itself. The usual method adopted in order to augment the force available is to arrange iron 'cheeks' on the outside of each single-phase arc chute, which are magnetised by a coil built into the structure for that purpose. This intensifies the field within the chute and assists the interruption process. This is also, clearly, a self-compensating system, in that the heavier the current, the stronger the field and therefore the greater force to overcome the arc's desire to stay short.

The use of the magnetic circuit complicates the contact structure, as it is not acceptable to retain the coils in circuit continuously and arrangements have to be made to insert them into the circuit during the opening of the circuit-breaker. This is usually done by splitting the arc runner, a metal strip that runs up from each contact to guide the arc into the required path, so that the coil is automatically inserted into the current circuit as the arc runs up towards the chute.

Another type of chute, which extends the arc into offset slots and uses the magnetic effect of induced flux in embedded iron plates to force the current into the slots, is illustrated in Fig. 4.16(b).

The heavier the fault current, the more effective now will be the air circuit-breaker, until it either becomes so effective that the arc runs straight through the arc chute and re-establishes itself as a nice short arc outside the chute, or the pressure inside the chute below the plates, created by the arc, becomes so great that the arc is prevented from rising into the plates. In either case, this obviously constitutes the upper limit of breaking capacity for a given arc chute, which needs to be a respectable margin above the designed rating.

At the other end of the scale, there may still be a current where the magnetic field is insufficient to move the arc into the chute with the alacrity that the designer is looking for. Left to its own devices the air circuit-breaker may have an arc duration of tens of current loops at currents of a few hundred amperes. This problem is overcome by fitting a small air piston below the arc chute which is driven by the moving contact system and forces a jet of air against the arc and drives it into the chute so that it is quickly extinguished. Figure 4.16 illustrates the features described above.

The penalty paid for the above method of arc extinction is the high energy that is developed in the arc chute by the combination of high current and high voltage, virtually in phase with one another. When interrupting maximum fault current, an air circuit-breaker rated typically at 40 kA and 12 kV would emit an impressive fire ball from its arc chute, if a large cooling baffle was not fitted to contain it. The result is that air circuit-breakers of heavy ratings are large, heavy and expensive pieces of plant.

They do, however, have advantages for certain applications, particularly where the system switching parameters are severe, as the high arc resistance makes the performance of the air circuit-breaker virtually independent of the circuit in which it is connected. Figure 4.17 is an oscillogram of a fault interruption by such a circuit-breaker and illustrates how the power-factor is changed and the TRV greatly modified as a result of the circuit-breaker arcing characteristics.

However, the use of the third technique can make for less energy generation by

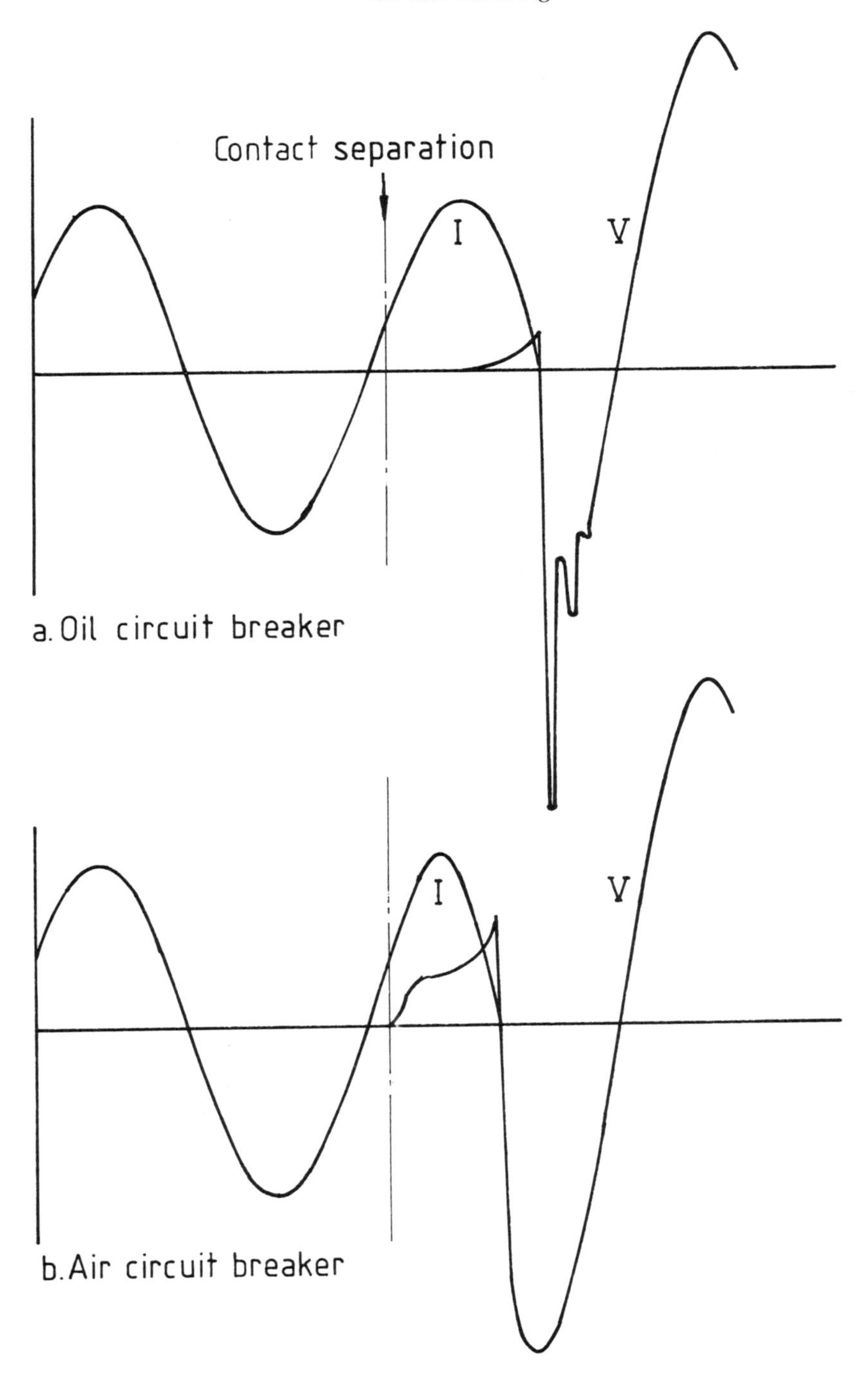

Fig. 4.17 Oscillogram of fault clearance by a magnetic air circuit-breaker.

using metal arc splitter plates in the arc chute, so that not only is the voltage gradient increased, but also a large number of anode and cathode voltage drops are introduced. A combination technique has been designed which uses an assembly of small arc chutes individually of the 'cooling and lengthening' type, but which are so arranged that the initial arc is split into a number of series arcs, each in its own mini arc chute (see Fig. 4.16(d)).

This is a more complicated design but one which is suited to the higher distribution voltages. It has the advantage, also, of not requiring the provision of extra components for the creation of a magnetic field in order to drive the arc into the chute. (See Figs. 4.18 and 4.19.)

In summary, the art of designing an effective air circuit-breaker requires the correct combination of chute shape, magnetic field development and air puffer to give a uniform arc extinction performance over the whole current range. The parameters that have to be balanced are many, including some not yet mentioned such as arc chute height, length and outlet area. The mechanical forces developed in a heavy duty air circuit-breaker are also considerable. Figure 4.20 is a typical British design of air circuit-breaker.

A final point concerning the characteristics of the air circuit-breaker is that the interrupting technique described is the only one that does not really depend on the presence of a current zero. If the arc voltage can be made to exceed the supply voltage, the arc will be extinguished whether the current is a.c. or d.c., so that it is suitable for the control and protection of d.c. circuits. Figure 4.21 shows the current and voltage waveforms when an air circuit-breaker interrupts a d.c. circuit containing inductance.

Fig. 4.18 Multi-chute 'Solenarc' air circuit-breaker. (*Courtesy Merlin Gerin.*)

Fig. 4.19 'Belledonne' air circuit-breaker switchboard. (*Courtesy Merlin Gerin.*)

Fig. 4.20 British magnetic air circuit-breaker. (*Courtesy GEC Installation Equipment Ltd.*)

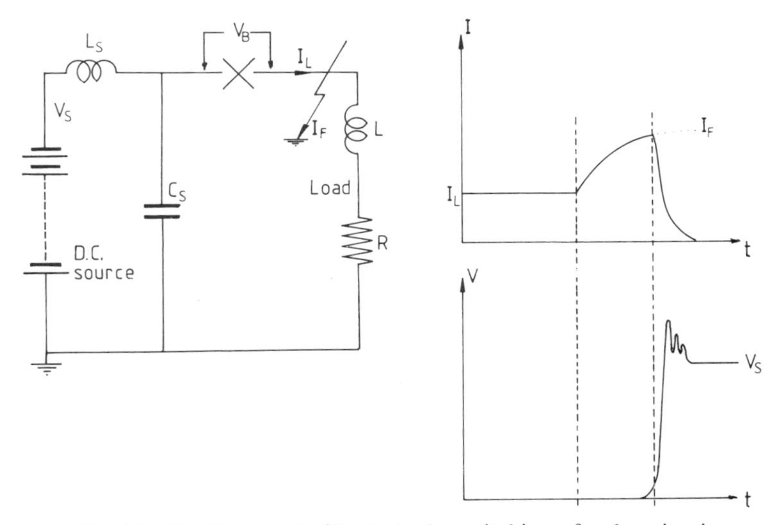

Fig. 4.21 Oscillogram to illustrate the switching of a d.c. circuit.

SF_6 CIRCUIT-BREAKERS

Historically, the SF_6 circuit-breaker was first introduced at transmission voltages as a simpler alternative to the air blast circuit-breakers that were used at those voltages in the 1960s. The original circuit-breakers, designed by the American Westinghouse Co., were of the two-pressure type. In this type the contacts and arc control devices were immersed in a metal tank full of SF_6, which served the dual purpose of an insulating and arc extinguishing medium. The gas used for arc extinction was drawn out of the tank, compressed and pumped into a separate small reservoir. When the circuit-breaker was opened, a valve discharged the compressed gas through nozzles round the contacts and extinguished the arc.

After a few years of development work, high voltage circuit-breakers began to appear in which the compression of the gas took place as the contacts opened, the energy required being supplied by the operating mechanism. Soon after that, in the late 1960s, the same principle began to be applied to distribution circuit-breakers, the French manufacturers being in the forefront of this development.

Since that time, more development work has been done and there are now a number of designs available in the world, and other interrupting principles have been introduced. This has been necessary partly because of the patent activities of the earlier workers, and partly because new ways of applying this remarkable gas have been discovered.

SF_6 'PUFFER TYPE' CIRCUIT-BREAKERS

The first technique to be applied to SF_6 circuit-breakers for distribution systems used the self-generated gas blast principle, usually referred to as the 'puffer' system. The contact system is arranged so that the action of opening the circuit-breaker starts to compress a quantity of SF_6 gas before the arcing contacts are opened. When the contacts do part, an arc is drawn and when the current zero is reached the flow of gas rapidly cools the remaining plasma and sweeps away the arc products, leading to a very rapid increase in the dielectric strength of the gap and a successful clearance. Figure 4.22 shows a section through a typical puffer type SF_6 circuit-breaker and Fig. 4.23 shows this circuit-breaker in three different positions during the opening stroke, to illustrate the puffer action.

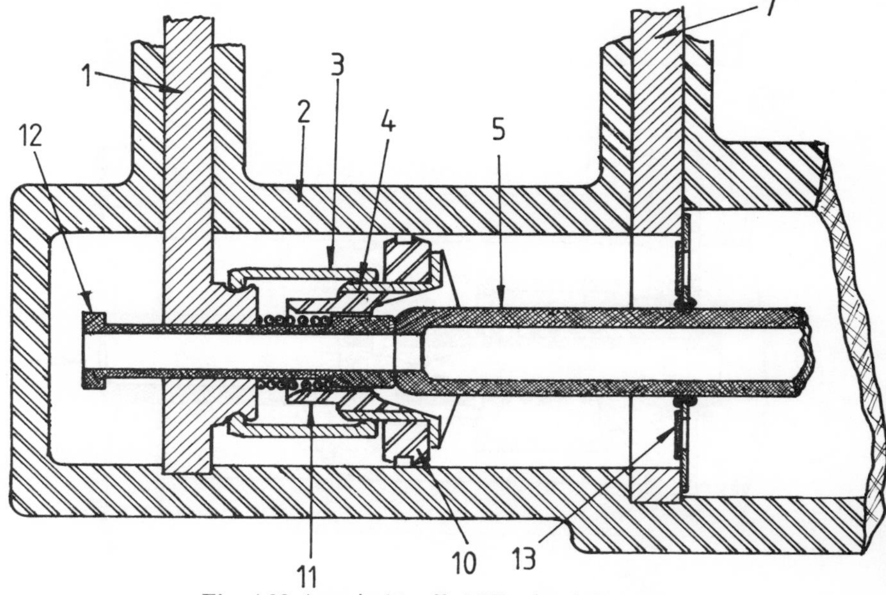

Fig. 4.22 A typical 'puffer' SF_6 circuit-breaker.

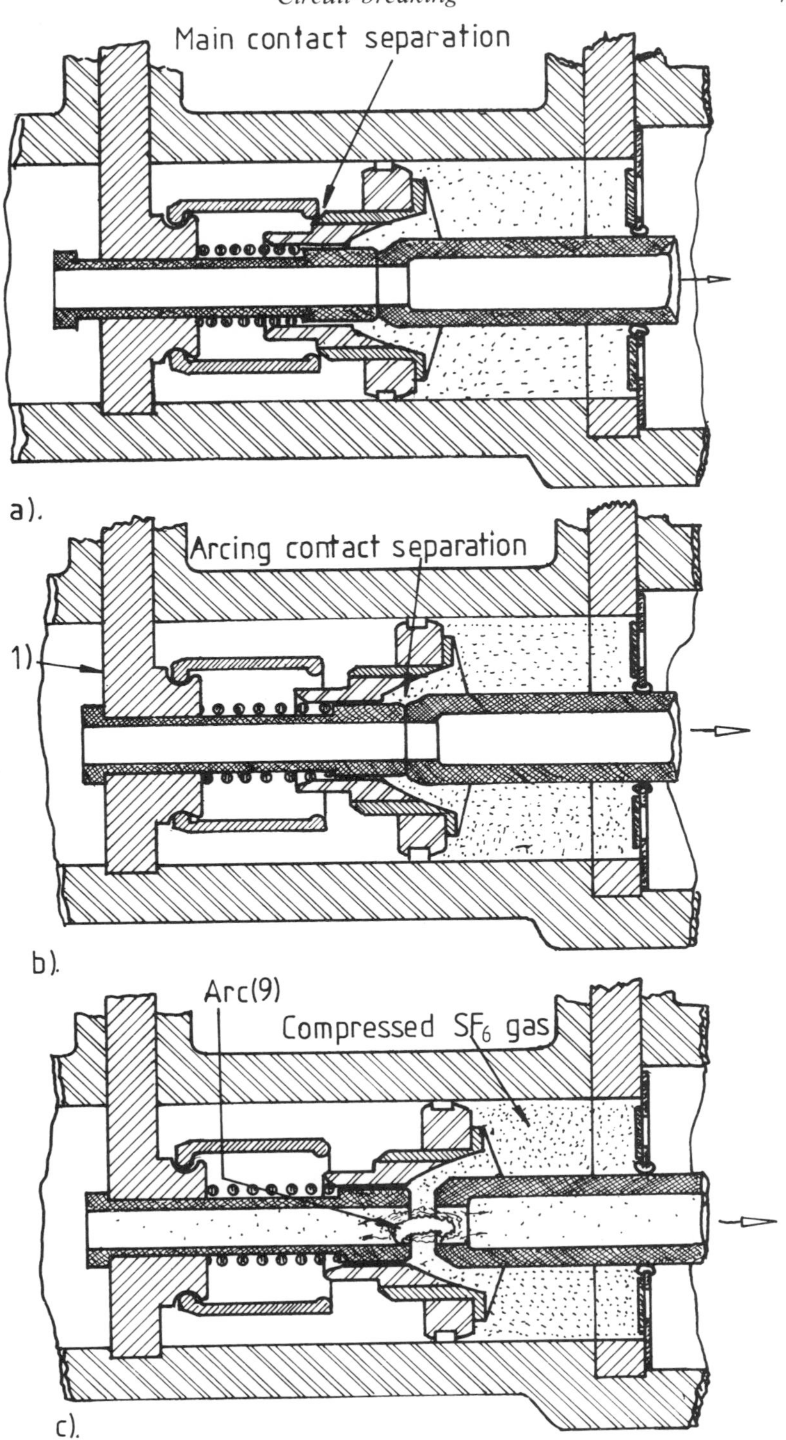

Fig. 4.23 Opening sequence of a 'puffer' SF_6 circuit-breaker.

Figure 4.22 indicates the circuit-breaker body (2) which is totally gas-tight, and through the walls of which two conductors (1) and (7) carry the current into and out of the circuit-breaker. When the circuit-breaker is closed, as shown, the majority of the current is carried through the main contacts, fixed (3) and moving (4). The moving contact (4) is part of a structure which also carries the moving arcing contact (5), the gas guide (11) and the piston (10).

The inside of the circuit-breaker contains SF_6 gas at a pressure between 0.5 bar and 1.5 bar (gauge) depending on the breaking current rating. When the circuit-breaker begins to open, the contact system starts to move to the right (Fig. 4.23(a)), and the gas trapped to the right of the piston begins to rise in pressure. The only escape for the compressed gas at this time is the very small annular space between the gas guide and the fixed arcing contact (6), which is able to follow the moving contact under the pressure of the spring shown, until the collar at the left (12), strikes the fixed contact support (1) (Fig. 4.23(b)). Before this happens a gap has been established between the main contacts, and all the current is now flowing through the arcing contacts.

As soon as the fixed arcing contact reaches its stop, an arc commences to burn (9), as indicated in Fig. 4.23 (c). The moving contact system continues to pressurise the gas, which can now escape through the space between the arcing contacts and exhaust into the lower pressure areas to the left and right of the contact area. At high fault levels, the pressure in the nozzle area between the arcing contacts can rise to a level which will prevent the gas from flowing through the exhaust tubes. This has the favourable effect of conserving the gas until it can be useful, at the current zero period, and increasing the pressure of the gas, as energy continues to be fed in from the external mechanism which is driving the moving contact structure. This nozzle-blocking action is a feature of puffer circuit-breakers.

As the current zero approaches and the ionisation of the plasma reduces, as described earlier in this chapter, gas flow is re-established and clearance will occur, unless the contacts opened so close to a current zero that the gap is insufficient to withstand the TRV. In that case one of the other phases, in a three-phase circuit-breaker, will be the first phase to clear. The minimum arc duration for this type of SF_6 circuit-breaker in the first phase to clear is about 3 ms. The other phases will follow at their next current zeros, as discussed in Chapter 2.

After current flow has ceased, the moving contacts continue to open until the piston is practically at the end of the cylinder creating a substantial gap between the contacts in clean SF_6 to withstand any voltage that the system is likely to produce. At the end of the cylinder is a valve plate (13) which lifts when the circuit-breaker is closed again and allows virtually unrestricted gas flow back into the space under the piston. Figure 4.24 illustrates a circuit-breaker employing the contact system just described.

Other designs, which behave in the same way, introduce variations on this theme. Some only exhaust in one direction, a 'single-blast' version of the puffer, and other arrangements move the cylinder over a fixed piston, but the effect is the same. Figure 4.25 is a section through a continental design which operates in a similar way to the UK design.

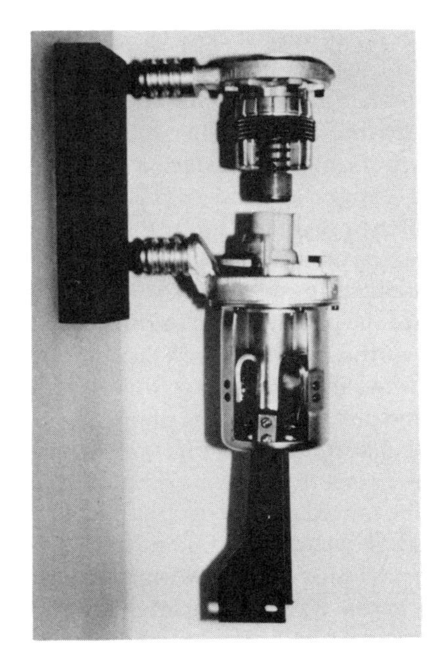

Fig. 4.24 The internal parts of a 'puffer' type SF_6 circuit-breaker. (*Courtesy Yorkshire Switchgear Ltd.*)

Fig. 4.25 Another type of 'puffer' circuit-breaker. (*Courtesy Merlin-Gerin.*)

SELF-EXTINGUISHING SF_6 CIRCUIT-BREAKER

In the 'competition' between the SF_6 circuit-breaker and the vacuum circuit-breaker, to which reference will be made later, one of the claims made in favour of the latter is the low operating mechanism energy required, because the vacuum interrupter has its interrupting capability 'built-in' by the vacuum. The types of SF_6 circuit-breaker described here and in the next section require less energy to be supplied from external sources as they make more use of the energy available from the arc itself.

The essential aim of any SF_6 circuit-breaker is to achieve relative movement between the gas and the arc. In the puffer type, this is done by compressing the gas from external sources, but the introduction of a magnetic field allows control to be exercised over the arc itself and thus relative movement can take place in other ways.

One design retains the nozzle of the puffer type, but the pressure is raised by thermal means rather than by mechanical energy. Figure 4.26 shows a section through a sealed, gas-filled circuit-breaker which has a fixed contact (3), mounted on the conductor (1), which is surrounded by a coil (2). One end of this coil is connected to the fixed contact support (1), and the other end is connected to an arcing ring (4) which is nominally insulated from the contact support (1). As the circuit-breaker is opened and the moving contact (5) is withdrawn from the fixed contact (3) an arc is created which is quickly transferred to the arcing ring (4) as the moving contact tip passes through it. This connects the coil into the circuit and a magnetic field is created in the contact zone which causes the arc (9) to rotate.

The purpose of rotating the arc in this design of circuit-breaker is to bring the intense heat of the arc column into contact with as much of the gas as possible, thus raising its temperature and, consequently, its pressure. When the current zero approaches, the gas flow will be established through the hollow moving contact and deionisation of the arc column will follow.

As with all circuit-breakers that rely on the energy of the arc for causing its own extinction, the designer faces a problem of balancing the lack of energy at low currents and a possible excess at high currents. The same applies to this design, and evidence of this is seen in the small puffer piston (6) which is used to help low current arcs to be extinguished, in a similar way to that used in airbreak circuit-breakers. The other end of the spectrum will result in excessive ionisation being produced in the arcing zone, with consequent difficulty in clearing away the charge carriers in the few microseconds available after current zero.

However, the arrangement will handle a very useful range of fault currents, and makes for a lighter operating mechanism than in the case of the classical puffer arrangement. Figure 4.27 is another view of the circuit-breaker described.

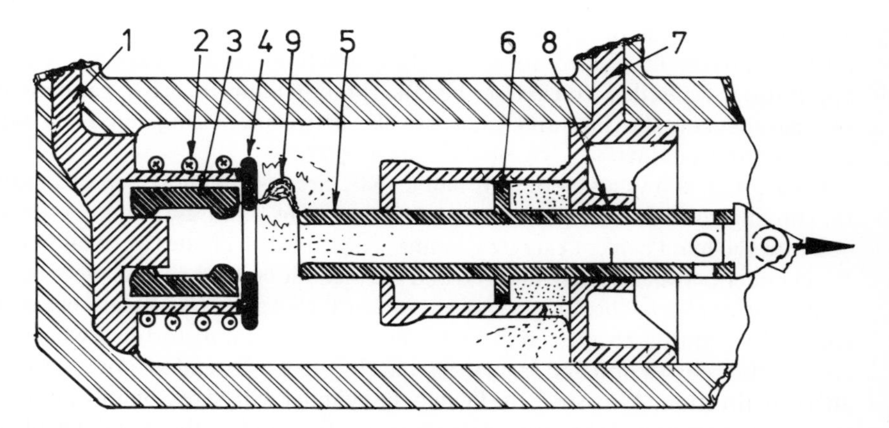

Fig. 4.26 A self-extinguishing SF_6 circuit-breaker.

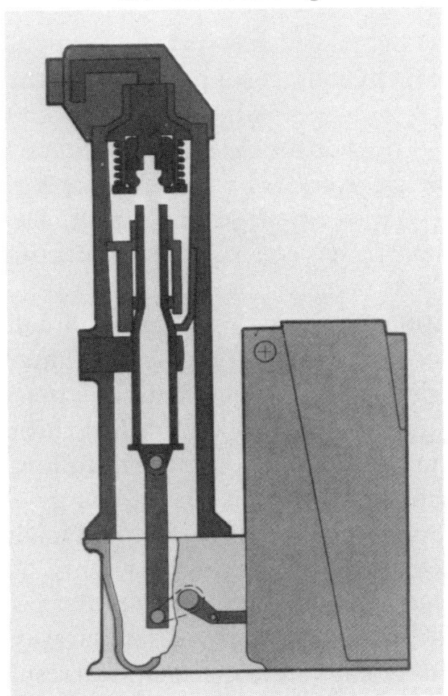

Fig. 4.27 A self-extinguishing circuit-breaker. (*Courtesy Calor-Emag Elektrizitäts-Aktiengellschaft.*)

ROTATING ARC SF_6 CIRCUIT-BREAKER

The same magnetic system that is used in the device just described, to move the arc for the purpose of raising the gas pressure, can also be used to create the relative movement needed to cause arc extinction. Some years ago patent applications were filed which described the rotation of an arc in SF_6 gas between two parallel contact rings and, alternatively, between one ring and a coaxially arranged rod contact.

Figure 4.28 shows a section through an SF_6 circuit-breaker which withdraws the moving contact through an arcing ring (as in Fig. 4.26), and thus connects a coil into the main current carrying circuit. This coil then creates an axial field in the cylinder between the contact rod and the metallic coil former. As the arc path is at right angles to this flux, the usual motor laws apply and the arc begins to rotate. In addition to the axial field there will be local distortions where the current flow passes into and from the arc itself, which will tend to move the arc in an axial direction. Also, since the field in the cylinder will be reasonably uniform, the speed of motion of the arc root on the central rod and on the cylinder wall will tend to be similar, and therefore the inner arc root will rotate faster than the outer one and the arc will tend to take a spiral shape.

High speed cine-photographs of the arc, taken looking in the axial direction during short-circuit tests, confirm this (Fig. 4.29). Figure 4.30 is a sketch of such an arc, showing the direction of the Lorenz forces on the arc when it has taken up its spiral form. It will be seen that these are acting in such a way as to drive those parts of the arc that are almost parallel to the cylindrical wall towards that

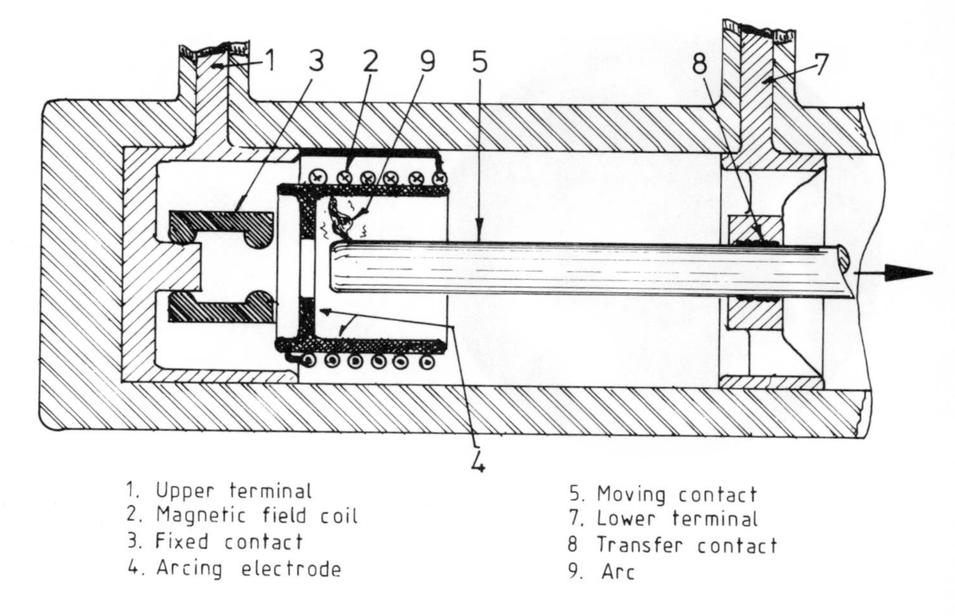

Fig. 4.28 A rotating arc circuit-breaker.

wall. However, there will be induced currents flowing in the arcing tube wall, in the same direction as the current in the arc. Thus there will be a repulsion force between the plasma and the wall. Eventually the balance of forces will allow the arc to touch the wall, and part of the arc will be short-circuited. This is illustrated at F in Fig. 4.30, where the part of the arc between E and F is cut out.

This action has the advantage that it keeps the arc length short and thus keeps down the arc voltage and consequently the arc energy. Also, the amount of ionised gas in the arcing zone is reduced. A possible drawback is a limitation to the voltage levels at which such a device can be rated without making a very large contact system.

An essential requisite in the design of such an SF_6 circuit-breaker is to choose materials and proportion the components in such a way that there is a significant phase shift between the flux and the current ensuring a strong action upon the arc plasma right up to current zero. Otherwise the arc would tend to slow down and reduce the relative motion just at the point when it is most needed. Again, such circuit-breakers need the careful attention to design required by all circuit-breakers that develop their arc extinguishing capability from the current in the arc itself, so that adequate effort is available at low currents without leading to excess forces at the high currents. The required compromise has been achieved satisfactorily by a number of manufacturers who market a range of designs of rotating arc circuit-breakers for voltages up to 36 kV and breaking currents up to 25 kA. Figures 4.31–4.33 illustrate some of these.

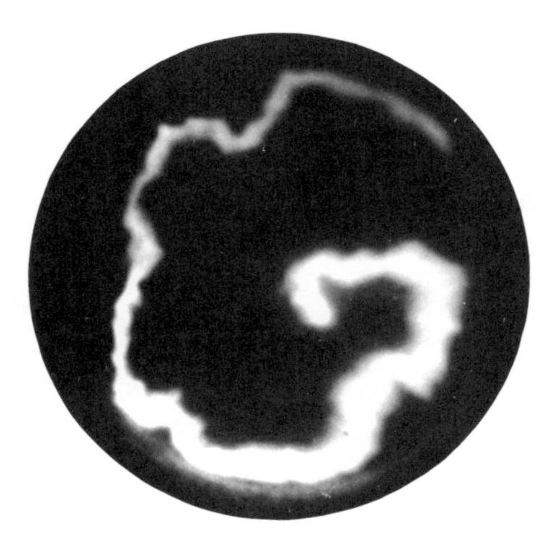

Cylindrical electrode
A
B
Successive arc positions
C
D
E
Contact
F
The arrows indicate direction of Lorenz forces

Fig. 4.29 The spiral arc in a rotating arc circuit-breaker. (*Courtesy South Wales Switchgear Ltd.*)

Fig. 4.30 Showing the magnetic forces on the spiral arc.

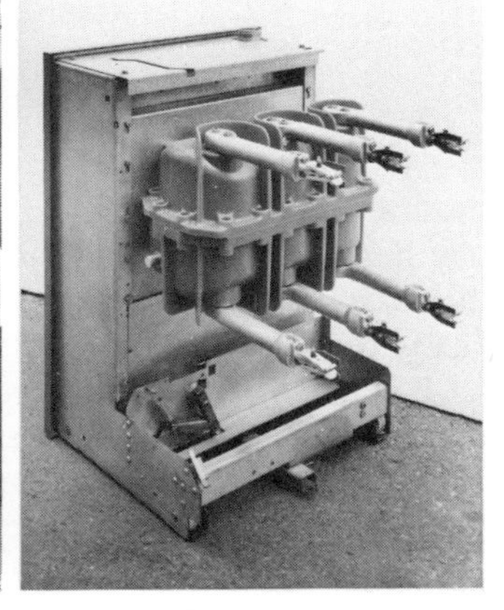

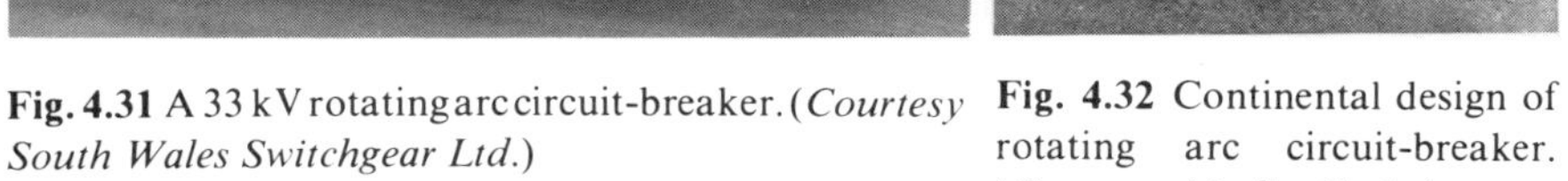

Fig. 4.31 A 33 kV rotating arc circuit-breaker. (*Courtesy South Wales Switchgear Ltd.*)

Fig. 4.32 Continental design of rotating arc circuit-breaker. (*Courtesy Merlin-Gerin.*)

VACUUM CIRCUIT-BREAKERS

With the exception of the SF_6 puffer circuit-breaker, all the circuit-breakers described so far have been dependent on the energy of the arc and its associated current to develop the arc extinguishing effort. Whilst this makes for circuit-breakers that do not need the input of large amounts of energy from the operating mechanisms, it poses a challenge for the designer in finding the best compromise of performance over the wide breaking current range from magnetising currents to heavy short-circuit currents.

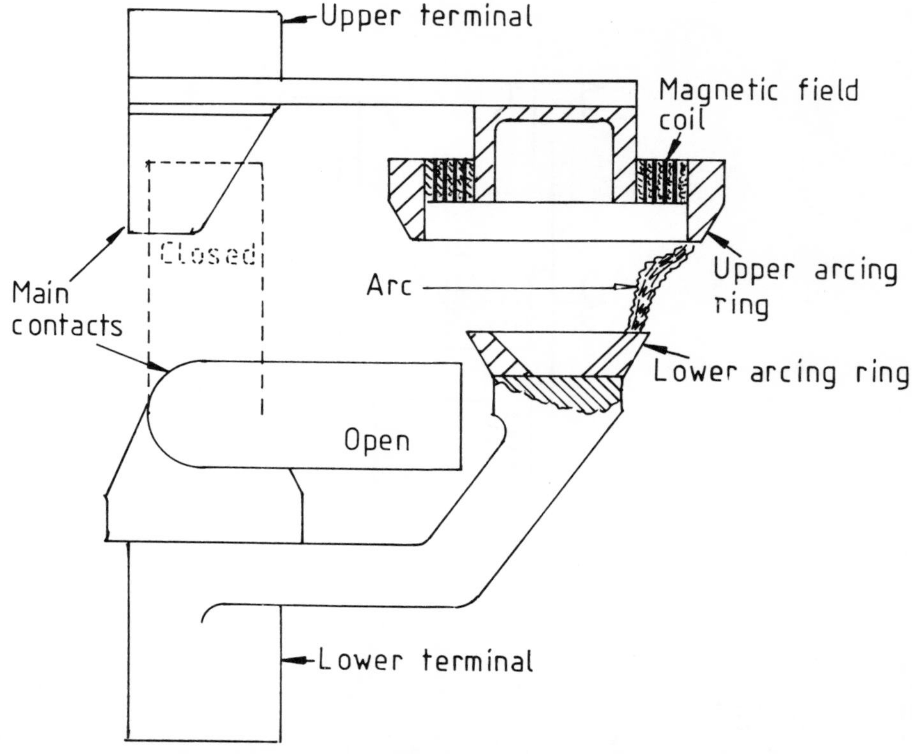

Fig. 4.33 Rotating arc circuit-breaker with parallel ring electrodes.

The vacuum interrupter seems to be the ultimate in switching devices that are independent of the operating system, as its breaking capability is dependent only on the material and geometry of the contact structure and the quality of the vacuum. In practice this is not quite true, as care has to be taken to ensure the correct travel and force characteristics of the operating mechanism to ensure the availability of the potential rating of the interrupter.

Figure 4.34 shows a combined section through typical vacuum interrupters with full insulating envelopes, and Fig. 4.35 is a section through an alternative design which has the central screens exposed. The left-hand half of Fig. 4.34 interrupter resembles the ones made in the UK, whilst the right-hand half is typical of American and Japanese devices. Figure 4.35 illustrates a type of vacuum interrupter made on the Continent.

The earliest attempts to make circuit interrupting devices using contacts in a vacuum date back to the 1920s in America. The conclusion from these early tests was that they were sufficient to display the technological promise of interruption in vacuum, but that it was necessary to await the development of ultra-high-vacuum technology and gas-free metals before sustained performance could be achieved. In the 1950s the Jennings Company in America made a range of specialist switches for high voltage use in such areas as the control of capacitor

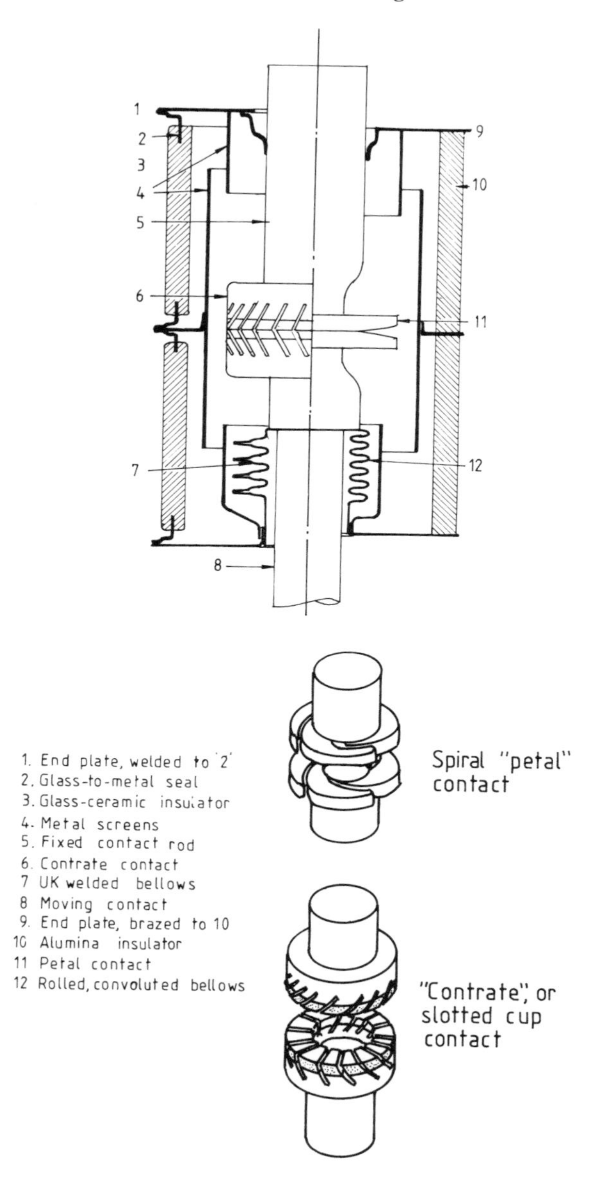

Fig. 4.34 UK and USA vacuum interrupters.

banks. The extremely high rate of dielectric strength recovery made the vacuum switch a natural choice for restrike-free performance when switching capacitance currents at high voltages.

It was not until the 1960s that the technology reached the point at which reliable vacuum interrupters capable of respectable short-circuit breaking current ratings could be made economically. Even today, the manufacture of vacuum interrupters requires sophisticated manufacturing techniques and specially processed materials which lead to the interrupter being an expensive article. To some extent the rest of the circuit-breaker can be simpler than some of the other types, but the interrupters in the average circuit-breaker account for about half its cost, depending on rating.

The development of the vacuum interrupters as we see them today took place separately in the UK and America. Most of the other designs in the world owe their origins to one or other of these sources of original work. As the markets on the two sides of the Atlantic vary quite substantially in detail, the objectives of the investigators were somewhat different, and this is reflected in the eventual products, initially brought out by the General Electric Company of America and the forerunners of the Vacuum Interrupter Company of the UK.

The essential requirement for the enclosure for a vacuum interrupter is that it be vacuum-tight(!), a good insulator and contain a system of screens which prevent the condensation of metal vapour on the inside insulating surfaces. The annular clearances between the contacts and the screens also play a part in determining the performance of the interrupter. The problems in making the envelope, apart from using gas-free metals for the internal parts, lie in the area of finding materials with the correct characteristics and suitable for joining in a vacuum-tight manner to insulating components at an economic cost.

For long life they need to be corrosion resistant to a high degree, as nothing must impair the gas-tightness during long years in service, so stainless steels are a

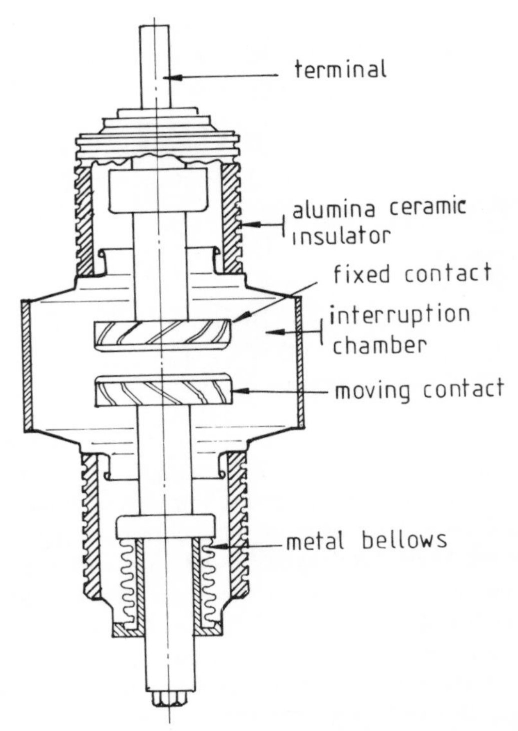

Fig. 4.35 A German type of vacuum interrupter.

natural choice. To ensure that the insulating body is totally impervious to gas diffusion, all the manufacturers use glass or ceramic type materials. The method of making the insulator-to-metal joint is dependent on the choice of insulator material, the glassy materials lending themselves to the melting in of a metal ring in the manner well established in lamp and valve manufacture. The alternative is high alumina ceramic which can be metallised on the end faces and the metal parts are then brazed on.

Once the envelope has been satisfactorily made, the moving contact has to be attached so that it can move freely without putting the vacuum at risk. This requires the use of metallic bellows, as no other technique has the gas-tightness required. Fortunately, the dielectric strength of the vacuum is so high that only a small contact gap is required for service voltages in the distribution range, and this facilitates the design of the bellows. The majority, in fact all except the UK design, use a convoluted bellows which is made by pressing or rolling a thin-walled stainless steel tube as is shown in section in Fig. 4.35 and the right-hand half of Fig. 4.34. The left-hand half of Fig. 4.34 shows the alternative design, made by welding the alternate inside and outside perimeters of a series of pressed discs. The biggest difference in the performance of these bellows is shown by comparing the ratio of permitted travel to relaxed length, which has a relationship to the fatigue life of what is basically the only mechanically stressed component in the vacuum interrupter.

One of the characteristics of pure metals in a vacuum environment is their propensity for 'cold welding'. Depending on the material, if two contacts are pressed together it will be found that quite a force will be required to separate them. Even after choosing a material with a low welding characteristic it is not practical to use sliding contacts in a vacuum interrupter.

Taking account of some of the other points raised in the earlier discussion of the vacuum arc, the requirements for a good contact material for use in a vacuum interrupter are as follows:

1. Low specific resistance.
2. Low arc erosion.
3. Good arc stability.
4. Low weld strength.
5. Excellent purity and low gas content.

Different manufacturers choose different compromises between these factors, depending on the chosen market place. For example, in the USA there is a requirement for switchgear with heavy current and high fault ratings. Typically, a common circuit-breaker in USA at 15 kV has a current rating of 1200 A and a breaking capacity of 750 MVA. (As explained in Chapter 19, the USA still rates the breaking capacity of circuit-breakers in terms of MVA (the product of breaking current, rated voltage and phase factor, $\sqrt{3}$) and rated voltage. In addition the circuit-breaker has to be shown to maintain its MVA rating over a range of voltage, and the rating quoted above has to be demonstrated as a breaking current capability of 28 kA at 15 kV, and 36 kA at 11.5 kV.) This fact created a need to concentrate on low resistance and low weld strength. The fairly

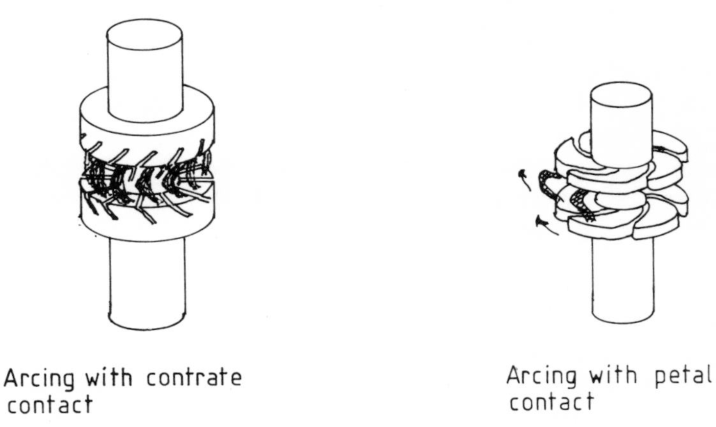

Fig. 4.36 Magnetic arc control in a vacuum interrupter.

general use, already, of surge arresters on loads with high inductance made the achievement of low chopping currents (i.e. good arc stability) of less importance and the general acceptance of regular maintenance meant that low arc erosion was also less important in operational terms.

Conversely, the UK market was predominantly modest in fault and current rating, surge arresters were not in common use, and pressure existed to extend the intervals between maintenance visits, so vacuum interrupters with different characteristics were produced. These gave priority to good arc stability and low arc erosion and accepted a higher contact resistance as part of the price to be paid.

The two contact materials originally used in the vacuum interrupters developed on the two sides of the Atlantic were an alloy of copper and bismuth in the USA and a composite material containing approximately equal proportions of copper and chromium in the UK interrupter. The former has a conductivity not much less than pure copper and a good cold weld resistance, but a current chopping propensity rather higher than can be tolerated without paying attention to the switching duty for which it is to be applied. The copper/chrome material has a higher resistance, a good cold weld resistance, low arc erosion and a chopping current comparable with that of the traditional oil circuit-breaker.

In addition to the material of the contacts, an important factor in their performance is their geometry, or shape, which plays a vital part in the magnetic control of the vacuum arc. As was mentioned in the description of the vacuum arc, if the contacts are plain, i.e. there is no particular control of the arc, then the arc changes from the diffuse form with a low arc voltage to a constricted form with a high arc voltage and high heat generation at the arc roots. The two forms of contact geometry used in the earliest vacuum interrupters are illustrated in Fig. 4.34. Figure 4.36 also illustrates the effect on the vacuum arc of the magnetic field generated by the passing of the current through the slotted contact surfaces.

In the USA pattern, known colloquially as the 'petal' contact, the current flows radially outward along the contact finger-like extensions and then adds a circumferential component of travel. The effect on the arc is to cause it to motor round the periphery of the contact at high speed. There is also a component of the flux forcing the arc to bow outwards away from the contact axis. With this design the arc changes into the constricted form when the limiting current is reached, but the arc plasma reverts to the diffuse form when the current again falls below that limiting value, the release of metal vapour and the local heating at the arc roots being kept at a sufficiently low level by virtue of the rapid movement of the arc roots during the constricted phase.

In contrast, the arc control created by the 'contrate' design of contact retains the arc substantially between the faces of the contacts and prevents the arc changing to the constricted form, the arc continuing to burn throughout the current loop in the form of an intense tube of plasma, which cools to the normal diffuse arc with many parallel arcs and fast moving roots as zero is approached.

In both cases the condensation of the unused metal vapour as the arc current reduces occurs so rapidly that as the current approaches within 100 A of current zero the arc column is reduced to a single arc which disappears at current zero. Dielectric recovery is very fast, with the normal dielectric strength returning in microseconds. The contrate contact shows two differences in performance compared with the petal contact that have a significant effect on the rate of contact wear. The continued diffuse arc has a much lower arc voltage and hence arc energy than the constricted arc, so less material is vaporised, and also the arc control does not drive the arc outside the ring of the contact to the same extent as the petal arrangement, so that at least half of the metal vapour re-condenses on the contact surface and can be re-cycled.

The way in which the contacts are operated has some influence on the performance of the vacuum interrupter, and the characteristics of a suitable operating mechanism are as follows. It should provide:

1. Additional contact pressure to ensure consistent contact resistance and the ability to carry short-time through faults.
2. The ability to open consistently and quickly so as to establish the necessary contact gap in the requisite time.
3. The ability to close quickly without contact bounce and immediately establish enough contact pressure to withstand safely the effects of closing on to a fault.
4. A mechanical life, with little or no maintenance, to match that of the vacuum interrupter.

Whilst the forces required to move the contacts of the vacuum interrupter are low, the contact loads are high, typical loads for a short-circuit rating of 25 kA being in the 150–300 kgf region for each vacuum interrupter, depending on the design, so an assembly of three interrupters requires attention to be paid to the rigidity of the circuit-breaker structure. (Figures 4.37–4.39 illustrate typical vacuum circuit-breakers.)

Having now surveyed the range of arc interruption technologies in use in distribution switchgear, it is interesting to look at two diagrams. One is based on manufacturers' published figures for contact life on their circuit-breakers and compares the contact endurance in terms of the number of current interruptions

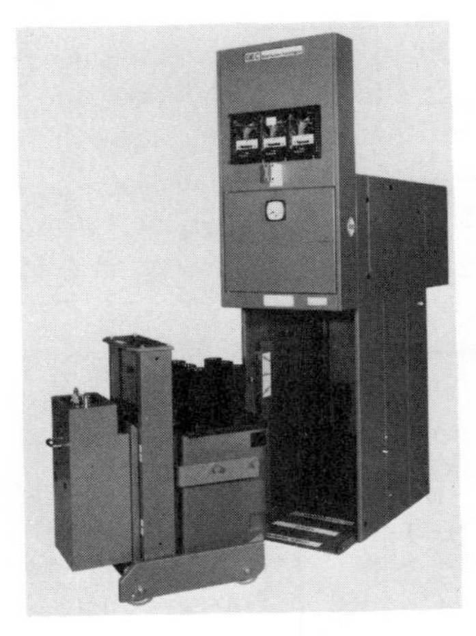

Fig. 4.37 British vacuum circuit-breaker. (*Courtesy GEC Distribution Switchgear Ltd.*)

Fig. 4.38 Vacuum circuit-breaker with axial magnetic field for arc control. (*Courtesy Holec Nederland BV.*)

Fig. 4.39 American vacuum circuit-breaker. (*Courtesy General Electric, USA.*)

permitted between contact changes at various proportions of the 100% fault rating for the types of circuit-breaker described in this chapter. This is Fig. 4.40, to logarithmic scales, and the only explanation necessary, in addition to matters already raised in connection with the different constructions, is to mention that the band marked 'magnetic air' relates to the type used in the USA for distribution purposes in large numbers.

In the UK the medium voltage (usually 11 or 6.6 kV) air circuit-breaker forms only a tiny fraction of the circuit-breakers used in distribution systems, and the majority of those in use are installed in power stations for very heavy duty. They are correspondingly expensive and not comparable with the lighter duty

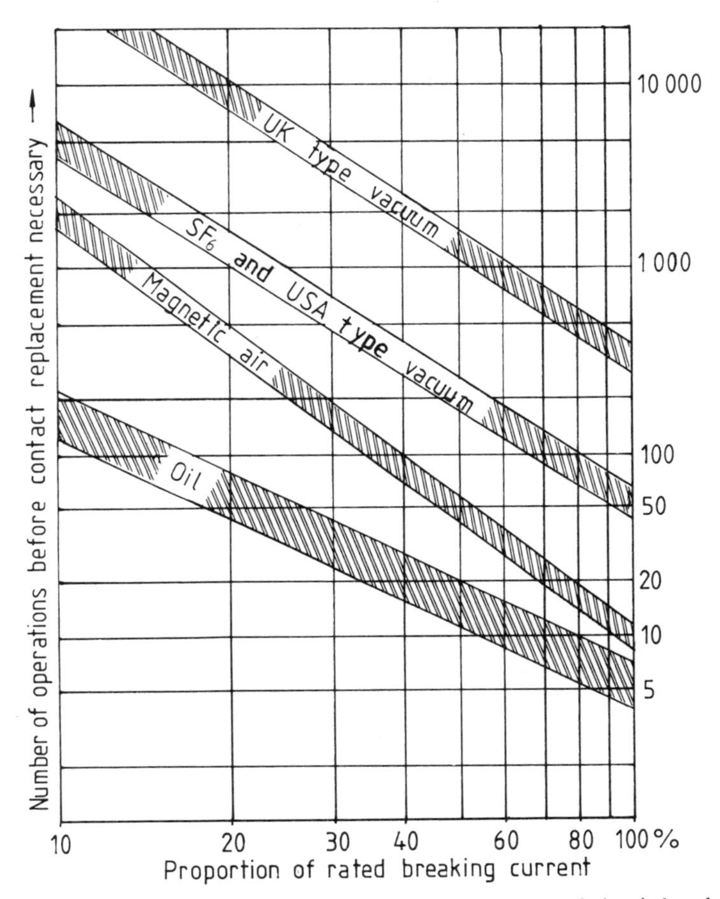

Fig. 4.40 Relative contact endurance of different types of circuit-breaker.

switchgear used in utility distribution systems in America and Europe. Today the vacuum circuit-breaker is rapidly taking the lion's share of the USA market, but there are many thousands of the air-break type in service, as indeed there are many thousands of bulk oil circuit-breakers still in service in the UK.

The other diagram, Fig. 4.41, compares the 50 Hz breakdown voltages of the principal insulating media that have been discussed in this chapter. It is interesting to note that SF_6 equates to the same dielectric strength as vacuum or oil at very modest pressures.

During the early years of the introduction of the 'new technology' circuit-breaking methods of SF_6 and vacuum there was much discussion about the relative merits of the two systems. Now, it is becoming accepted that they are two very good media for arc interruption and the choice of one or the other for a particular application often depends on factors that have little connection with the arc interrupting duty.

As an example, the SF_6 circuit-breaker is a better choice for a country that has

Table 4.2 Comparison of circuit-breaker characteristics with different media

Characteristic	*Bulk oil*	*Minimum oil*
1. Switching small inductive currents	Oil is a good insulator and arc extinction is more efficient than in air-break circuit-breakers, giving shorter arc duration at low currents and a greater degree of current chopping. The current chopped will be less than 10 A and will give rise to measurable over-voltages in highly inductive circuits. However, many years of experience have shown that these over-voltages are not sufficient to cause damage to system insulation.	
2. Switching capacitance currents	Bulk oil circuit-breakers employing two breaks per phase at 12 kV have a build-up of dielectric strength across each pole sufficient to ensure restrike-free capacitance current interruption.	Minimum oil circuit-breakers are inherently of the single-break type, and unless designed with oil jet action (as many are), they are unlikely to be free from restrikes at voltages of 12 kV and above.
3. Mechanical behaviour	The international standards require 1000 no-load operations to be performed without attention and with negligible wear. Regular lubrication at this sort of operation interval is envisaged in the design. Mechanism energies depend on the fault and can be quite high at fault currents of 25 kA. Powers of 10 kW are not uncommon.	
4. Pressure development	The dissociation of the oil into hydrogen and hydrocarbons by the electric arc generates a very high pressure in the arc control device, and this contributes to arc extinction. In the bulk oil circuit-breaker a proportion of this pressure is transmitted to the metal tank but the existence of an adequate air cushion in the top-plate of the breaker helps to keep the tank pressure to modest levels and cylindrical design makes it a simple matter to contain this residual pressure rise safely.	Pressure in the arc control device will be generated as for bulk oil design, and will similarly contribute to arc extinction. The pressure will be transmitted to the container, which usually takes the form of a cylinder of glass-reinforced insulating material. The air cushion is relatively smaller than in the bulk oil design but the smaller diameter of the tube assists in the creation of an adequately strong pressure container.
5. Exhaust gas	Modest quantities of inflammable exhaust gases emissions are produced during arcing and arrangements are made for these to be passed through cooling baffles before escaping from the circuit-breaker vents. These baffles also serve to separate the gases from the oil so that oil loss is minimised.	
6. Mechanical reaction on the foundations	Heavy	Heavy
7. Noise generation	Moderate	Moderate
8. Fire risk	The use of oil as the interrupting medium and the emission of inflammable gases (hydrogen, acetylene, methane, etc.) during operation must constitute a fire risk. In practice, reputable designs rarely give rise to any fire unless some serious malfunction occurs. The *quantity* of oil is not a factor either in the liability to fire or in the amount of damage created in the event of a fire. In environments where a fire could have serious consequences, enclosed substations with fire-fighting installations may be desirable.	
9. Suitability for industrial systems	Oil circuit-breakers are very suitable for industrial systems, and they have a long record of satisfactory service in such applications. The only drawback, if the frequency of operation is high (e.g. daily, or more), is the need for frequent oil checks and, rather less frequently, changes of contacts and arc extinguishing devices. Obviously the minimum oil circuit-breaker is even more subject to the need for frequent oil changes.	

Air-break	*Vacuum*	*Sulphur hexafluoride*
At low currents the a.c.b. has a gentle extinguishing action and arcs for several half-cycles. This results in negligible current chopping and negligible voltage surges.	The v.c.b. practically always interrupts at the first current zero, irrespective of current value. Arc stability is a function of the contact material. Chromium/copper alloys allow current chopping at levels compatible with oil circuit-breakers, but other materials may make it necessary to apply surge protection to inductive systems.	The current-chopping behaviour of a g.c.b. is somewhat dependent on the means of arc extinction employed, but the level of chopping is generally of the same order as the o.c.b.
For the same reason that a.c.b.s are gentle at extinguishing inductive currents they are prone to restrike when interrupting capacitance currents and have a limited capacity for such duty.	The dielectric recovery of the vacuum gap is extremely fast, giving restrike-free interruption of capacitance currents, usually up to the full rated current of the interrupter.	The electro-negative characteristic of SF_6 gas deionises the contact gap quickly and restrike-free capacitance switching is assured.
The mechanical attributes of the air circuit-breaker are similar to those of the oil circuit-breakers.	The short travel and low energy requirements assist the designer to construct a robust mechanism capable of matching the maintenance-free life of the vacuum interrupter. At least 10000 maintenance free operations are usual.	'Puffer' type circuit-breakers need more energy than rotating arc types and the requirement obviously increases with short-circuit rating. The mechanisms are designed for minimum maintenance and intervals of 10000 operations are normal.
The rapid establishment of a heavy arc in the arc chute generates high pressure and shock waves which have to be withstood by the mechanical construction. This leads to large and heavy arc chutes which are accordingly expensive.	The metallic vapour produced during arcing increases the vapour density in the contact region synchronously with the current. There is no general pressure rise inside the interrupter, the envelope of which only has to withstand external atmospheric pressure.	Internal pressure will be developed during fault interruption, being greater in the 'puffer' type device. As this has a lower standing pressure to begin with, the end result is usually that the two designs are very similar in maximum pressure requirements. In either case the maximum pressure is well within the strength of the container.
Large quantities of hot ionised air are discharged from the arc chutes, creating a need for insulated cooling and ducting arrangements for safe discharge.	The vacuum interrupter is permanently sealed and all metal vapour produced during arcing condenses immediately. There is no emission of any sort.	The SF_6 circuit-breaker is totally sealed so there is no gas emission. Some of the gas is dissociated into other sulphur and fluorine compounds. The gaseous products of arcing are absorbed by special filters inside the circuit-breaker.
Very heavy	Negligible	Slight
Heavy	Negligible	Slight
While no oil is used and no inflammable gases are emitted, the hot exhaust gases produced during fault operation constitute a very minor fire risk.	The fire risk is negligible as no flammable materials are used and no gases are emitted.	As for vacuum switchgear.
Industrial systems often have onerous environments for the switchgear installation and add frequent operating duty as well. This leads to a need for frequent maintenance attention, particularly in respect of insulation surfaces exposed to the atmosphere.	Industrial situations with frequent operation provide the application which makes the best use of the advantages of vacuum switchgear. The annual costs of vacuum switchgear in this environment are less than the other types.	It is in the industrial environment that the ability to operate frequently without maintenance is of the greatest advantage. SF_6 switchgear fits the requirement well, with the proviso that the mechanical requirements should be studied if the closing energy is high.

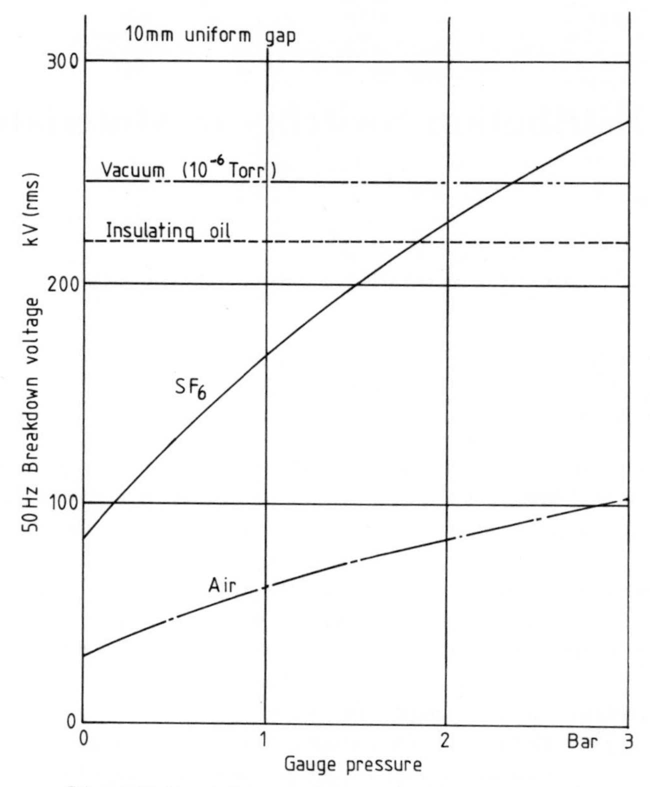

Fig. 4.41 Breakdown voltages of different arc media.

no indigenous source of vacuum interrupters, as the locally made content can be significantly higher than if a vacuum design was chosen for local manufacture. The capital investment in a facility for the manufacture of vacuum interrupters is only feasible if there is a local demand of substantial magnitude, so that the throughput of the factory is commensurate with the size of the investment.

Table 4.2 summarises some of the comparative characteristics of the different types of circuit-breaker described in this chapter.

CHAPTER 5

Distribution Switchgear Materials

The process of selecting materials for use in switchgear is subject to the usual engineering factors, as follows:

1. Adequate technical properties to achieve the desired objective.
2. Ease of working (forming, welding, cutting, etc.).
3. Economic cost.

The desired objective, from the designer's point of view, is to produce an item of switchgear that fulfils all its rating parameters and is attractive to the potential user in terms of price, size, ease of use and appearance. Over the years the availability of new materials, or new forms of old materials, together with new ways of manipulating them, has changed the appearance even of switchgear – that most conservative of items of electrical plant.

One of the most important characteristics of any material used in switchgear is its long term stability and reliability. This means its ability to retain its characteristics for long periods of time – periods of many years – even in hostile environmental conditions. It is obviously difficult to obtain information on this aspect of the performance of a new material and much effort has been, and is being, expended on trying to find representative ageing test procedures. The switchgear business is well known for the cautious attitude of its customers, who would like to see evidence of 20 years' satisfactory service experience before they buy!

ENCLOSURES AND SUPPORTS

Most items of switchgear have some form of enclosure associated with them, and that enclosure usually requires to be at earth potential for the safety of passers-by. The enclosure may totally surround the switchgear, as in the case of indoor metal-enclosed switchgear equipment; or it may protect certain parts of it as in outdoor open-terminal equipment, where the insulated terminals are exposed, but all the remaining components are enclosed.

The main reason for the enclosure is to prevent unauthorised operation and tampering, to provide some protection from the environment and to protect personnel from harm. In recent years the term 'environmental protection' has had to include vandalism, or deliberate interference and tampering with the equipment.

Another use for structural materials is as supports for the components of the switchgear. Typical of this is the framework on which outdoor switchgear is usually mounted in order to lift the terminals to a height normally out of reach of operating personnel.

Steel is still the most common choice of material for these purposes, giving as it does a good combination of mechanical strength with economy. Enclosures in the form of cabinets or cubicles are usually made of steel sheet in the thickness range of 2–3 mm depending on the size of the enclosure. One factor in the choice of sheet thickness nowadays is related to the risk of 'burn-through' in the unlikely event of an internal arcing fault to earth, where the enclosure would form one of the arc electrodes. This factor might outweigh purely mechanical considerations.

Rolled steel sections are finding less and less use in modern distribution switchgear, being substituted by components folded from sheet steel.

The advantages of steel for these applications are its mechanical strength and economic cost. It is easily available in a wide range of sizes and shapes.

Its disadvantages include the fact that standard mild steel sheet is magnetic and needs corrosion protection. The use of a suitable stainless variety can avoid both these problems, but at a cost. The use of steel in switchgear therefore requires a suitable finishing process to improve durability substantially and give a reasonable appearance. Over the years the use of zinc as a sacrificial protection, either applied hot as in galvanising, sprayed on as a liquid or, for small parts, electrolytically deposited, has played an important role. For indoor use, the application of plastics powder coatings by electrostatic spray techniques over a phosphate underlayer has become the accepted long life treatment. For outdoor applications, zinc spray coatings under a paint layer are commonly used on distribution equipment. As the cost of stainless steel comes down, increasing use will probably be made of this material. A number of American 'Padmount' switch units are already made from stainless steel.

CONDUCTING METALS

The most important characteristic of the materials used as conductors in switchgear is to have a low specific resistance. Energy loss in switchgear is a revenue expense for the user, and excessive temperature rise due to high resistance conductors in a small enclosure will lead to a reduction in the life of associated insulation materials. The next most important characteristic is that the material must resist corrosion. Any corrosion in joint areas leads almost inevitably to an increase in resistance and this becomes a 'vicious circle' as the higher resistances lead to higher temperatures and they in turn lead to more corrosion. So the designer must choose a low resistance corrosion resistant material and use it within the safe temperature range corresponding to that material.

The material with the lowest resistance is silver, but this is ruled out in every switchgear application because of its cost. However, it does find extensive application in switchgear equipment as a thin coating, and in small, high current units, in the form of small contact inserts. The coatings are almost always applied by electroplating and typical thicknesses are in the range of 0.02–0.05 mm. The greater thickness coating is used when sliding motion takes place

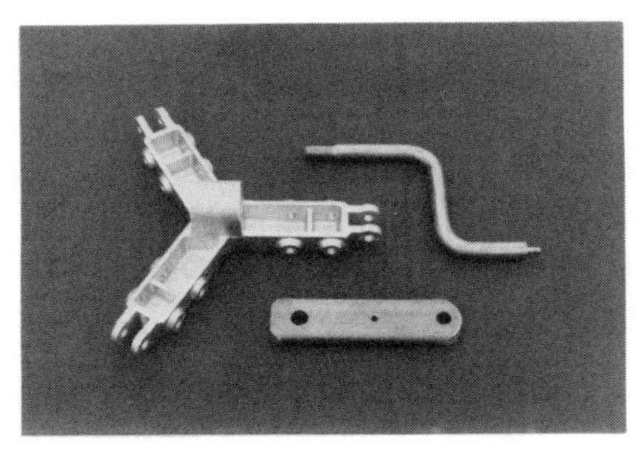

Fig. 5.1 Parts made from copper and aluminium. (*Courtesy Yorkshire Switchgear Ltd.*)

between the plated surfaces. The virtue of the silver coating is that it allows contacts to be used at significantly higher temperatures than for the cheaper base metal, as even when silver oxide forms on the surfaces, the resistance of any joint does not increase significantly.

The two materials usually considered for conductors of any size are, of course, copper and aluminium, which also have low resistivity. Copper has the lower resistivity, and also resists oxidation better than aluminium, particularly at temperatures below about 80° C. Aluminium has a much smaller specific gravity and although larger cross sections of conductor are needed to give the same resistance as the equivalent copper, the weight and therefore the cost of the material often show to advantage. Typical parts are shown in Fig. 5.1.

As used in switchgear, copper can take many forms. Rods and bars are perhaps the most common, but in the past some of the complex shapes required for some of the conducting parts were created by casting. This was never very satisfactory, as the castings are prone to 'blow-holes' and these often only came to light after considerable, expensive machining operations had been performed. Nowadays, advances in the technique of hot-stamping have led to extended use of this process for those components that cannot be made from rolled or drawn sections. The usual grade of copper used is the high conductivity variety. In its natural form this is very soft, so a degree of work-hardening through rolling or drawing is desirable. It follows that subsequent manufacturing processes should not apply significant amounts of heat to the material, as for example in a brazing process, as this will soften the copper again.

Cast and hot-pressed materials are usually soft by the nature of the forming process, and if the application is such that hardness is important, i.e. a fairly highly stressed component, then the use of a copper alloy is sometimes preferred. Chrome-copper is one such alloy which can attain a reasonable compromise between an increase in strength and hardness without too great a loss in conductivity. Typical hot-pressed components are shown in Figs. 5.2 and 5.3 and a diecast brass cable gland in Fig. 5.4.

Aluminium in its pure state is also a very soft material and therefore it is frequently used as an alloy, silicon being a common alloying agent, to give it greater strength and stiffness. Its use in switchgear is usually restricted to conductors, typically busbars, and also it is more frequently used in low voltage equipment than at medium voltages (i.e. 1–36 kV).

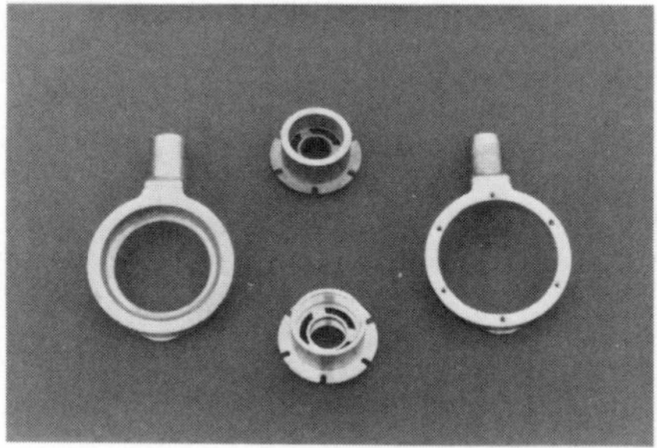

Fig. 5.2 Hot copper pressings – shown as received and after machining. (*Courtesy Yorkshire Switchgear Ltd.*)

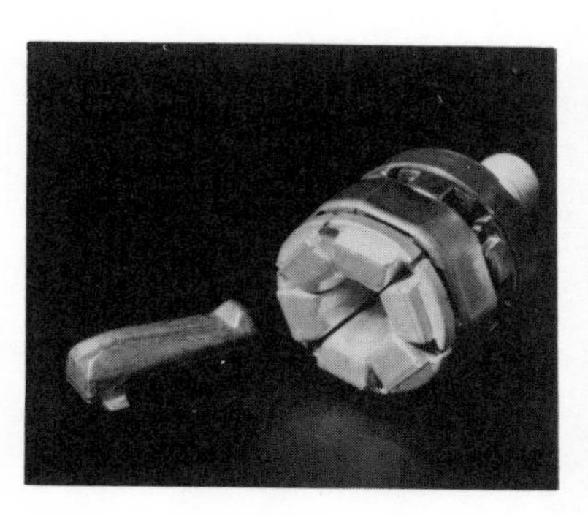

Fig. 5.3 Hot brass pressing – contact finger – as received and as a finished assembly. (*Courtesy GEC Distribution Switchgear Ltd.*)

Fig. 5.4 Diecast brass cable gland. (*Courtesy GEC Distribution Switchgear Ltd.*)

Aluminium oxide is a good insulator, so it is important to follow good jointing practices when aluminium conductors are being bolted together. A recommended procedure is to coat the mating surfaces with vaseline, to exclude the air from the surface and then to remove the oxide layer by scratch brushing through the vaseline. Before making the joint, the contaminated vaseline is wiped off and the surface immediately recoated. Then the bolting operation should follow without delay. In some low voltage busbar systems, particularly in the USA, the aluminium conductors are welded together to overcome this problem.

Another problem arises if aluminium conductors are used in association with copper, as the electrochemical potentials of these two elements are widely separated. If a copper-to-aluminium joint is allowed to become moist, then electrochemical corrosion will begin to take place. To avoid this phenomenon, it is customary to coat such bimetal junctions with a suitable material to exclude moisture. Alternatively, jointing methods which avoid bolted faces of dissimilar materials can be used. For some applications, friction welding is a suitable process, a typical example being a bimetallic cable ferrule connecting an aluminium cable conductor to a copper conductor on the switchgear (Fig. 5.5). It is now possible to electro-plate aluminium with either copper or silver by using a special process, and this is also used to simplify the bimetallic jointing procedures.

Another group of conducting materials, used in circuit-breakers in particular, are those containing tungsten, usually in association with copper. The presence of the tungsten substantially increases the resistance of the circuit-breaker contacts to arc erosion, and it is used in the form of contact tips for this reason. The conductivity is low and for that reason tungsten-tipped arcing contacts are frequently used in conjunction with a parallel main contact system, which would use low resistance materials, usually silver-plated copper. The design of the contact structure would then ensure that the main contacts separated before the arcing contacts so that the circuit-breaking duty was restricted to the tipped contacts. The copper-tungsten material is difficult to machine, and it tends to be brittle, so the contact tips are usually made as a sintered component and attached by soldering or brazing. Typical components are shown in Fig. 5.6.

Finally, mention should be made of the use of special purpose copper alloys. One such is beryllium-copper which can be used for lightly stressed springs and is of particular value when a non-magnetic spring material is desired.

Typical characteristics of these materials are listed in Table 5.1.

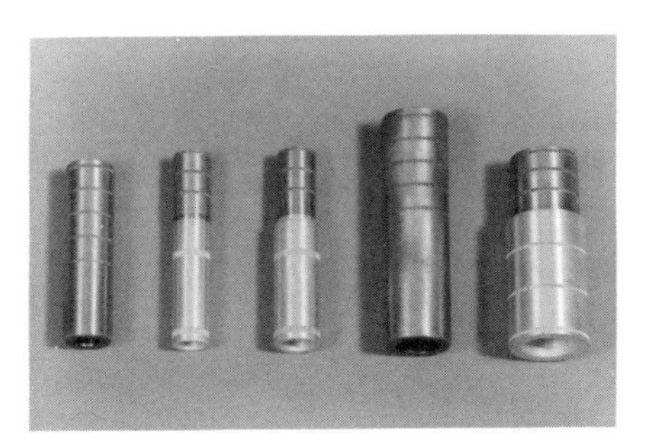

Fig. 5.5 Plain copper and copper/aluminium friction welded bimetal cable ferrules. (*Courtesy BICC.*)

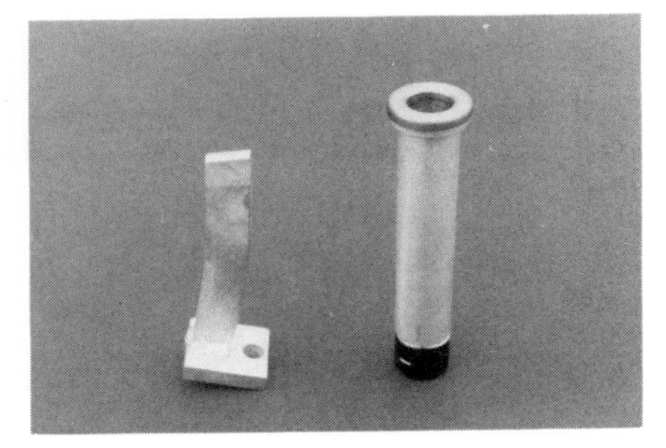

Fig. 5.6 Contact parts with copper-tungsten tips. (*Courtesy Yorkshire Switchgear Ltd.*)

INSULATING MATERIALS

A wide range of materials are used for the purpose of insulating live parts of switchgear from the earthed enclosure that usually surrounds them. These materials can be solid, liquid or gaseous. They are used to fill spaces, cover conductors, support live parts and extinguish arcs depending on their physical form and properties.

Air is probably the most commonly used gaseous insulation material as it has the virtue of being readily and cheaply available. It is composed of approximately 80% nitrogen and 20% oxygen, so its physical, chemical and electrical properties are close to those of nitrogen. However, whilst it is a very good insulator when it is clean and dry, it is usually subject to contamination by

Table 5.1 Typical characteristics of conducting materials

Material	1	2	3	4	5	6	7	8	9
Aluminium	2.70	2.80	3.90	2.00	0.895	658	1.35	1.90	0.70
Brass 60Cu/40Zn	8.50	9.00	1.60	1.10	0.418	890	3.10	4.70	0.90
Copper – HDHC	8.90	1.80	3.90	3.80	0.389	1083	1.10	3.40	1.25
Iron (wrought)	7.80	11.00	5.50	0.60	0.481	1530	2.40	2.90	1.70
Phosphor-bronze	8.80	16.60	—	0.80	0.335	1000	1.50	—	0.70
Silver	10.50	1.60	4.10	4.00	0.234	960	—	—	—
Tungsten-copper	13.20	5.10	—	0.20	0.167	3380	—	—	—

Explanation of column headings:
1. Density – g/cm^3
2. Specific resistance – ohm-cm $\times 10^{-6}$
3. Temperature coefficient of resistance – $1/°C \times 10^{-3}$
4. Thermal conductivity – W/cm/°C
5. Specific heat – Ws/g/°C
6. Melting point – °C
7. Elastic stress limit – $kg/cm^2 \times 10^3$
8. Ultimate tensile strength – $kg/cm^2 \times 10^3$
9. Young's modulus – $kg/cm^2 \times 10^6$

dust and moisture. At low voltages, where mechanical clearances are often more than adequate for the purpose of insulation in air, this is not a problem, but at higher distribution voltages precautions need to be taken to allow for the reduction in the insulation value when the air is polluted. If these precautions amount to hermetically sealing the switchgear enclosures to maintain the quality of the insulation, then it may prove more economical to use a filling medium of greater dielectric strength. The reduction in space and the quantity of filling medium will allow a more expensive material to be used.

A recent addition to the gases used to fill switchgear enclosures is sulphur hexafluoride (SF_6). Its value as an arc extinguishing medium has already been discussed, but it is also used for its insulating property. It has a dielectric strength approximately three times that of air, and where the reliability of air is not deemed adequate, then SF_6 is increasingly being used as an insulating filler. The characteristics of these gases are listed in Table 5.2 and their electrical characteristics are discussed in more detail in Chapter 6. Chapters 10 and 11 make reference to the growing use of fully sealed SF_6 filled metal-enclosed switchgear, for both primary and secondary switchgear applications.

A range of other fluids have been tried or considered for use in switchgear as

Table 5.2 Typical characteristics of insulating gases

Characteristic	*Unit of measurement*	N_2	H_2	SF_6	*Air*
Molecular weight	Gram-mol	28.016	2.016	146.07	28.96
Critical point:					
Temperature	°C	−146.95	−235.95	45.55	−140.73
Pressure	bar	34.1	13.2	37.8	37.7
Density	kg/m^3	310	31	730	328
Density at 0°C and 1 bar	kg/m^3	1.25	0.089	6.139	1.293
Specific heat (C_p)*	$Ws/kg/° \times 10^3$	1.04	14.22	0.91	1.00
Specific heat (C_v)*	$Ws/kg/°C \times 10^3$	0.745	10.06	0.691	0.72
Thermal conductivity	$W/cm/°C \times 10^{-4}$	2.4	16.8	1.4	2.4
Viscosity	$Poise \times 10^{-7}$	1580	835	1450	1708
Adiabatic constant (C_p/C_v)		1.4	1.41	1.3	1.4
Sound velocity, 1 bar and 30°C	m/s	355	1330	138.5	350

* C_P= Specific heat at constant pressure.
C_V= Specific heat at constant volume.

filling media, but the only other one which merits mention in this chapter is insulating oil. Reference is often made to low and high viscosity oils, but the type which is in common use in distribution switchgear is the low viscosity type in accordance with BS 148. This is a hydrocarbon oil, often referred to as 'transformer oil', because of the vast quantities that have been used over the years to insulate medium and high voltage transformers. When used in the bulk oil circuit-breaker, it has the advantage of serving the dual role of an insulator as well as the arc extinguishing medium. It has a dielectric strength almost three times that of SF_6 at atmospheric pressure, but is coming under increasing scrutiny from a safety point of view because of its flammability.

Finally, mention should be made of another rather special insulating medium – vacuum. It is very expensive both to make and contain, so its use in distribution switchgear is restricted to circuit interrupting devices either as vacuum contactors (mainly load break, although some overload capability is available), or for power circuit-breaking applications. This application has been covered in Chapter 4. As a dielectric it is probably the best available, but the realisation of the potential levels requires considerable attention to the surface condition of the electrodes. In the usual application, opening of the contacts on no-load leads to roughening of the surface due to the breaking of the cold welding which takes place, and the practical levels that can be achieved are well below the potential, although still very adequate for the purpose.

Solid insulating materials include many of natural origin, such as mica, asbestos and slate, or those derived from natural materials such as porcelain, paper and shellac varnishes, but their usage is declining. There have been many changes to the range of solid insulating materials used by switchgear manufacturers in the last 20 years or so. Most of the newer materials are based on petrochemicals and include a wide range of polymers, resins and films. For mechanical strength, glass fibre reinforcement is often used, a particular example being the insulating tubes which form the enclosures for minimum oil circuit-breakers.

One principal use of solid insulation is in the form of bushings to insulate conductors where they pass through the walls of enclosures. Not long ago, porcelain tubes or wound, resin impregnated paper were the common types, but today the predominant material used for this purpose in distribution switchgear is cast resin. The resin is usually epoxy, but other polymer resins have begun to compete with this in recent years. These casting resins are characterised by usually being a two-part mix of a resin and a hardener, or polymerising agent. The resin usually contains substantial quantities of mineral filler, partly to reduce the cost, but mostly to reduce stresses due to shrinkage during curing. One major advantage of the resin casting technique is the almost limitless range of shapes which can be made, and the ease with which conductors of various shapes can be embedded in it.

The range of plastics materials now available to the switchgear designer includes thermoplastic materials suitable for injection moulding, thermosetting materials for compression and transfer moulding and two-part casting resins for pouring into moulds, usually at atmospheric pressure. The thermoplastic materials are mostly used for small components required in quantity, such as fuse holders, indicating lamps, control and selector switch components and other such low voltage auxiliary products.

The thermoset materials include glass-fibre filled polyester 'dough', usually referred to as DMC (dough moulding compound). Early versions of this material were of variable electrical quality, usually due to inadequate 'wetting' of the glass fibres by the resin, leading to channels of electrical weakness. Continuous research has overcome this problem and also led to the ability to produce ever larger components. Single components weighing as much as 25 kg and containing moulded-in conductors and fixing inserts are now regularly produced. This material has a high resistance to tracking and is economical to produce, provided the quantities are sufficient to justify the rather high tooling costs. The components are mechanically strong and light in weight, see Fig. 5.7.

Fig. 5.7 Insulation components made from DMC. (*Courtesy Yorkshire Switchgear Ltd.*)

The main casting resin used until recently for medium voltage distribution switchgear insulation components has been epoxy. The resin is usually mixed with a filler, frequently silica, which improves its characteristics in a number of ways, e.g. increasing the mechanical strength, resistance to tracking, etc. This practice also reduces the shrinkage which occurs during the curing process and during cooling from the processing temperature. One of the problems with designing components for the epoxy resin casting process is to eliminate, or at least reduce to harmless levels, the internal stresses arising from the resin contracting around inserts and embedded conductors. Obviously these problems are greatest when the conductors are large and the insulation thickness is small. Care has to be taken to avoid trapping the resin by changes in section of the embedded metalwork, which would prevent the small movements of the resin during curing and cooling. In some cases it is advantageous to apply cushioning layers to the conductors so that strain can be absorbed.

The designer of modern medium voltage switchgear components by casting or moulding must ensure the avoidance of internal stresses as just described, and also the inclusion of air pockets or 'bubbles' in the moulding. The electrical stress present in the insulation, which has a permittivity greater than the trapped air, will lead to ionisation in the air pockets and eventual damage to the resin. This phenomenon is referred to as 'internal discharge' and measuring techniques are available for detecting it at very low levels. It is common practice therefore for every important insulator made of these materials to be given a routine discharge test.

The full curing of cast resin components used to occupy a mould for 24 h or so, and it was common practice to pour the resin into the mould in a heated vacuum chamber, after having first drawn a vacuum in the resin mixing chamber, to eliminate the possibility of internal voids. Then the filled mould would be cured in an oven. This made the process suitable only for products required in small numbers, or a number of tools would be required, thus increasing the costs of production. The pouring and curing temperature of these resins is in the region of 130°C which results in a considerable input of energy for each moulding.

Now, new production techniques and new materials have dramatically shortened production cycle times. One new technique is to apply air pressure to the resin in the tool after pouring, which has the effect of shortening the 'gel' time, i.e. the time for the moulding to set to the point that it can be handled (with care). Typically moulds can be re-used after half an hour in some instances, the 'gelled' casting being transferred to an oven for curing. Any critical dimensions or faces that must be flat need to be attached to curing fixtures during this period.

Casting resins from the polyurethane family have been used for some time in low voltage applications, but in recent years their use has been extended to medium voltage applications. It is important to select formulations that have the requisite electrical characteristics, and which maintain those characteristics after

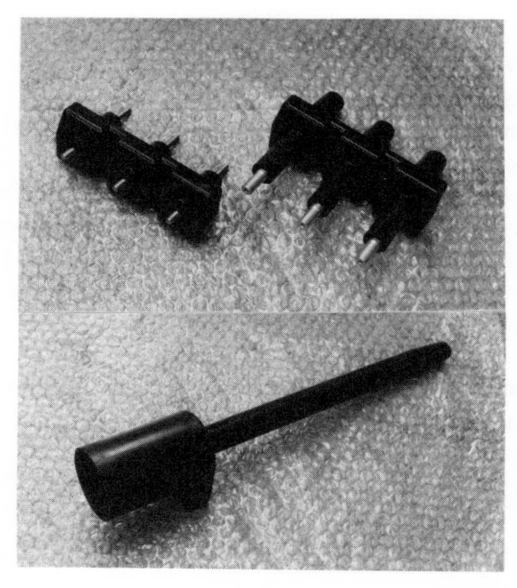

Fig. 5.8 Cast resin bushings. (*Courtesy Yorkshire Switchgear Ltd.*)

Fig. 5.9 Cast resin SF_6 circuit-breaker container. (*Courtesy Yorkshire Switchgear Ltd.*)

Fig. 5.10 Cast resin voltage transformer. (*Courtesy Yorkshire Switchgear Ltd.*)

long periods of exposure to heat and high humidity. When these requirements are met, polyurethane materials offer advantages in cost and tool turn round, and avoid the need for vacuum pouring in all except the most exacting applications. Much of the heat required for curing the casting is generated from the exotherm created by the chemical reaction, and this considerably reduces the energy input into the moulding. Typical components are shown in Figs. 5.8, 5.9 and 5.10.

In summarising the most important characteristics for solid insulating materials for use in distribution switchgear, the following items head the list:

1. Long term stability and retention of electrical characteristics, under arduous environmetal conditions.
2. No measurable internal discharges at working voltage. (Usually a small margin above working voltage is used as a criterion.)
3. High tracking index, i.e. it must be very resistant to the formation of surface breakdown tracks, even when under high electrical stress which may arise under damp, polluted conditions.
4. Good mechanical strength is often necessary because insulators are usually subject to mechanical stress.
5. Suitable for simple, low cost manufacturing processes.

Typical characteristics of these materials are listed in Table 5.3.

Table 5.3 Typical characteristics of solid insulating materials

Material	1	2	3	4	5	6	7	8	9
Bulk materials									
Porcelain	2.7–2.8	875	20	1500	0.0	0.800	30	5–6.5	17–25
Glass (pyrex)	2.2–2.6	790	11	550	0.0	0.6–0.9	10–20	4–6	1.5–6
Epoxy and chalk	1.5	—	—	120	0.2	0.060	14–18	4–5	15–50
Epoxy and silica	1.5–2.3	—	—	120	<0.2	0.08–0.16	14–16	3–5	10–30
Dough moulding comp'd	1.7–1.9	—	—	140	0.2–1.0	0.120	10–15	3.0	15–20
Polyurethane resin	1.5–1.55	1000	6–7	110	0.2	0.04	10	5–6	10
Sheet materials									
SRBP*	1.3–1.4	—	2.75	105	0.2–10	0.09–0.15	15–32	4–7	20–100
SRBF**	1.3–1.4	—	3.20	105	0.2–10	0.07–0.08	7–18	5–6	100–300
Glass filled epoxy	1.6–1.9	—	5.50	120	0.2–0.8	0.11–0.20	10–28	5–7	3–40
Glass filled polyester	1.2–2.0	1400	—	120	0.2–2.0	0.08–0.20	8–20	4–6	10–40
Thermoplastics									
Nylon	1.09–1.14	1500	2.80	50	0.2–6.0	0.02–0.03	20–30	4–7.5	10–120
PVC	1.3–1.5	1000	1.50	70	0.2–0.8	0.02–0.04	15–50	3–4	15–40
Polyethylene	0.92–0.94	2250	3.50	40	0–0.2	0.02–0.05	20–60	2.3	0.2–1.0
Polytetrafluorethylene	2.11–2.3	1000	1.60	200	0.0	0.003	20–40	2.0	0.5

* Synthetic resin bonded paper
** Synthetic resin bonded fabric

Explanation of column headings:

1. Density – g/cm^3
2. Specific heat – Ws/kg/°C
3. Thermal conductivity – W/cm/°C$\times 10^3$
4. Limiting temperature – °C
5. Water absorption – % at 20°C
6. Young's modulus – kg/cm$^2 \times 10^6$
7. Breakdown strength 50 Hz – kV/mm
8. Dielectric constant – 50 Hz, 20°C
9. Power factor – tan $\partial \times 10^{-3}$

CHAPTER 6

Insulation Aspects of Switchgear Design

BREAKDOWN OF SOLID INSULATION

Compared with the mechanisms of breakdown in liquids and gases, the breakdown by puncture of a solid dielectric is a complex process.

The factors which affect the voltage level at which breakdown of a solid material occurs are:

(a) The nature of the material.
(b) The duration of the voltage application.
(c) The thickness of the material.
(d) The shape of the electrode system.
(e) The waveform of the voltage applied.
(f) The temperature.
(g) Mechanical stress.

It is to be expected that the type of material and its molecular structure will affect the development of a conducting channel through the insulation, because of the electronic nature of that conduction. Attempts to measure or predict an 'intrinsic electric strength' based only on the molecular structure have been largely unsuccessful because of the difficulties of eliminating the other factors. Approximations to such an intrinsic property fall into the range from 1–10 MV/cm. The stresses actually used in practice fall in the range of 1–100 kV/cm, which illustrates that the other factors tend to be dominant. An important aspect of the nature of the material is its homogeneity. This includes both the uniformity of its physical structure and the presence of impurities or voids – and the content of any voids. Most often this will be air, but could be gaseous products from the manufacture of the material.

Since some of the factors listed above are time-dependent, it is not surprising to find that the breakdown strength is a function of the time of application. Figure 6.1 shows qualitatively the way in which this takes place. For very short times, in the microsecond range, the mechanism is mostly electronic and gives results that are closest to the intrinsic value.

As the time of application extends to the seconds region, results become erratic, as streamer formation due to ionisation of the air adjacent to the electrodes becomes a factor in the breakdown, and this is a variable factor in the short time application of voltage. As the time extends into minutes, the effect of streamer formation becomes more consistent.

These streamers cause no lasting damage in this period, but serve to extend the area of material being stressed and increase the level of voltage gradient at the tip of each discharge streamer which are mobile and seek out weaknesses in the

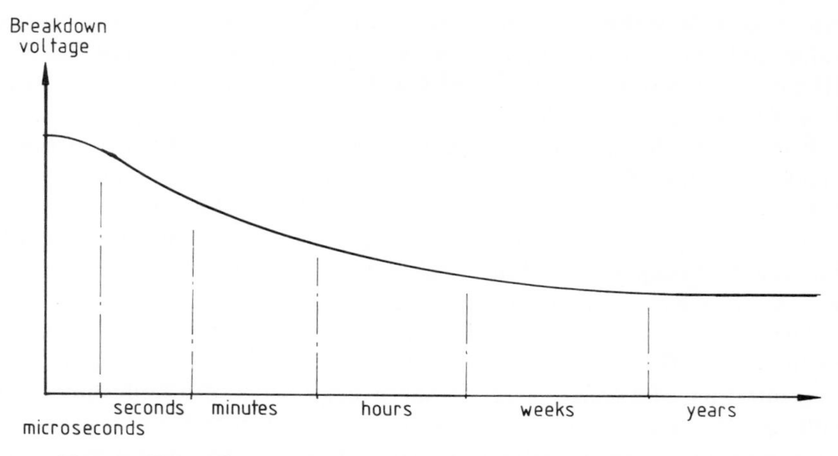

Fig. 6.1 Illustrating the time dependence of solid insulation breakdown.

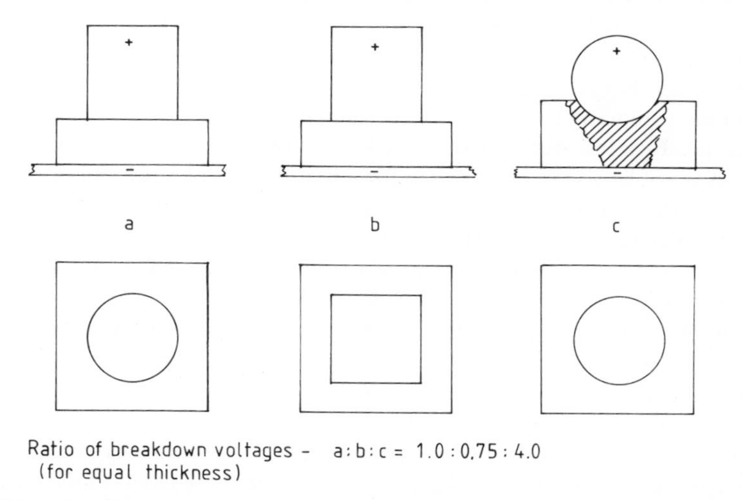

Fig. 6.2 Effect of electrode shape on breakdown voltage of solid insulation.

insulation. This reduces the breakdown value even further.

As the time of application lengthens, thermal factors begin to play a part. In addition to the electronic conductivity of the material, ionic mobility can also occur, leading to internal heating. As in all atomic behaviour, activity is increased with rising temperature and a 'run-away' condition is established with breakdown occurring at even lower values of voltage stress. Finally, as the period of application becomes hours, days, and eventually years, other factors such as the ionisation of internal voids leading to material damage, and deterioration of the surface adjacent to the electrodes, which may be aggravated by the environmental conditions, begin to affect the level of breakdown.

For the above reasons it is important to establish consistent procedures when testing the voltage strength of materials, by raising the voltage at a consistent rate, and, if the withstand level is being checked, to keep the withstand voltage value applied for a consistent time. Standard procedures for testing with 50/60 Hz a.c. require a time of one minute for withstand test and suggest values such as 1 kV/s for the rate of application.

Because of the questions of homogeneity, and the thermal nature of breakdown, the thickness of the material sample is also a factor in the breakdown level. Thin samples are statistically likely to be more homogeneous than thick ones and will cool more easily. Tests show that the relationship between the thickness (t) and the breakdown voltage (V_B) can be expressed by: $V_B = K \times t^N$ where the exponent (N) varies from 0.5 to 0.75.

The shape of the electrodes plays a part in determining the breakdown voltage to the extent that it determines the formation of streamers because of the concentration of voltage stress at sharp edges. Figure 6.2 illustrates three different situations. In the first the stress is determined by the sharpness of the edge of the face of the cylindrical electrode, in the second this is aggravated by the greater effect of the corners of the rectangular electrode, while in the third case a uniform field condition is created by using a sphere. The ratio of breakdown strength for the three cases is shown, and clearly shows the effect of the surface discharge.

From the foregoing the effect of the voltage waveform can be forecast. D.C. voltages produce less heat because the losses are purely resistive and therefore d.c. breakdown levels are higher than for a.c. Impulse voltage breakdown is also higher than the 50 Hz breakdown because streamer and thermal effects play little part in the behaviour. 50/60 Hz a.c. generates capacitive current losses in the dielectric which contributes to the thermal effects, and gives the lowest breakdown levels for otherwise equal conditions.

The question of thermal effects has already been mentioned, so it follows that the effect of high ambient temperatures must be to reduce the breakdown strength. In materials that soften as the temperature rises it is to be expected that the effect will be even more marked.

It is less understandable for the electrical breakdown strength to be dependent on mechanical stress, but as this stress puts the molecules under strain, a mechanism can be supposed which facilitates electron formation and reduces the voltage stress necessary to cause breakdown. Not all materials exhibit this relationship, the hard and inorganic substances being unaffected, but some of the polymeric materials show a reduction of breakdown strength under mechanical stress. As a group, insulating materials are not particularly strong, so mechanical stress should be avoided anyway, but this is a further reason for doing so.

BREAKDOWN OF GASES

The basic process that leads to breakdown in a gas has been discussed in relation to phenomena in the electric arc, in Chapter 4. When the breakdown is initiated only by the voltage gradient in the gas between the electrodes the principal factors governing the level of breakdown are the physical properties of the gas, the electrode spacing, the shape of the electrodes and the waveform of the applied voltage.

The properties of the gas which affect the dielectric strength are basically molecular, and relate to the energies required to liberate an electron in an inelastic collision, the mass and hence the acceleration of the atoms and molecules when ionised, and whether the atoms have the electro-negative property of electron attachment.

Smoothly curved electrodes, e.g. spheres, create a uniform voltage distribution between them. When the voltage has been raised to a level that causes cumulative ionisation of the gas – the 'electron avalanche' – the spark propagation is rapid, as all the gas is more or less equally on the brink of breakdown.

When either electrode has sharp edges or a sharp point, the voltage stress at that point will be high, and diminish towards the other electrode, if that has a large radius. If both electrodes are sharp, the stress is higher at both ends and lower in the middle. Ionisation thus occurs first at the point, and propagates outwards as the voltage is raised further. The breakdown is preceded by streamers from the electrodes and assisted by the high stresses at the tips of the streamers. This ionisation of the gas before breakdown is called *corona*.

The formation of corona is more pronounced in electro-negative gases as the body of the gas retards the electron formation through electron capture. As the gas pressure increases, the voltage gradient required to accelerate an electron to the velocity needed to release further electrons by collision increases, because the distance available for that acceleration is reduced. Thus the breakdown strength of any gap in a gas increases with pressure. In an electron-attaching gas the pressure also inhibits the corona formation with the result that the breakdown strength actually reduces as the pressure increases up to a critical pressure –

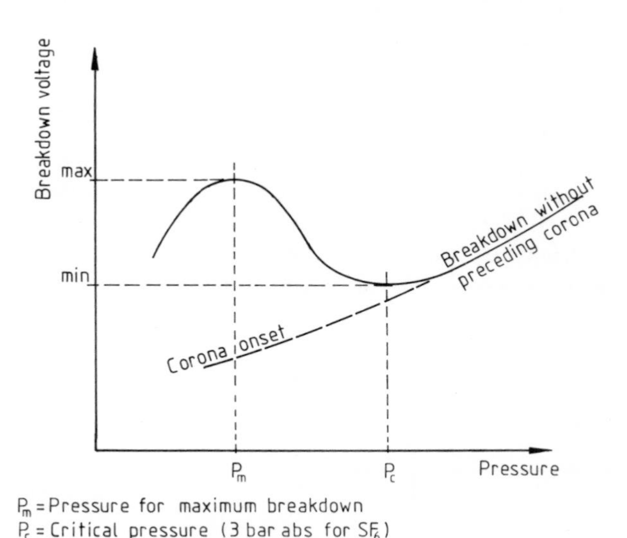

Fig. 6.3 Breakdown in a non-uniform gap in an 'attaching' gas.

about 3 bar abs for SF_6 – and then the breakdown curve becomes a continuation of the corona inception curve (see Fig. 6.3).

Since time is a factor in the development of the breakdown path in gases it is to be expected that if the voltage application is only brief, the value required to cause breakdown will be higher. Thus the test applied to simulate the effect of lightning, the impulse voltage test, results in higher breakdown values than a test at power frequency.

The combination of these factors and the effect on the breakdown voltage of gaps in a gas of particular interest to switchgear designers is shown in Figs. 6.4 and 6.5. The gas is SF_6 and Fig. 6.4 illustrates the breakdown strength for a uniform field, in a sphere gap, for both power frequency and impulse voltages.

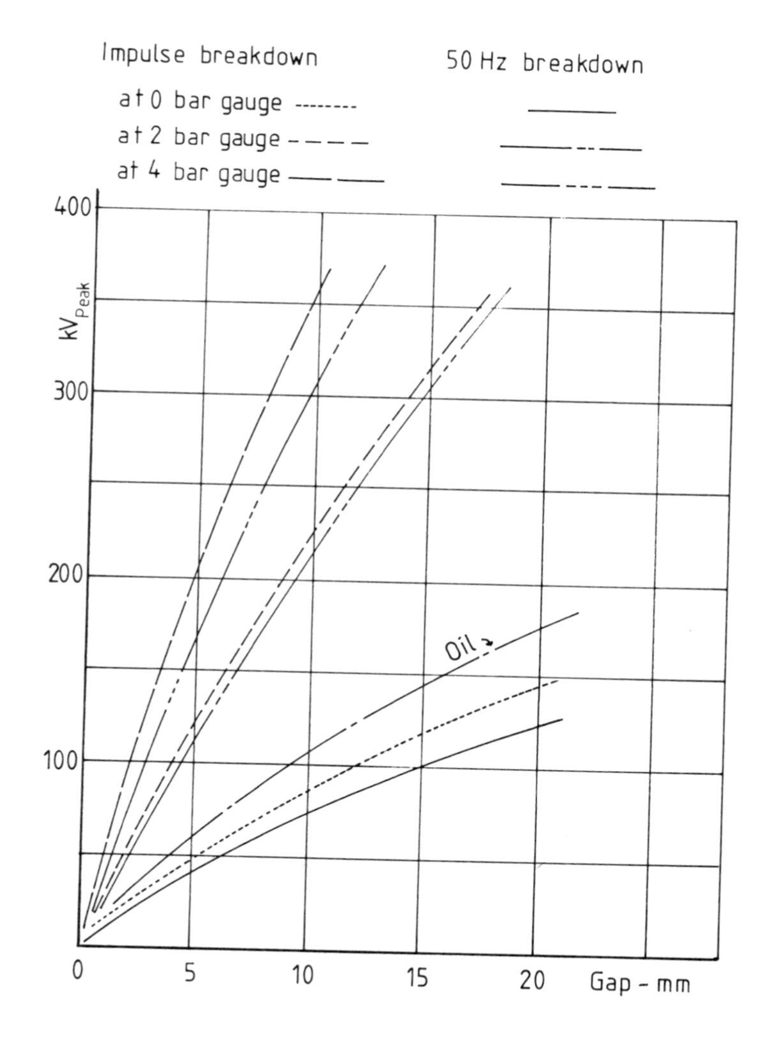

Fig. 6.4 Breakdown in a uniform gap in SF_6 and oil.

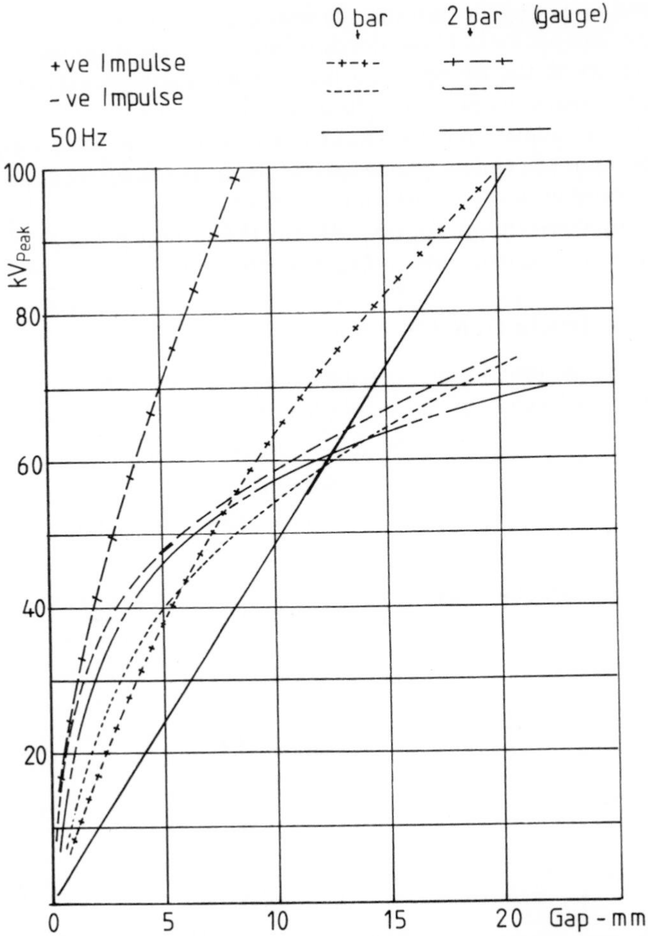

Fig. 6.5 Breakdown in a non-uniform gap in SF_6 and oil.

The effect of creating a non-uniform field, by substituting a point for one of the spheres, is shown in Fig. 6.5. The values are significantly reduced, and there is also a pronounced polarity effect displayed by the impulse tests. The effect of the corona discharge in artificially increasing the 50 Hz breakdown values at atmospheric pressure is also marked, the values for 2 bar gauge being lower than at atmospheric pressure at the wider spacing of the electrodes. These figures conceal the fact that unacceptable corona begins to occur well below the flashover point in the non-uniform gap at pressures below about 3 bar gauge in SF_6, as shown in Fig. 6.3.

BREAKDOWN IN LIQUID DIELECTRICS

The mechanism of breakdown in liquid insulating materials resembles that in gases. A conducting channel has to be produced within the liquid, and the energy of ionisation caused by the action of the voltage gradient on free electrons creates microscopic gas pockets. These develop into channels of ionised gas and the electron density at the tip creates a high voltage gradient which continues to propagate the spark channel until flashover occurs.

Therefore, it follows that the breakdown of a gap in an insulating liquid exhibits the same dependence on electrode shape as shown by gases. The most important liquid used in the distribution switchgear in service today is mineral insulating oil conforming to BS 148 or IEC Standard 296. Figures for breakdown in this oil are shown in Figs 6.4 and 6.6 for comparison.

COMPOSITE DIELECTRICS

In practical electrical plant, it is unusual to use insulating materials in isolation so that the breakdown data can be applied directly in the design process. It is

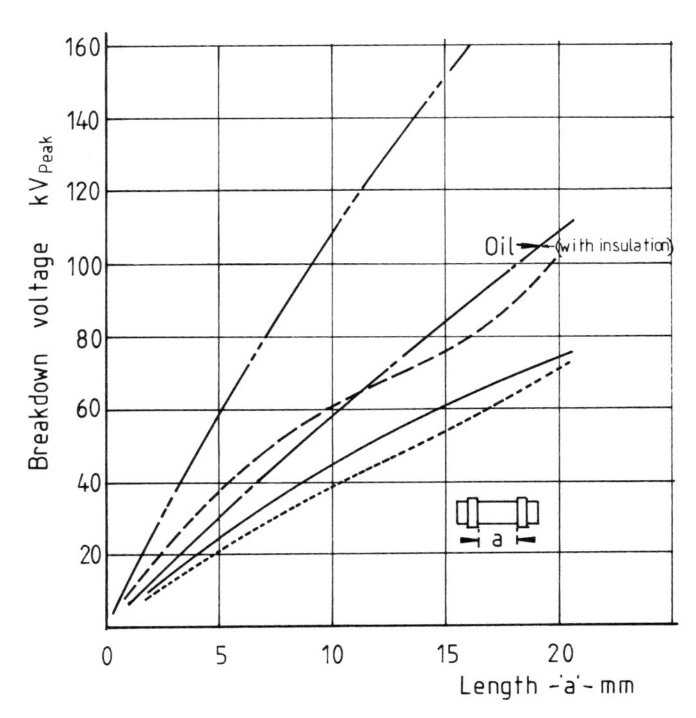

Fig. 6.6 Effect of solid insulation on the breakdown in SF_6.

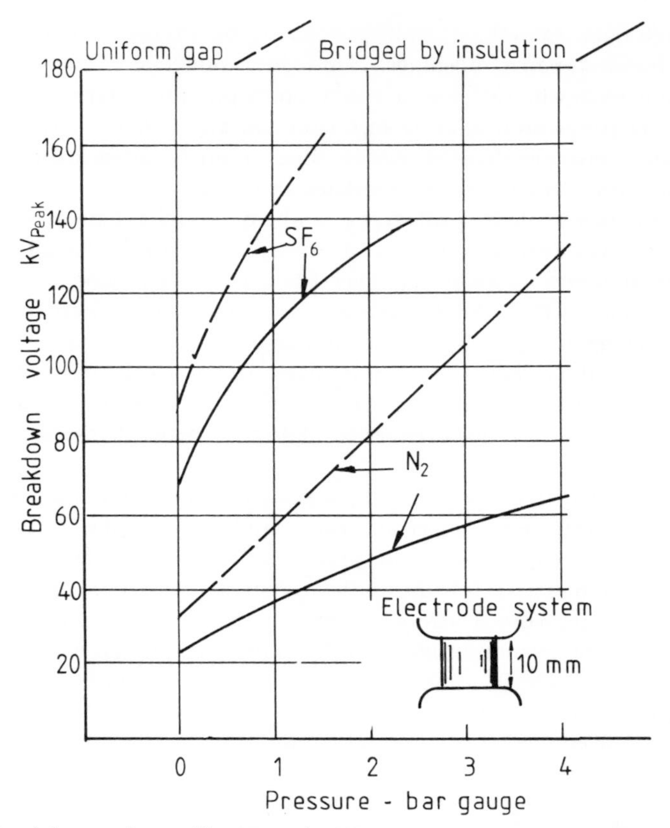

Fig. 6.7 Breakdown of a uniform gap in SF_6 and N_2 at 50 Hz showing the effect of bridging the gap by solid insulation.

necessary therefore to obtain an understanding of the behaviour of insulating materials in combination.

One typical case often met with in switchgear construction is the use of solid insulation between two conductors, the whole then immersed in a liquid or gaseous dielectric. The effect of an example of this arrangement is shown in Fig. 6.6, where a tubular piece of solid insulation has two electrodes wrapped round it at different spacings and the breakdown strength at 50 Hz is then measured in SF_6 gas at different pressures. Figure 6.6 also shows the corresponding results for insulating oil. A similar case of a reasonably uniform-field gap, bridged by solid insulation, is compared in Fig. 6.7 with the results for the unbridged gap. It will be seen that a similar reduction in the gap strength occurs in SF_6 and N_2.

Another type of 'composite' occurs when arcing takes place in an oil circuit-breaker and free carbon particles are created by the release of the hydrogen used in the extinction process. This carbon consists of very fine particles, many of which remain in suspension, and others fall onto whatever insulation surfaces there are. There is also a strong attraction of these light carbon particles towards areas of high field strength, typically the ends of buried foils in condenser type synthetic bonded paper bushings. This does not cause any measurable deterioration, even when allowed to reach totally unrepresentative proportions, but carbon coatings on insulating surfaces connecting phase conductors have been known to give problems in service. Modern oil circuit-breakers avoid the use of insulation that can be contaminated in this way.

From data such as these, prototypes can be built and exhaustively tested to ensure that adequate margins of safety are designed in to allow for all the factors and long term effects to which reference has briefly been made.

Of more concern than the deliberate use of composite dielectrics described above is the unconscious use of them due to the accidental inclusion of contaminants in the principal insulation material. One of the most common of these is air, included unwittingly in materials such as cast resin, or occurring as part of the design in the joints between insulators, or between insulators and the adjacent conductors.

The appendix to this chapter illustrates a classical condition of an air space in solid insulation between two plane electrodes, and evaluates the voltage stress in the void as a function of the total voltage stress and the ratio of the void thickness to the insulation thickness. It is shown that the stress in the void approximates to K times the average voltage stress, where K is the dielectric constant (specific permittivity) of the insulation material. For epoxy resin a typical figure for K would be 4, so the voltage stress on the air in the void would be four times the average stress, and this could quite possibly lead to ionisation of that air. Fortunately, test methods for detecting the presence of voids by measuring the occurrence of such discharges are easy to apply as a routine measure in the manufacture of such insulators.

LONG TERM RELIABILITY

Switchgear is installed to protect the distribution system in the event of an electrical fault occurring. It follows that switchgear should have a long term reliability to ensure that it does not itself add significantly to the probability of a fault occurring. The information in the possession of the designer enables him to produce a design prototype which has ample margin to avoid insulation breakdown when in new condition. Quality control, routine tests and diagnostic procedures ensure that production models correspond in their characteristics with the prototype. However, the one factor that is impossible to check on a new design, particularly if new insulation materials are incorporated, is how it will behave during its service life.

The best indicator of this is based on the known performance of similar designs, and there is no substitute for service experience, particularly if that experience already extends over a typical life span under severe climatic conditions. It is therefore no surprise to find that one of the most frequently asked questions by potential customers to a switchgear manufacturer relates to the location and period of service of his previous installations.

Manufacturers, therefore, are continuously evaluating insulation materials under conditions intended to represent, and indeed surpassing in severity, the worst service conditions in which their product is likely to serve. By devising accelerated life tests, usually by combining temperature and humidity cycles of shorter period than in service, it is possible to obtain indications of behaviour in periods of months rather than waiting years! The use of insulators of known service reliability as a comparison is a good guide to the performance of the test objects.

Attempts to shorten further the test period by artificially polluting insulation surfaces are being proposed, but it is felt in many quarters that the ratio of test period to intended life span cannot sensibly be reduced below a figure of 15 to 20, and that the introduction of artificial test conditions could prove misleading. Natural pollution, by dust, chemical particles, and in coastal sites, salt, occur in service, but it must be expected that any serious pollution of this sort would be removed in the course of maintenance.

In using insulation in the design of switchgear, factors of safety must be applied based on the maximum voltage stresses for which the equipment is intended. These need to take into account the statistical variation and the effect of the various factors discussed in this chapter, together with allowances for deterioration which experience dictates will occur during the lifetime of the switchgear. The stresses used after this evaluation will be far below the intrinsic breakdown strength.

APPENDIX: INSULATION ASPECTS OF SWITCHGEAR DESIGN. CALCULATION OF THE ELECTRICAL STRESS IN AN AIR-FILLED VOID IN SOLID INSULATION

Figure 6.8 shows a thin void in solid insulation of thickness t, the length being assumed sufficient for end effects to be ignored. If K is the relative permittivity of the solid material, and K_0 is absolute permittivity, then the capacitance of a part of the void with walls of unit area is:

$$C_v = (K \times K_0)/t_v$$

Similar calculation derives the capacitances of the two portions of the insulation between the void and the electrodes. Consequently the total capacitance of the three unit capacitors in series is:

$$K \times K_0/(t_1 + t_2 + K \times t_v)$$

As the voltage is shared in the ratio of the capacitances it can be shown that

$$V_v = V \times K \times t_v/(t - t_v(K - 1))$$

It then follows that the stress in the void E_v is given by:

$$E_v = E_{av} \times K/(1 + (K - 1)\, t_v/t)$$

where E_{av} is the average stress, $= V/t$ kV/cm. In practice $t_v << t$, so to a first approximation:

$$E_v = K \times E_{av}$$

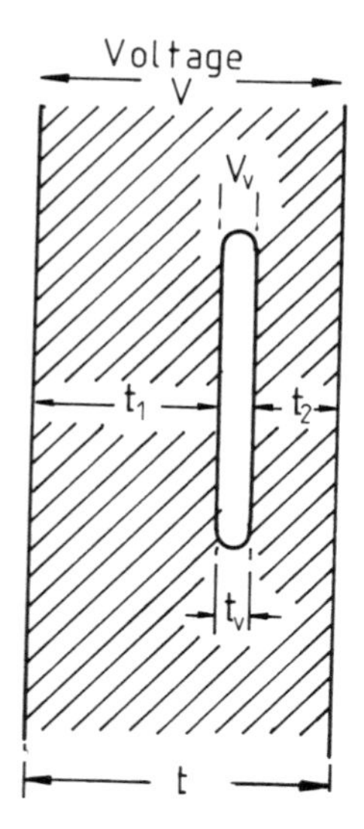

Fig. 6.8 Electric stress in an air-filled void in solid insulation.

CHAPTER 7

Mechanical Aspects of Switchgear Design

When the electrical aspects of switchgear were examined in Chapters 2 – 4, including the factors affecting arc interruption, it became clear just how much of a circuit-breaker's life is devoted to dealing with transient phenomena. This is also true of the mechanical performance.

Much of the design work associated with switchgear relates to its mechanical requirements, yet they occupy an infinitesimal part of its service life. Most of the time switchgear is static, either open or closed. Except for a relatively small number of equipments which have an active controlling role, e.g. switching arc furnaces or rolling mill motors, switchgear is operated very few times in a year. Yet when it is required to open, particularly if there is a fault condition in the system, there must be no hesitation, and its timing and speed of operation must be within its operating tolerances.

Even when it does operate, this is soon over, so the usual mechanical engineering problems of dealing with fatigue, wear, etc., arise in only a limited way. Resistance to corrosion, low friction and high reliability are the principal objectives of the switchgear designer. Ideally the mechanical parts of switching equipment should be able to remain in service for long periods – years – without requiring maintenance, a subject that is dealt with in more detail in Chapter 15.

INFLUENCE OF OPERATING TIMES

Time, then, is a very important parameter in mechanical performance, and the size of the time intervals with which the switchgear designer has to cope are worthy of a little examination. The second is too large a unit to use and the millisecond is the more usual unit of time in this field. One loop of power frequency current lasts for 10 ms at 50 Hz or 8.33 ms at 60 Hz. Fault currents can cause heavy damage in periods considerably less than one second, so the whole process of detecting the fault, initiating the opening of the circuit-breaker, releasing the opening mechanism, accelerating the contact system to its optimum speed and extinguishing the arc, must all be capable of happening in a few current loops. If we ignore the first element, which is adjusted by the user to suit the system's overall protective plan, the rest of the activity typically lasts between 3 and 5 cycles, i.e. from 25 ms to 42 ms at 60 Hz or from 30 to 50 ms at 50 Hz. Typically, this time is divided up as follows:

	50 Hz *ms*	*60 Hz* *ms*
Time from trip initiation to mechanism release	10	10
Time from contact release to contacts parting	15/25	15/25
Arcing time*	13	11
Making a total of	38/48	36/46

* For a modern vacuum or SF_6 circuit-breaker; oil or air circuit-breakers have longer times at fault currents below about 60% of rating.

This indicates that special measures are usually required, partly in the electro-mechanical release area, and partly in the contact acceleration region, to meet the 3 cycle rating, when that is the desired aim of the designer. It also indicates that this problem is aggravated for the American designer, who benefits marginally from the shorter time between current zeros, but has to be quicker off the mark than his European counterpart when circuit-breaker times are measured in cycles, since the mechanical parts do not recognise the frequency difference! As the 3 cycle rating is American in origin, they must be used to meeting this requirement.

CIRCUIT-BREAKER OPERATING MECHANISMS

Circuit-breaker operating mechanisms work on the principle that the closing effort, from whatever source, both closes the circuit-breaker and charges the opening springs. Thus, when the circuit-breaker is closed there is a bias towards the open position, and the circuit-breaker is held closed by a light prop or catch. Release of this, usually by a low power solenoid but also manually, allows the circuit-breaker to open. There are circuit-breakers that differ from this but they are few in number. Some of these arrange for the opening springs to be charged first, thus reducing the energy requirement for the closing of the circuit-breaker, and others use pneumatic or hydraulic mechanisms, in which case the closing and opening operations are usually separately driven from the same stored energy.

The prime mover, or source of closing power, is generally a spring, compressed either manually or by a motor driven device, or a solenoid. Many years ago, direct manual closing of circuit-breakers was common at low fault levels, typically up to 12.5 kA at 11 kV or 22 kA at 3.3 kV, but today direct manual closing of any switching device is hardly ever permitted when the system is energised, because of the possibility of closing on to a fault condition. With a direct manual mechanism, unless the operator uses what would normally appear to him to be excessive force, the chances of him closing the device safely are poor. With increasing fault levels, and recognition of possible risks, all equipment that is designed to switch a live circuit, that is to say, everything except a disconnector, should have a power closing mechanism.

As mentioned, circuit-breakers are normally biassed to the open position by springs. It is not common, or possibly even desirable, to treat switches in the same way. In the event of a fault passing through a switch, it is designed to withstand safely the thermal and dynamic effects of the fault current for a limited time. However, with the passing of years, and allowing for inadequate maintenance, there always exists a slim possibility that the electromagnetic forces might shock a circuit-breaker type trip mechanism into allowing the switch to open, with probable disastrous consequences. So switch mechanisms, although power operated, are of the simplest type, in which a spring is manually charged and then, as the end of the charging stroke is reached, the spring is automatically released, either to open or close the switch. This type of mechanism is usually referred to as an *independent manual mechanism*.

Other common types of closing mechanism include the spring mechanism in which the springs are first charged and held in that condition by a latch. At a convenient time they can be released, either manually or by a solenoid release. The latter enables the operation to be initiated remotely. For full remote operation, it is arranged for the charging operation to be performed by an electric motor.

The other common power source used for the closing of distribution circuit-breakers is the solenoid. In this type a large and powerful coil surrounds a cylindrical armature and a fixed pole piece. Operation is usually through a low voltage contactor from a d.c. source. The provision of this d.c. source is not popular with most users, which explains the popularity of the spring mechanism. The source is generally a secondary battery and unless this already exists because of some other emergency power requirement, is expensive to install just for circuit-breaker closing. Keeping it in good order is also a possible problem. However, if remote operation of the switchgear under all possible conditions, including electrical power failure, is a requirement, then the solenoid mechanism has its virtues.

Reference has already been made to the time element in circuit-breaker operation, and it is interesting to explore the accelerations, masses and forces that can be needed to meet the requirement for short operating times. In order to establish an adequate contact gap within the arcing time of a modern circuit-breaker, contact speeds in the order of one metre per second are required. Consider the best case, a vacuum interrupter that at 11 kV will have a maximum gap of some 8 – 10 mm. It really needs to have opened that gap in 8 – 10 ms, which means an average speed of 1 mm/ms or 1 m/s. Oil circuit-breakers perhaps need speeds of two or three times this for the same purpose. Depending on the type of arcing contacts used, the distance during which the moving contact system can be accelerated to reach this speed is quite short.

Tulip type contacts, discussed in Chapters 8 and 9, have a contact engagement in the region of 20 – 25 mm, but detract from the opening forces because of the friction they impose on the moving contacts. Butt contacts, on the other hand, have very short travel, but add to the opening forces by the contact loading

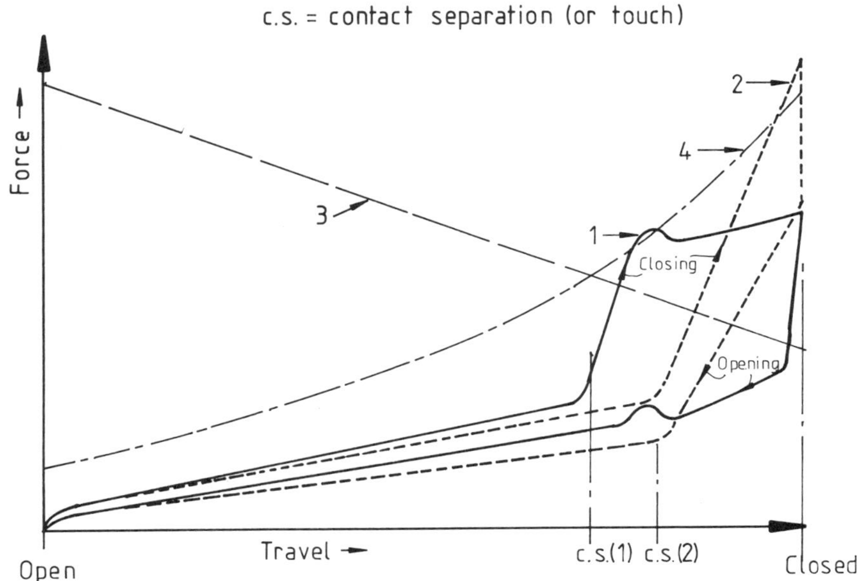

Fig. 7.1 Circuit-breaker force/travel diagram.

springs. If the circuit-breaker is a heavy current design having main contacts to increase its normal current rating which must break before the arcing contacts part, then a little longer travel is available. From this simple example it can be seen that the design of an operating mechanism is no easy task.

Figure 7.1 shows in diagrammatic form the travel/force relationship for two hypothetical circuit-breakers. One is a vacuum circuit-breaker with butt contacts, and the other is for an SF_6 design with wiping contacts. The first part of the closing travel represents the gradually increasing load presented by the charging of the opening springs and then the contacts meet.

The additional load is steep and high in the butt contact case (2), and lower and more gradual in the version with wiping contacts (1). In the latter case, though, the load comes on sooner. The lower curve in each case is the condition for opening slowly, with the opening springs now providing the driving force. The difference between the closing and opening curves represents the friction losses. The area under the top curves is a measure of the work done. Taken together with the time in which the action takes place an estimate of the power requirements can be made.

To set against this force and power requirement is the way in which force and power are developed by the energy source. The solenoid develops a low force when the air gap is large, i.e. at the beginning of its travel, and this force increases non-linearly as the air gap diminishes. The shape of this curve is shown on Fig. 7.1, and at first sight gives a good match with the load requirement. The spring, on the other hand, develops its peak force at the beginning when it is fully charged and this falls as it is discharged.

For this reason circuit-breaker spring mechanisms use substantially preloaded springs, which gives a flatter characteristic, but requires bigger springs, since only part of their work content is used. This fact is important for maintenance fitters, because the preloaded springs need to handled with circumspection, if the mechanism is dismantled for any reason. This characteristic is also suggested in Fig. 7.1, and appears to be quite wrong for the job.

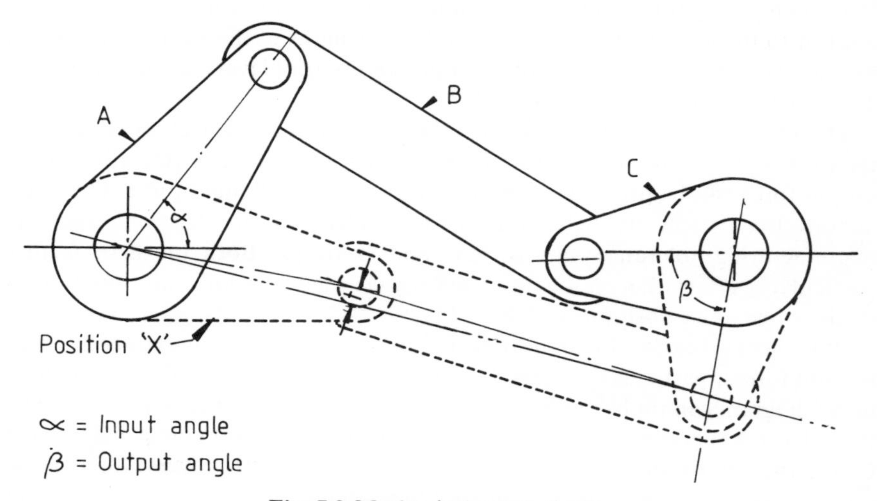

Fig. 7.2 Mechanical toggle linkage.

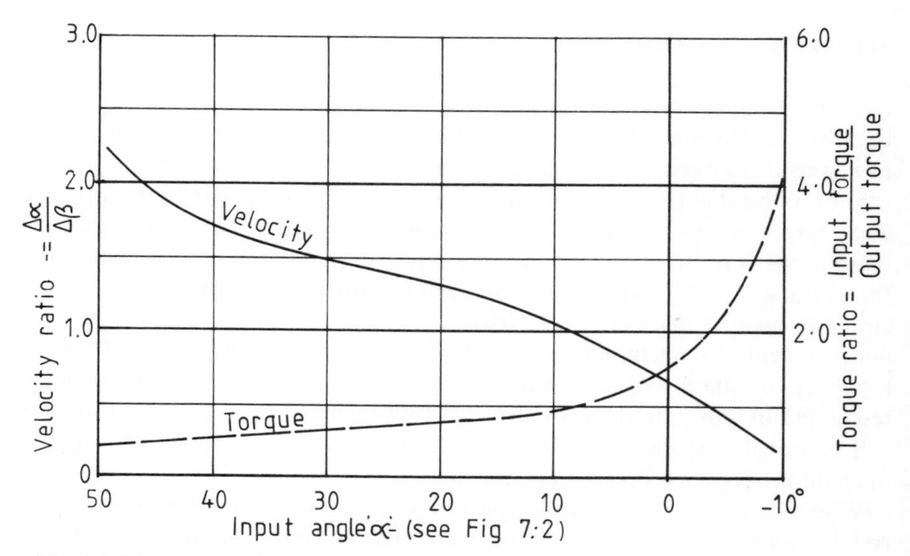

Fig. 7.3 Torque/travel and velocity-ratio/travel relationship for the toggle linkage.

THE 'TOGGLE' LINKAGE

However, between the energy source and the circuit-breaker moving assembly lies the operating mechanism itself, with its catches, spring loading cams, releases and various links. This has to be designed to match the characteristic of the power source to the load requirements of the circuit-breaker.

A simple mechanical linkage that forms a principal part of most circuit-breaker mechanisms is illustrated in Fig. 7.2, and its force and travel characteristics are shown in Fig. 7.3. This is the 'toggle linkage' and it can be seen that a constant torque applied to the input lever A results in an increasing torque applied to the lever C. By varying the relative lengths of the levers and links, as well as the angular movement of the input and output levers, a wide variety of 'transfer functions' can be produced.

The principal trade-off that occurs is between force and travel, or velocity, at the output. It will be seen that in the position X in Fig. 7.2, a virtually infinite output force can be produced with no travel, for a small input force. The other factors that affect its behaviour as a mechanism are friction, and the manufacturing tolerances in its pins and bearings. Both of these factors particularly affect the geometry of the toggle when the links are nearly in a straight line as in position X in Fig. 7.2.

With heavy bearing loads near to the toggle point, and consequently large friction forces, particularly on the bearing between links A and B, the distance t in the diagram cannot be made too small or the linkage will become irreversible. If the lever A represents the closing input to a circuit-breaker connected to lever C, and the latter wants to open, it would be impossible if the reverse force of the opening springs acting from C could not break the knuckle joint between B and A, even though they were not geometrically in a straight line.

MECHANISM DESIGN

Figure 7.4 illustrates the use of the toggle linkage to develop a simple circuit-breaker mechanism which contains the main elements required. In (a) the mechanism is at rest and the circuit-breaker is open. The lower end of link (2) cannot move because the trip roller T is resting on the release prop (6). Therefore, when the closing force F is applied to the rod (7), an increasing torque is transmitted to the circuit-breaker closing lever (4). When the rod (7) reaches the end of its travel, a closing latch (5) drops behind the drive roller C, locking the circuit-breaker closed (see Fig. 7.4(b)).

Movement of the trip release prop (6) by the solenoid (not shown) allows the lever (1) to rotate and the circuit-breaker to open, links (2) and (3) being restrained by other mechanism elements (not included to avoid confusion). It will be noted that the close roller C rolls across the drive rod (7) during the opening process (see Fig. 7.4(c)).

When the closing springs are recharged, or the solenoid is de-energised, drive rod (7) moves to the left and the lever (1), which is spring biassed in an anticlockwise direction, rotates until the trip prop can reset under its roller.

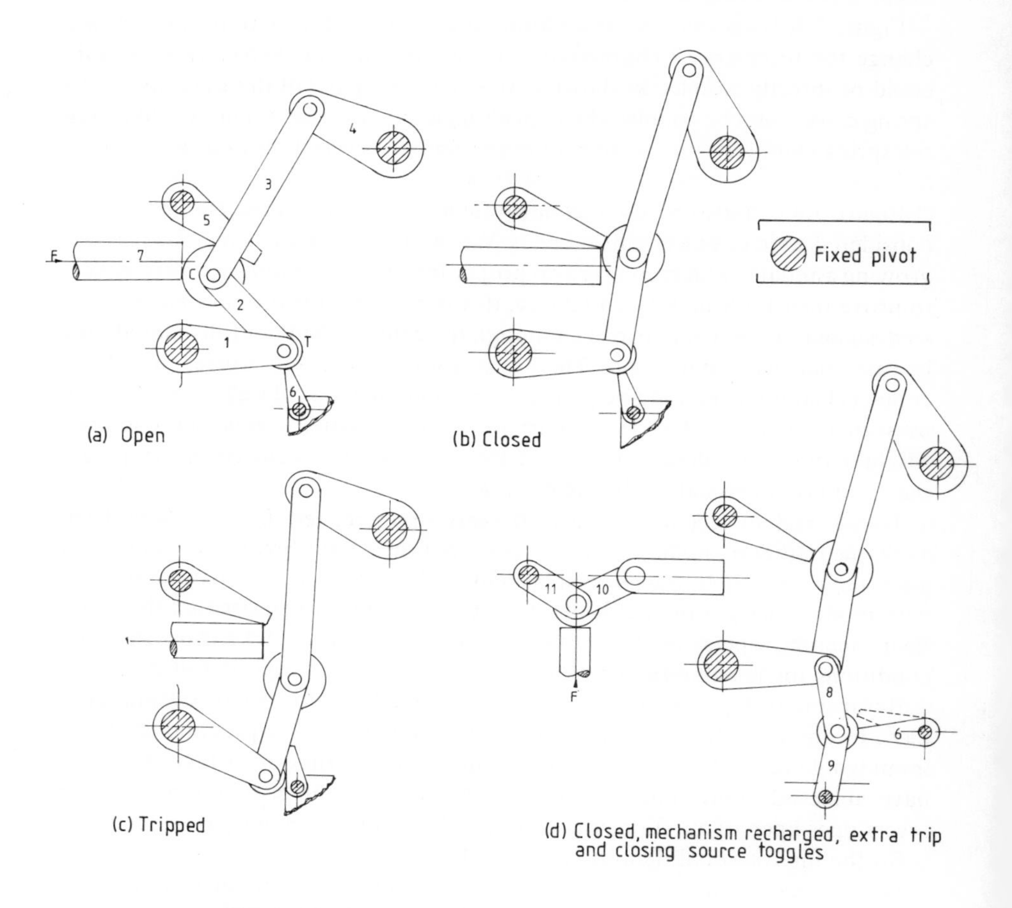

Fig. 7.4 Elements of circuit-breaker mechanism design.

Components in the spring charging system will ensure that no attempt to reclose the circuit-breaker can take place until this resetting has occurred. At the same time, the links (2) and (3) regain the positions shown in Fig. 7.4(a), the roller C returning to its position under the latch (5).

In the linkage shown so far the load on the trip prop (6) with the circuit-breaker closed will be quite high, and a correspondingly high force will have to be generated by the trip coil in order to move it from under the roller T because of the friction generated. Figure 7.4(d) shows how the addition of a further toggle linkage can be used to reduce considerably the tripping forces required. The force required to support the circuit-breaker reaction forces on the links (8) and (9) when they are nearly on toggle are quite small, and a low energy tripping source can be used. Obviously a price must be paid for this benefit and it is in the form of an increased opening time, because the links (8) and (9) have to be accelerated out of the way before the links (2) and (3) can drop, allowing the circuit-breaker to open.

Figure 7.4(d) also shows the addition of a toggle to the drive rod, which will change the force/travel characteristic of the mechanism. Typically a solenoid could be directly coupled as shown in the first three parts of this diagram, and a spring drive could be coupled through a toggle as shown here. This would match the spring characteristic to the load requirement, as was discussed earlier in the chapter.

Figure 7.4 can also be used to illustrate a feature that in the past used to be considered a virtue in a circuit-breaker. When the fault capacity of networks was growing annually, and there were no proper facilities for testing circuit-breakers to prove their fault-making capability, it was desirable for the mechanism to be so designed that it would trip even though the contacts had not fully closed, just in case that event happened. There was always a risk, of course, that if the contacts had not achieved adequate engagement, they would weld because of the heat generated by the heavy current in the partially engaged contacts. Manufacturers also designed inertia devices that would accelerate the tripping if the circuit-breaker stalled during closing.

Today, and in fact for the last 50 years or more, test facilities have been available so that manufacturers can prove that the circuit-breaker will close and latch against its rated short-circuit current peak. Furthermore, it is arguable that with modern designs of circuit-breaker, it is important to ensure that they start their opening operation from the fully closed position, to ensure that the conditions for meeting the breaking current rating are properly fulfilled.

Returning to Fig. 7.4(a) and (b), it can be seen that if the trip operation takes place at any time during the closing stroke, the linkage collapses and the force input would cease. However, the moving masses inside the circuit-breaker would have achieved some momentum, so where they might have stopped before reversing is open to question, and conceivably might have made passing contact.

So the international definition of 'trip-free' has changed over the years and now only means a mechanism in which the opening operation can prevail even if the closing command is maintained, and a note is added to the effect that it may be necessary for the circuit-breaker momentarily to reach the closed position to ensure proper breaking of the current which may have been established.

Another feature that may be required of the operating mechanism is an auto-reclose facility. The purpose of this is discussed in Chapters 10 and 11 and the time interval during which the contacts must remain open is very short, of the order of 0.3 s.

With a solenoid mechanism, the armature normally resets as soon as the current supply to the coil is cut off. Therefore it only requires the closing current to be reconnected for reclosure to occur. The other feature that is needed is to ensure that the mechanism links have reset before the solenoid is re-energised.

However, to achieve the required 'dead time' between arc clearance and contact reclosure with a spring mechanism, it is necessary to recharge the springs as soon as the circuit-breaker has been closed, so that it is ready to reclose as soon as it has opened. Figure 7.4(d) shows that this simple mechanism allows the drive rod (7) to be retracted by recharging the closing springs without affecting the latching of the circuit-breaker or its opening functions.

Figure 7.1 shows the mechanical forces of circuit-breakers under no-load and virtually static conditions. Under short-circuit conditions some of the forces are substantially increased by electromagnetic effects. This is particularly true of wiping contacts, as discussed in Chapter 9, and the increased energy required to close the circuit-breaker fully home against its rated peak making current is considerable.

This is a condition that is probably never experienced by the majority of circuit-breakers and even those that do operate under such conditions may only do so once or twice in a lifetime. The rest of the time, excess energy is being pumped in every time the circuit-breaker is closed. The situation is also aggravated by the tolerances that have to be applied to the operating mechanism due to the variations in its own construction and in the source of energy. The largest variation is probably in the terminal voltage supplied to a solenoid from a storage battery.

In testing the circuit-breaker, allowance must obviously be made for the worst condition of that storage battery, i.e. a partially discharged condition. In service, the most likely condition is that the battery will be on continuous float charge, so that the cell voltage will be above nominal. Further, with the relatively heavy current taken by a closing solenoid, there is a difference in the on-load terminal voltage at the coil, depending on whether the circuit-breaker concerned is close to the battery or far away. The IEC and BS standards require the short-circuit making tests to be done with 80% of the rated voltage at the coil terminals – and it will be realised that this is only 64% of the rated power. On the other hand, the mechanical endurance test has to be performed with 95% terminal voltage and this is representative of the likely service condition.

The designer has to include in the circuit-breaker means of absorbing this excess energy, or mechanical damage could ensue. Air or oil dashpots are one solution, and the use of friction as an energy absorber is also useful, particularly in a circuit-breaker with wiping contacts. In this case the allowance for a degree of override when closing may be sufficient precaution.

This is also a reason for preferring spring mechanisms, because, although springs are subject to manufacturing tolerances, it is possible to design their use so that adjustment can compensate for these and the only force variation then is the presence or absence of short-circuit forces.

Much ingenuity has been shown over the years in the design of switchgear mechanisms and the development of new types of circuit-breaker has created new fields for study. There are other mechanical devices that are added to these mechanisms, such as no-volt or under-voltage releases. These are electrically operated releases that trip the circuit-breaker if the supply voltage disappears, or prevent the circuit-breaker from being closed if there is no supply voltage.

The scope for mechanical ingenuity comes when the action of the device has to be defeated in certain circumstances, for example if a circuit-breaker fitted with an under-voltage release has to be prevented from tripping out if it is closed on to a fault. Under those conditions the fault current would cause sufficient voltage drop to actuate the under-voltage release, yet the probable presence of an operator at the circuit-breaker location might make it undesirable for it to trip.

Elsewhere, the scope for novel design can occur in the field of interlocking to prevent certain undesirable or dangerous sequences of actions. This subject is also discussed in Chapter 15.

CHAPTER 8

Thermal Aspects of Switchgear Design

Switchgear has only two continuous rating duties as it sits quietly in the distribution system. First, it has continuously to withstand the system voltage applied to its insulation, and second it has to carry its rated current. In both cases these continuous duties must not lead to any deterioration in the equipment. Thermal design requires an understanding of the factors affecting both sides of the heating equation, in achieving a balance of heat input and cooling to arrive at an operating temperature that does not overstress any of the materials in the design.

AMBIENT TEMPERATURE

Since it is the final steady state temperature of the conductors in relation to their material and surface finish and the material of adjacent insulation that is the limiting factor in a design, it is important first to establish the design parameters for the maximum ambient temperature. This is specified in international standards as an average temperature over 24 h of 35° C with occasional peaks of 40° C.

However, international standards are based on average conditions in general use, so a manufacturer must also look to his intended markets to see if they are different from this average. Many tropical countries have national standards that vary the international recommendation, the figures ranging up to 15° C higher. Also, outdoor equipment may have to take account of the fact that it could be subjected to tropical sunlight for hours at a time, with the corresponding increase in external temperature that would create.

Once the design parameter of maximum ambient temperature has been set, the permitted temperature rise is obtained by subtracting that figure from the maximum safe operating temperature for the conductor and its associated insulation. Guidance for these safe operating temperatures is also available from the international standards, both for conductors and for insulation. For switchgear, these are contained in IEC Standards 694 and 56, and repeated in British Standards BS 5227 and BS 5311, and are summarised in Table 8.1. As for most of the information in these product standards it is based on a consensus of international experience. In this case it is based on service experience and research concerning the temperatures at which deterioration resulting from long term exposure to high temperatures begins to show.

Switchgear is looked upon by many users as a 'necessary evil' and a non-revenue earning capital expenditure, so losses due to the resistance of the conductors in the switchgear are seen as adding to that revenue loss. The

Table 8.1 Recommended temperature limits for contacts and material. (Based on Table V of IEC Standard 694)

Item	*Maximum operating temperature, °C*		
	in air	*in SF_6*	*in oil*
Contacts			
Bare copper	75	90	80
Silver- or nickel-coated	105	105	90
Tin-coated	90	90	90
Connections, bolted or equivalent			
Bare copper	90	105	100
Silver- or nickel-coated	115	115	100
Tin-coated	105	105	100
Oil in oil switching devices (near the surface)			90
Insulation materials, and metal parts in contact with insulation materials*		In any medium	
Class Y**		90	
Class A		100	
Class B		130	
Class F		155	
Class H		180	

* The maximum temperature limit for any metal or insulated part in contact with oil is 100° C.

** Insulation classification:

Class Y, cotton, silk or paper and combination thereof, unimpregnated.
Class A, as Class Y, but suitably impregnated or immersed in oil.
Class B, mica, glass fibre or asbestos, etc., with suitable bonding agent.
Class F, as Class B, but with higher temperature bonding agents.
Class H, silicone elastomer or Class B materials bonded with silicone resins.

reduction of these losses means increasing the quantity of conducting material and hence the cost. Both capital cost and revenue cost are resented, but at the time of purchase capital cost has the greater effect on immediate cash flow and gains priority. The result is that the greatest pressure on the designer is to keep just under the permitted temperature rise limits, knowing that there is an element of safety margin in the recommended figures.

HEAT GENERATION

The following discussion concentrates on the creation of heat due to the passage of power frequency alternating current through conductors in three-phase distribution switchgear. It tends to relate mostly to indoor metal-enclosed switchgear, since that brings more factors into the 'heat balance equation'. Direct current produces less heat than alternating current for equal values, equal meaning average d.c. equivalent to rms a.c. Outdoor equipment is dealt with in exactly the same way, with allowances for the greater range of ambient temperatures and for solar heating.

If R is the resistance of a unit length of conductor carrying a current of I amperes, the power loss is equal to I^2R watts. The value of I is known (the rated normal current), and the problem is to determine R. The d.c. value of the resistance can readily be calculated from a knowledge of the cross section of the conductor and its material. However, because of the change in direction of current flow 100/120 times per second, there are other influences to be taken into account when calculating the resistance to alternating current.

There are three influences and all increase the heat generation:

1. *Skin effect*. This arises due to the magnetic field set up by the change of current flow within the conductor, which has the effect of forcing the channels of current towards the outside, or 'skin'.
2. *Proximity effect*. This arises from the magnetic field set up in a conductor by currents flowing in adjacent conductors, which has a similar result of increasing the resistance by concentrating current flow in part of the conductor cross section.
3. *Iron effect*. If a conductor is close to surrounding steel work, and especially if it passes through a hole in the steel enclosure, as in a bushing, eddy currents will be created in that steel, leading to heating of the steel components.

SKIN EFFECT

The increase in resistivity due to the skin effect depends on the system frequency, the resistivity of the conductor material, the size of the conductor and its shape. The effect becomes more pronounced when the frequency is high, the cross-sectional area (c.s.a.) is large and the resistivity is low. The influence of the alternating field causes the current to concentrate towards the outside of the conductor, and this gives rise to the concept of a penetration depth, which may be defined as that depth of conductor which would carry all the current at the surface current density. This figure for copper at 50 Hz is about 6 mm. (For iron it is 0.5 mm.)

An approximation to the ratio between the a.c. and d.c. resistance for round bars is given by the following equation for values of A greater than 5 cm^2, where r is the resistivity (microhm cm), f is the frequency (Hz) and A is the c.s.a. (cm^2).

$$\text{Ratio } (R) = 0.3 + 0.056\sqrt{\left(\frac{A \times f}{r}\right)}$$

This formula is also suitable for square bars, but the use of rectangular bars reduces the effect, which explains the use of multiple strips of copper in heavy current busbars, for example. The effect is small when the c.s.a. is less than 10 cm^2 and will double the resistance for a solid round conductor of about 35 cm^2.

The use of bars with a 'shape factor' (ratio of width to thickness) of five will reduce the resistivity by some 10 – 15%, see Fig. 8.1.

PROXIMITY EFFECT

The proximity effect is important in compact switchgear where the three phase conductors are enclosed in a common chamber with relatively small spacings. The current in the adjacent phases creates an alternating magnetic field which reacts with the current in other conductors and leads to uneven current distribution, which manifests itself as an increase in resistance. The effect is not as great as the skin effect but in a 12 kV, 50 Hz circuit-breaker with average busbar spacings of 15 cm and a c.s.a. of 7.5 cm², it would account for about 10% increase in resistance. This is multiplied by the 11% increase due to skin effect in this section of busbar. Further detail is given in Figs. 8.1 and 8.2.

Finally, after making allowance for these increases in the d.c. resistance of the conductors, correction has to be made for the temperature coefficient of the resistivity at the final working temperature of the conductors. This is assumed to be the maximum temperature allowed, at the beginning of calculation, with refinement later as the detail emerges.

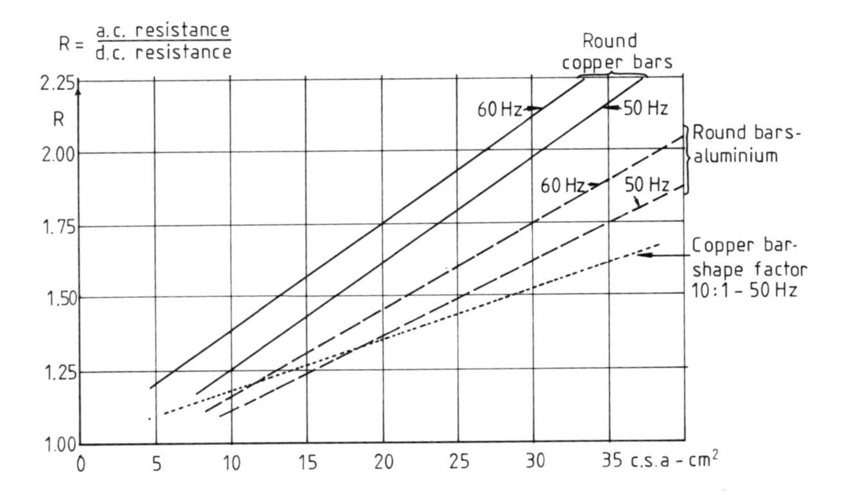

Fig. 8.1 Increase of conductor resistance due to skin effect.

IRON EFFECT

It is also possible to calculate the heating effect in steel sheets caused by a conductor passing through a hole. As a general rule the effect can be ignored for current ratings up to about 400 A, since the clearance needed to accommodate the insulation round the conductor is enough to keep the eddy currents in the steel to

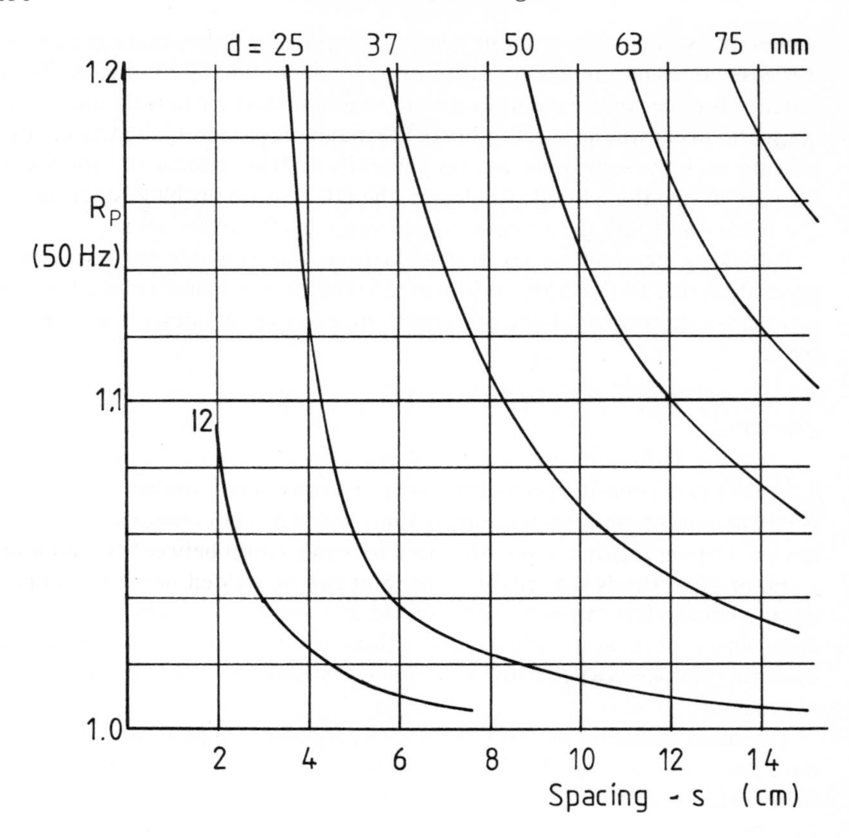

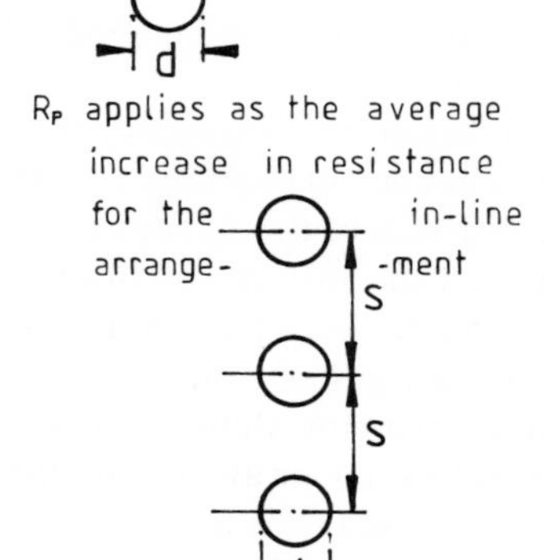

Fig. 8.2 Increase of conductor resistance due to proximity effect.

a level that will not cause undue heating. At currents above that value, however, it is wise not to permit a continuous iron circuit round any conductor. The usual practice is either to use a non-magnetic, stainless steel, or to weld inserts of non-magnetic steel into the sheet to break the magnetic circuit. If all three conductors pass through the same hole, as they generally do from cubicle to cubicle with the busbars of metal-enclosed switchgear, then there is no problem since the sum of the fields due to all the three-phase currents is effectively zero.

By taking account of the above factors it is possible to assess the heat generation due to the current flow in the conductors. Some conductors present problems due to their shape and some engineering adjustments will need to be made.

JOINTS

There are two kinds of conductor joint to be considered, the permanent or semi-permanent joint, and the 'temporary joint' made by the contacts of the switching device. Depending on the possible need to break a joint between two conductors, a choice of methods is available. The joint can be welded or brazed, depending on the material; it can be tinned, riveted and soldered, or it can be bolted. The application of heat to most conducting materials is usually deprecated in medium voltage switchgear, and such methods are labour intensive and expensive.

The most common material for the conductors is copper, and it is used in a hard drawn or at least half-hard condition; heating it, certainly if to brazing temperatures, reduces its ability to withstand mechanical stress. So the most common jointing method is by bolting, or sometimes clamping, the conductors together. Clamping takes up more space but it avoids putting holes in the conductors, and is most frequently used for the joints between conductors on outdoor switchgear, particularly where these are tubular, and embody flexible expansion joints.

The resistance of a bolted joint depends on the current distribution in the joint face, and the resistance of the joint face itself. The current distribution in the joint is not uniform and the current transfer tends to be concentrated towards the ends of the overlap. The joint resistance depends on the surface condition and the joint pressure. The surfaces must be clean and oxide free as far as practicable and this is easier to achieve with some materials than others. Aluminium oxidises quickly and its oxide is an insulator, so joints involving aluminium need more care than any other. The use of vaseline helps to keep air away from the metal and delays oxide formation. It also serves to seal the joint and prevent deterioration in service.

When bolt pressure is applied, the resistance initially reduces substantially as the oxide film is crushed and the area of contact round each bolt increases. Then a limit is reached beyond which further tightening causes very little reduction in resistance, so a properly designed joint does not need excessive tightening, and it is usual for a torque limit to be set by the manufacturer to prevent overtightening. Conduction is limited to an area only a little larger in diameter than the bolt head, although the addition of a stout washer under the bolt head makes for some improvement.

The bolt diameter obviously needs to be such that the necessary pressure to crush the oxide layer does not lead to overstressing of the bolt itself. This needs particular care when the joint is made by using a tapped hole in the conductor material itself. Where such a thread is used and the hole is blind, a stud is preferable to a bolt as there is no risk of it bottoming in the hole before it is tight. Because of the uneven current distribution in the joint there is no advantage in a long overlap and more than the distance necessary to fit two bolts in line is a waste of material, unless they are needed for mechanical reasons.

CONTACT DESIGN

The 'temporary' joints used in switchgear for the contacts of switching devices and for withdrawable circuit-breakers in metal-enclosed units fall into two groups: butt contacts and sliding or wiping contacts. Each group has many variations and the different characteristics of these types has a bearing on the design of the mechanical drives. It has been mentioned in an earlier chapter that SF_6 circuit-breakers need wiping contacts but vacuum circuit-breakers must have butt. Figure 8.3 shows some different types of contact and also illustrates some pitfalls to avoid in good design. High pressure line or point contact usually gives better and more consistent performance than trying to create an area of contact. Table 8.2 summarises some of the differences between the two groups of contact design.

TRANSFER CONTACTS

Another area of contact design that is worthy of mention is that of providing for current transfer to a moving contact. This is an essential requirement in any single-break circuit-breaker and many varieties have been developed. The simplest form is a bolted connection through a flexible conductor, usually laminated from a sufficient number of thin strips for the current rating, and to give adequate flexibility. If these are arranged so that the movement is controlled and does not impose any frictional forces, then a contact system with low mechanical and low electrical resistance can be achieved. Other types of contact relying on sliding or rolling contacts are illustrated in Fig. 8.4, all aimed at achieving a space saving, economical current transfer without introducing mechanical resistance to motion.

One of the recent additions to the methods of performing this function is the Multilam(TM) type of sliding transfer contact. It is made of a silver-plated copper alloy, and is punched out and shaped in such a way that there exist a large number of parallel, individual contact bridges between the two surfaces. Because of the large number of transfer points it does not require much contact pressure on each point, so it provides a very compact contact system with a high current rating and low friction (Fig. 8.5).

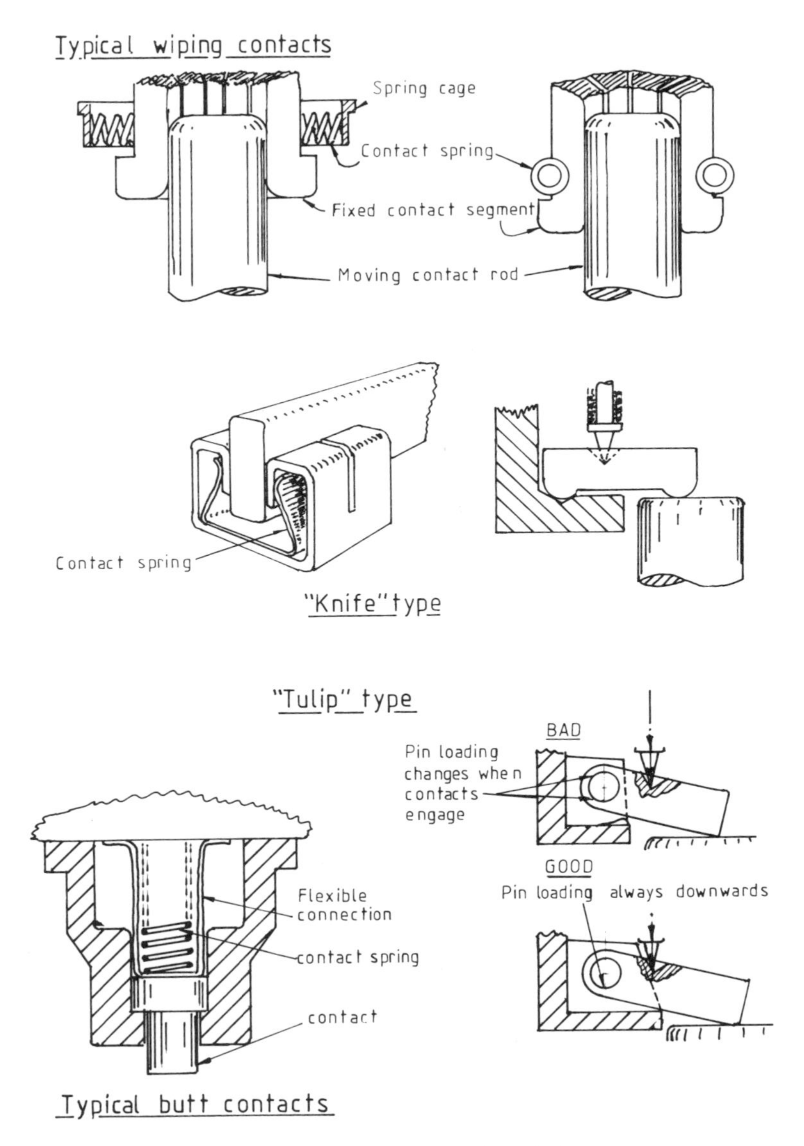

Fig. 8.3 Types of contact used in switchgear.

HEAT DISSIPATION

For heat to be transmitted from a body, it is axiomatic that it be at a higher temperature than its surroundings, and that the greater the temperature difference, the greater will be the rate of heat dissipation. As already mentioned, there are limits to the actual operating temperature of switchgear parts, therefore there are limits to the maximum temperature rise allowed. Some of these limits are listed in Table 8.1, based on British Standard recommendations.

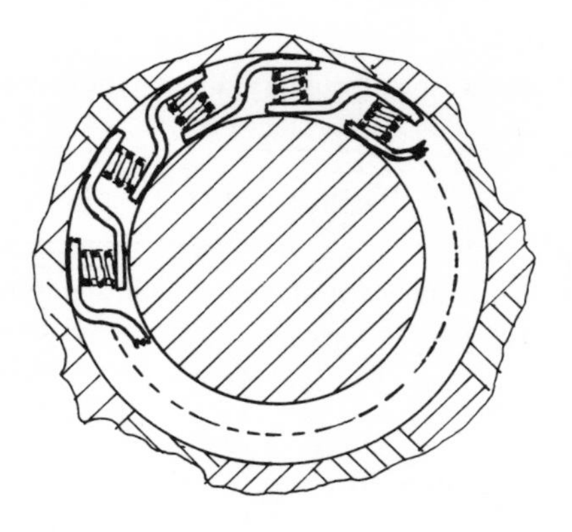

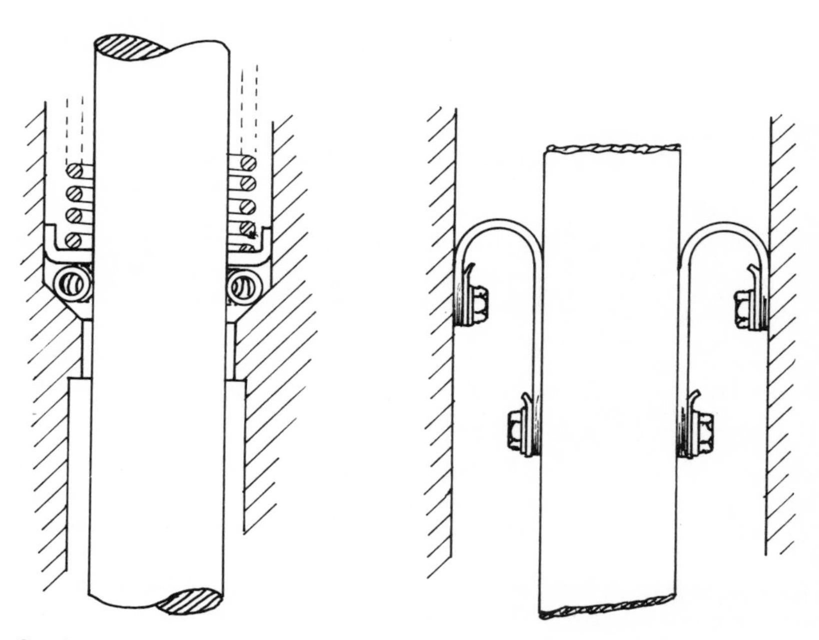

Fig. 8.4 Types of transfer contacts used in switchgear.

Table 8.2 Comparison of wiping and butt type contacts

Item	*Wiping contact*	*Butt contact*
Contact resistance	The wiping action is good for oxide removal	Insignificant wiping action, relies on high pressure for oxide penetration
Contact burning when interrupting fault currents.	The normal current carrying surfaces are not usually affected by arcing	Contact surface has to be designed to withstand burning, but generally rely on point contact anyway
Current carrying capacity	It is easy to provide quite large numbers of contact segments	Difficult to provide many segments without an unduly large moving contact
Current transfer to contact block	Bridge type spring-loaded line or point contact or bolted flexes	Spring-loaded pivots which require careful design
Effect on opening time	Long because of long in-contact travel and friction	Short because of short wipe and extra acceleration forces
Opening forces	Larger force required to overcome friction, but a longer time is available	No friction, but less time available, so larger forces needed
Absorption of excess contact force	Good – friction absorbs energy well, and there can be adequate override	Poor – resilient contacts and small permissible override
Effect of electromagnetic forces	Contact pressure increases due to attraction forces	Careful design needed to avoid pressure loss due to electromagnetic forces
Energy required to overcome electromagnetic forces	Large due to contact grip and long travel	Small due to lack of friction and short travel

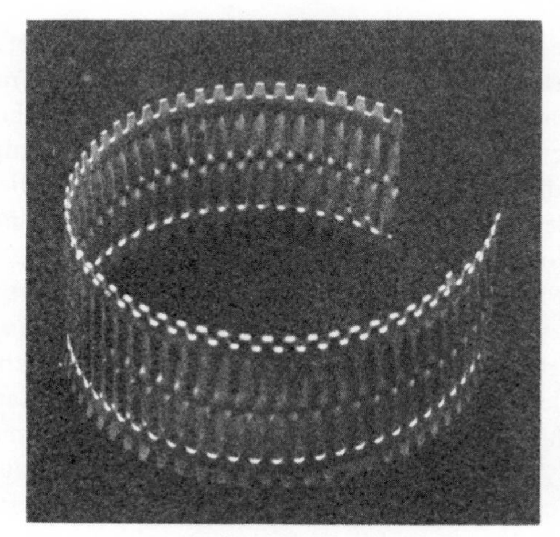

Fig. 8.5 A typical Multilam$^{(TM)}$ transfer contact.

There are three basic mechanisms of heat transfer:

1. *Conduction*. This is the mechanism whereby heat travels through a solid body by increasing the molecular energy and transferring the heat energy in this way through the material. Each material has a characteristic thermal conductivity, in the same way as it has an electrical conductivity. Materials with good electrical conductivity also tend to be good conductors of heat, silver being at one end of the scale and polystyrene foam at the other.
2. *Convection*. This is the mode of heat transfer in a liquid or gas. The layer of fluid next to the hot object is heated and becomes less dense, so it rises and is replaced by cooler fluid. The scientific representation of this process so that calculations can be made mathematically is too complex for easy use, and empirical relationships based on dimensionless groups of parameters have been developed which allow the behaviour of the heat transfer to be predicted.
3. *Radiation*. Also relevant to heat transmission in gases, although as the heat transfer is by electromagnetic radiation, like light or radio waves, it can be transmitted through a vacuum. This is an important factor in the cooling of vacuum interrupter contacts as convection is clearly not possible.

In any practical arrangement of switchgear, all three mechanisms play a part. Figure 8.6 gives an elementary picture of the mechanisms of heat transfer in a metal-enclosed switchgear cubicle. By breaking down the heat transfer problem into its separate parts it is possible to form a good idea of the temperature difference between the conductors and the surroundings required to carry away the heat being produced by the current flow. For those readers interested in a deeper understanding of the cooling processes and the derivation of the balance between input and output, further detail has been included as an appendix to this chapter.

The limitation of permissible temperature rise is caused either by the effect on contact or joint resistance, or by the effect of the heat produced on adjacent insulation. Chapter 18 describes how switchgear is tested to prove that the limits laid down in the standards are not exceeded, and the temperatures of contacts and joints can easily be measured by thermocouples. The temperature of embedded conductors is not so easy to establish, and the temperature gradient along those conductors needs to be calculated to predict the 'hot spot' temperature inside the insulation, and make sure that it is not likely to cause gradual degradation over the service life of the equipment.

In general there are many factors at work to ensure that the temperature rises achieved on the type test are hardly ever likely to occur in service. When a switchgear installation is being planned, the maximum current likely to flow in each circuit is calculated, using the worst possible combination of operating circumstances. The limits of temperature rise allow for a maximum ambient temperature rarely experienced in temperate climates, although often realistic enough for tropical applications. Then the next highest standard rating is chosen, and the combination of these factors usually ensures that a respectable margin exists between the maximum temperature rise likely to be experienced and that permitted.

SHORT-TIME RATING

One condition that can sometimes have an overriding influence on the design of conductors is the temperature rise which takes place when a fault current has to be carried for a time dictated by the protection system. In the UK, where extensive interconnected urban networks exist, and much old switchgear is still in service, the need to provide adequate time discrimination in the protection can lead to quite long time delays for those switching devices near the power source. This results in a requirement for most distribution switchgear in the UK to have a three second short-time rating. If the short-circuit rating exceeds the normal current rating by a factor of about 40, then the size of the conductors is likely to

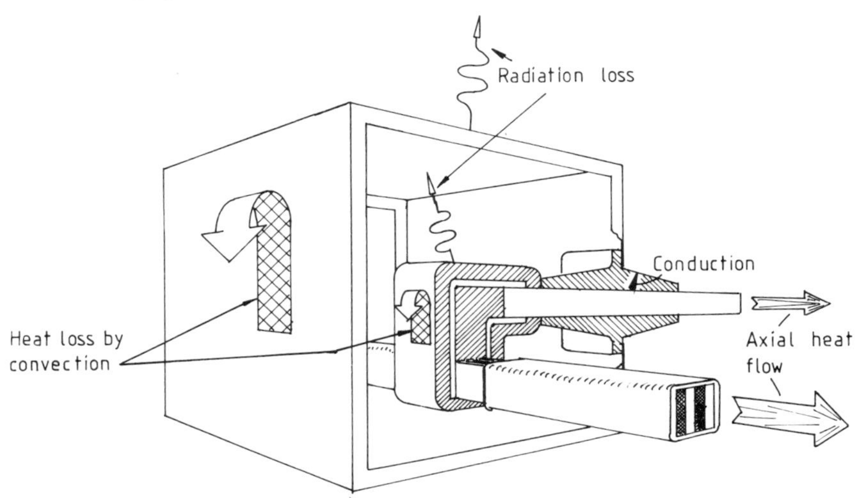

Fig. 8.6 Heat transfer at work.

be determined by its short-time current carrying capability, and the need to avoid temperatures that could damage insulation.

Since most conductor materials have a positive temperature coefficient, i.e. the resistivity increases with temperature, the temperature increase with time tends to be exponential in form. Figure 8.7 shows calculated figures for the temperature rise of copper and aluminium conductors as a function of current density and time. Curves for two different starting temperatures are given, and the graph ignores skin effect and proximity effects. The inclusion of these factors would add perhaps 15% to the resistivity, but the heat capacity of the conductor is not affected by the frequency.

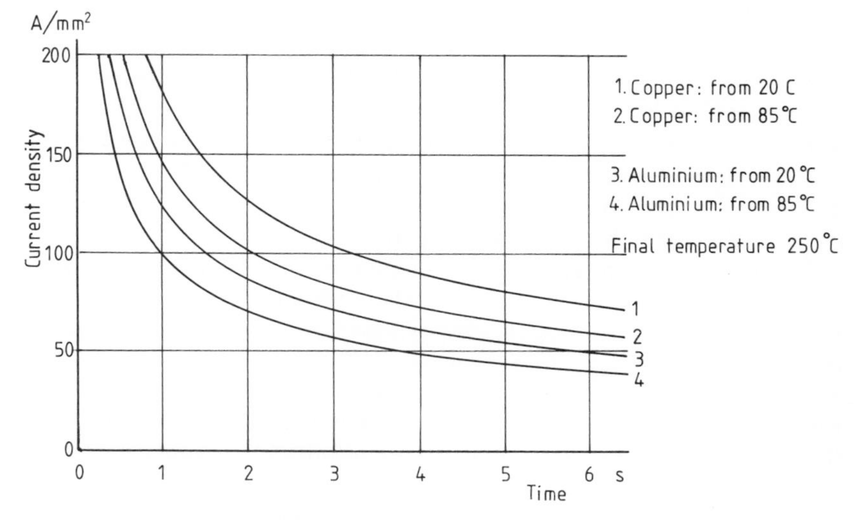

Fig. 8.7 Short-time rating graph.

CYCLIC RATING

One final point is worth mentioning, in view of the fact that the current flow through switchgear circuits tends to vary according to a usage pattern that is a factor of the season of year and the time of day. Whilst a capacity to cope with overload conditions is not part of the design practice for switchgear manufacturers, if the load peaks are of relatively short duration, many switchgear units can have a cyclic rating. This is not such a margin as it used to be in the past when bituminous compound and oil figured largely in the insulation of equipment, but it may be of use if there is an emergency need to take a heavy load for a short time. Because of the time it takes to raise the temperature of the conductors and the insulation materials, this temperature rises exponentially with a time constant that may be two or three hours in some cases. If the extra current is limited to a value less than this it is possible that the limiting temperature will not be exceeded. It is fairly easy to calculate by the manufacturer, and it could be of value in an emergency to have some knowledge of the cyclic rating capability of the equipment.

APPENDIX: HEAT LOSS CALCULATIONS

CONDUCTION

The practical calculation of heat transmission by conduction can be facilitated by using the concept of a 'heat resistance' which is analogous to the electrical resistance and can be manipulated in series and parallel arrangements in the same way.

It can be proved mathematically that

$$R_H = \frac{L}{k \times A}$$

where R_H = heat resistance – °C/watt
k = thermal conductivity – watts/cm/°C
A = area of conducting path – cm²
L = length of conducting path – cm

The driving potential that is analogous to the electrical potential is the temperature difference, θ (°C), and the equivalent for the rate of flow of quantity of electricity is Q, the rate of heat flow (watts). So as in Ohm's law: $\theta = Q \times R_H$ °C.

Where the conducting medium is regular in shape – concentric cylinders, for example, which often approximate conditions arising in practical cases such as the insulation around conductors – R_H can be calculated mathematically. Often,

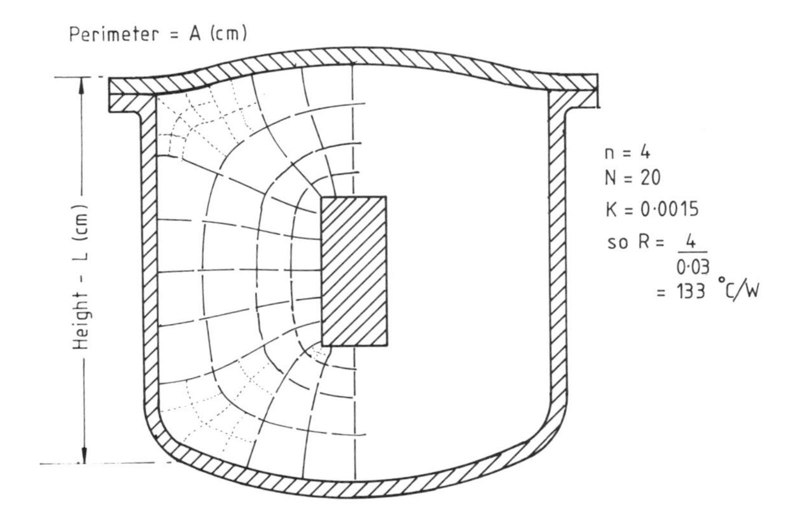

Fig. 8.8 Field plot for the calculation of heat resistance.

however, insulation does not arrange itself in regular mathematical shapes, and then field plotting techniques, such as the use of curvilinear squares, are needed to estimate graphically the field pattern in two-dimensional cases. Mathematical techniques such as relaxation methods or finite difference analysis, nowadays with computer assistance, can be used for more complex cases, and for three-dimensional models.

Figure 8.8 shows a typical field plot for a filled, insulated joint and shows the method of calculating the heat resistance. Then if the heat produced at the conductor is calculated from the current and a.c. resistance as Q (watts), the temperature rise $\theta = Q \times R_H$ °C.

CONVECTION

The dimensionless numbers referred to in Chapter 8 are associated with the names of known researchers in the fields of heat transfer and physics, i.e. Nusselt (N_{Nu}), Prandtl (N_{Pr}) and Grashof (N_{Gr}). The groups of parameters are as follows:

$$N_{Nu} = \frac{h \times d}{k} \qquad N_{Pr} = \frac{v \times d \times C_P}{k} \qquad N_{Gr} = \frac{g \times L^3 \times C_E \times \theta}{v^2}$$

where		
	h = the convection coefficient	W/cm²/°C
	k = thermal conductivity	W/cm/°C
	L = characteristic length	cm
	g = gravitational acceleration	981 cm/s
	C_P = specific heat of the fluid (constant pressure)	Ws/kg/°C
	θ = temperature difference	°C
	C_E = coefficient of cubic expansion	1/(°C+273)
	v = kinematic viscosity	cm²/s
	d = density of fluid	kg/cm³

These characteristics of the fluid in which the convection is taking place are to be expressed as the values that appertain to a temperature midway between the temperature of the fluid and that of the hot object. As the latter is an unknown, then a certain amount of iteration will be necessary to solve the problem.

Researchers have shown that the empirical relationship between these numbers, which expresses the basic equation for convected heat transfer, is:

$$N_{Nu} = K \times (N_{Gr})^m \times (N_{Pr})^n$$

where K, m and n are constants.

Typical values used in heat calculations are:

$K = 0.555$, $m = n = 1/4$ for $N_{Gr} \times N_{Pr}$ in the range 10^4 to 10^8
$K = 0.129$, $m = n = 1/3$ for $N_{Gr} = N_{Pr}$ in the range 10^8 to 10^{12}

As an example, using an intermediate temperature of 50° C and putting in the values for air including the value $C_P = 1000$ Ws/kg/°C, we calculate the values for the dimensionless groups as follows:

$N_{Pr} = 8.43$
$N_{Gr} = 0.965 \times L^3 \times \theta$
$N_{Nu} = h \times L/(2.3 \times 10^{-4})$

These figures give rise to the equation:

$$h = 2.156 \times 10^{-4}\ (\theta/L)^{1/4}\ \text{W/cm}^2/^\circ\text{C}$$

This equation gives good results for horizontal cylinders where the characteristic dimension L = the diameter of the cylinder.

Corresponding equations for vertical and horizontal plates can be deduced in this way, and other fluids can be introduced.

RADIATION

The heat loss due to radiation has been shown to be proportional to the difference between the fourth powers of the absolute temperature of the hot body and the temperature of its surroundings. If:

T_1 = the body temperature (Kelvin) and
T_2 = the temperature of the surroundings

$$h_r \text{ (radiation coefficient)} = K/\theta \times ((T_1/100)^4 - (T_2/100)^4)\ (\text{W/cm}^2/^\circ\text{C})$$

where K is a constant (=0.57 for a black body).

These radiation equations need to be multiplied by an emissivity factor which is a function of the surface finish. For copper, for example, this factor varies from 0.035 for polished material up to 0.75 for a black oxidised surface. Painting the radiating surfaces with matt black paint can make a worthwhile reduction in the temperature rise.

The total calculations require the combination of these three types of heat transmission, to work out the temperature drop from the conductor, through its insulation covering, to the chamber walls and then from those walls to the ambient surroundings.

EXAMPLE OF THE USE OF HEAT RESISTANCE IN CALCULATION

Figure 8.8 shows a simple example of the need to calculate the temperature drop between a single busbar situated in a metal chamber filled with insulation, in this case, bituminous compound. This material has a thermal conductivity of 0.0015 W/cm/°C.

On a cross-sectional drawing of the chamber, the flow lines and thermal equipotential lines can be drawn in freehand as curvilinear squares. These are 'squares' with curved sides having equal diameters and right-angle corners. With practice and trial-and-error, a set of these squares can be sketched in with sufficient accuracy for practical purposes, in simple two-dimensional systems.

Figure 8.8 shows this for the busbar chamber, and it is seen that there are four squares in series and 20 in parallel. Mathematically it can be shown that the heat resistance per unit length is now given by the formula

$$R_H = \frac{n}{N \times K} \ ^\circ\mathrm{C/W} \text{ per cm}$$

where n = number of squares in series
N = number of squares in parallel
K = thermal conductivity of the medium.

In this case, substituting the above figures gives $-R_H = 133^\circ \mathrm{C/W}$.

If we then take a 1 cm slice of the chamber the external area and the heat generation can be established. Call the circumference A cm and the height of the chamber L cm. Then if r is the a.c. resistance of a cm length of the conductor at the estimated final temperature, the heat generated at a load of 1200 A will be $1.2^2 \times r \mathrm{W/cm}$.

The heat loss from the walls of the chamber is calculated as described earlier in this appendix and gives the following equation:

$$\text{heat loss coefficient } H = A \times h_r \times e + h_c \times L \ \mathrm{W/cm^2/^\circ C}$$

where h_r = radiation coefficient
e = emissivity
h_c = convection coefficient.

Then T (temperature rise) = $T_1 + T_2$, where T_1 = conduction drop through the compound and T_2 = convection and radiation drop to the ambient temperature.

$T_1 = 1.2^2 \times r \times R_H$ and $T_2 = 1.2^2 \times r/H$. Therefore $T = 1.44 \times r \times (R_H + 1/H)^\circ \mathrm{C}$.

A process of iteration is likely before the exact temperature rise is established, using the calculated figure as a guide to a better estimate each time. Computer programs are available which will carry out this process, with the additional capability of analysing the field by mathematical methods.

CHAPTER 9

Electromagnetic Forces and their Effects

Any conductor carrying a current is surrounded by an electromagnetic field. Any other conductor also carrying a current and in close proximity to the first conductor is subject to a mechanical force, the magnitude of which is proportional to the size of the current and the spacing of the conductors. If the current is alternating, then the relative phase of that current is also a factor in determining the magnitude of the force, which is also alternating.

A simple relationship to examine first is that where a single conductor changes direction, in which case the current creating the field, and that reacting with it, are the same. Starting with Biot-Savart's Law, values for the force developed can be determined mathematically. The equations become quite complex when allowance for the finite size of the conductors is made, and workers in this field have developed graphs and charts from which the values can be read.

For discussion purposes, it is noted that the instantaneous value of the force is proportional to the product of the instantaneous values of the two currents and the distance between the conductors, at any one point. End effects are the problem areas, where conductors meet at angles, or at the ends of short conductors, and empirical solutions have to be used, checked by experimental results.

Figure 9.1 shows on a per-unit basis the shape of the force developed between two parallel conductors carrying an asymmetrical fault current, as would occur in switchgear busbars in the event of a phase-to-phase short circuit. Figure 9.2 is an

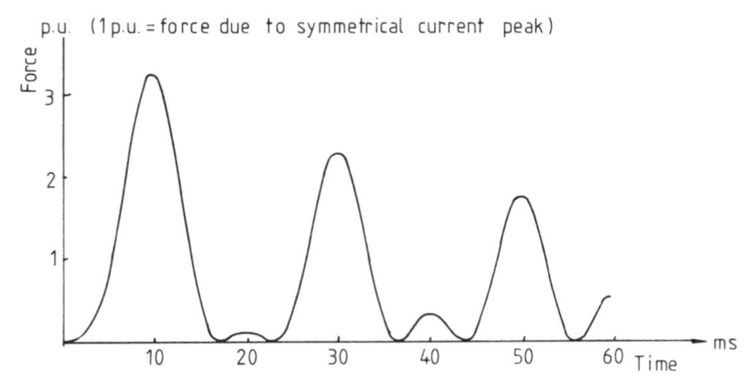

Fig. 9.1 Single-phase electromagnetic force between two parallel conductors.

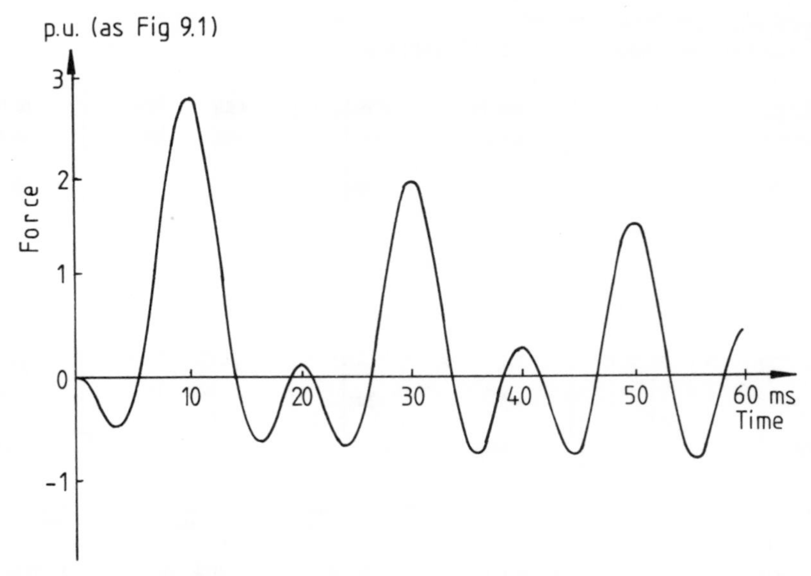

Fig. 9.2 Force on the central conductor of a planar three-phase system of conductors.

extension of this to a three-phase short circuit with the maximum asymmetry in the centre phase of a co-planar system of busbars, which is the most severe condition and the most common physical arrangement.

Two characteristics are immediately apparent: the frequency of the force graph is twice the system frequency, and the effect of the d.c. component in the fault current is accentuated due to the square law implicit in the relationship, particularly in the single-phase case.

The modern tendencies for switchgear to become more compact and for fault levels to increase, as electricity systems expand, both tend to increase the need to give careful consideration to the effect of these forces in the design of the cubicles for metal-enclosed switchgear. Factors in the design of certain types of circuit-breaker can be sensitive to the effects of these forces. For example, vertically isolated bulk oil circuit-breakers have six parallel conductors on quite small spacings, bridged by a moving contact structure at their lower ends (Fig. 9.3). This diagram illustrates the cantilever forces which affect the conductors and the acceleration forces applied to the moving contact system.

When these forces are added to the high pressures developed within the arc control devices which surround the contact system, when the circuit-breaker opens under fault conditions, the mechanical forces on the various conductors can be quite formidable. The efflux of gases from the side vents and the bore of the arc control pots partly balances and partly reinforces the electromagnetic forces, and the effect on the moving contact system under maximum asymmetrical conditions is to produce acceleration forces significantly higher that those on no-load. That is why oil dashpots are frequently found in oil circuit-breakers which control the travel of the moving contact structure over most, or at least the last part, of its movement.

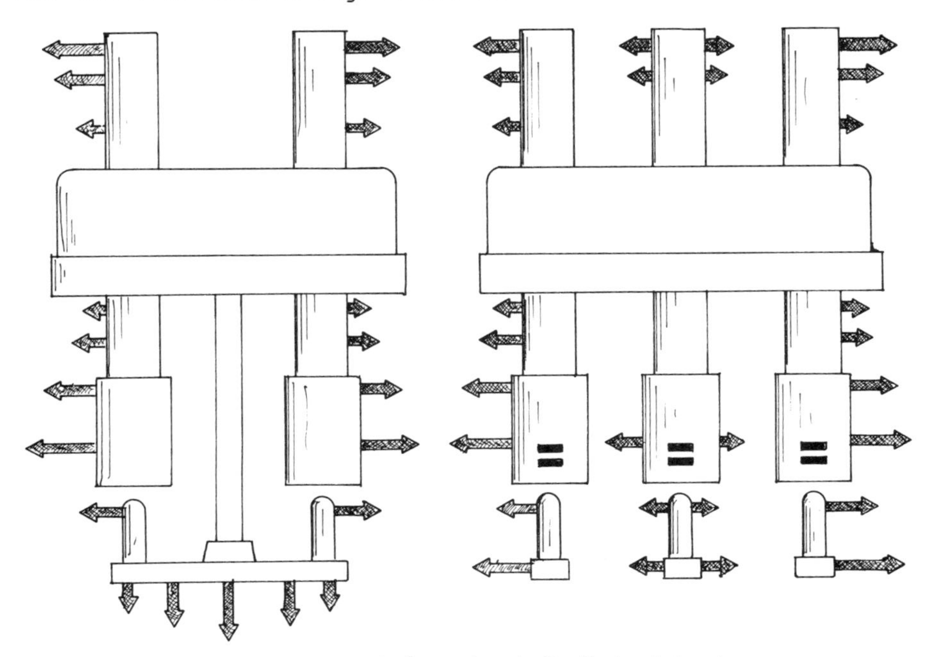

Arrows indicate direction and approximate magnitude of forces
Time course as shown in Figs 9.1 and 9.2

Fig. 9.3 Electromagnetic forces in a bulk-oil circuit-breaker.

In switching devices which depend to some degree on the magnetic control of the arc, and this is particularly true of vacuum circuit-breakers, the presence of other conductors in close proximity must be carefully considered in the circuit-breaker design. Again, this probably affects vertically isolated circuit-breakers to a greater degree than the horizontally isolated versions because of the need to bring the return conductors in a fairly tight loop to keep dimensions reasonably small.

The switchgear designer can also turn these forces to his advantage. In the oil circuit-breaker case, a faster contact speed at high fault currents tends to shorten arcing times and reduce the pressure in the arc control device if kept within limits. There are designs of low voltage circuit-breakers that use electromagnetic forces to overcome a special restraint feature built into the mechanism and which thus allows very fast opening to occur on heavy fault currents, even to the point that the arc voltage then developed has a significant effect in limiting the peak value of the fault current.

These so-called 'current-limiting' circuit-breakers can be useful in permitting the use of smaller conductor and cable sizes in the system design. This is because the size of the conductors in switchgear of low normal current rating can be increased in cross-sectional area beyond that needed for normal current carrying, due to the short-time rating requirements. The effect is similar to that of a current-limiting fuse.

The contact structure in all vacuum interrupters is designed to make good use of electromagnetic forces for arc control. If well executed, this can prevent the arc becoming constricted, which leads to increased energy loss, as explained in Chapter 4.

Also, special measures to create magnetic fields for control of arc movement are used to good effect in magnetic air circuit-breakers, and rotating-arc SF_6 circuit-breakers.

An unwelcome effect of electromagnetic forces in contact design is their tendency to reduce the contact pressure precisely when it is most needed. Even in the form of wiping contact, often referred to as the 'tulip' or 'rose' type because of the cylindrical arrangement of the contact segments, and which on face value seem to benefit from the magnetic effect of parallel currents flowing in the same direction, there are local forces tending to blow the contacts off the contact rod. This is known as the 'pinch effect', and arises because the current flow is concentrated in a small number of points and is not evenly distributed over the contact area.

However well made the contacts may be, in the microscopic region of mechanical contact it has to be assumed that flat areas only make contact in three points, line contacts only at two points and point contacts only at one. The mechanical pressure at each point is usually enhanced by this fact, and the avoidance of the area contact in modern designs stems from the designer's wish to know just where the points of contact are likely to be.

It is the concentration of the lines of current flow in the region of these points of contact that gives rise to the pinch effect, as illustrated in Fig. 9.4. This shows the effect applied both to tulip contacts and butt contacts, and also shows a butt contact design which tries to compensate for the effect by increasing the normal electromagnetic effect.

With the tulip type of contact, it is beneficial to have some reduction in the grip forces due to the close proximity of the parallel fingers, as otherwise there is an unwelcome increase in the resistance to the mechanical opening forces by the additional friction the contact grip causes, and this tends to delay the opening of the circuit-breaker.

This brief introduction to the issues raised by the effect of electromagnetic forces on switchgear design focuses therefore on the balance that has to be maintained between the many conflicting factors which have an influence on the designer's choice of such things as the contact structure, type of withdrawal and arcing medium.

One final issue arising from the presence of electromagnetic forces is the need to study the dynamic and transient mechanical behaviour of the conductors and their supports. An obvious area is the examination of the natural vibration frequency of the structure involved and its reaction to the forced vibrations generated during a short circuit. Any system with a resonant frequency close to 100 or 120 Hz is obviously undesirable. Fortunately, with the size and weight usually associated with switchgear conductors, resonant frequency is usually lower than these figures, but if supports are rigid and close, it is by no means impossible inadvertently to cause a resonant condition. An appendix to this chapter includes a nomogram, Fig. 9.6, which can be used in conjunction with Fig. 9.5 to predict the resonant

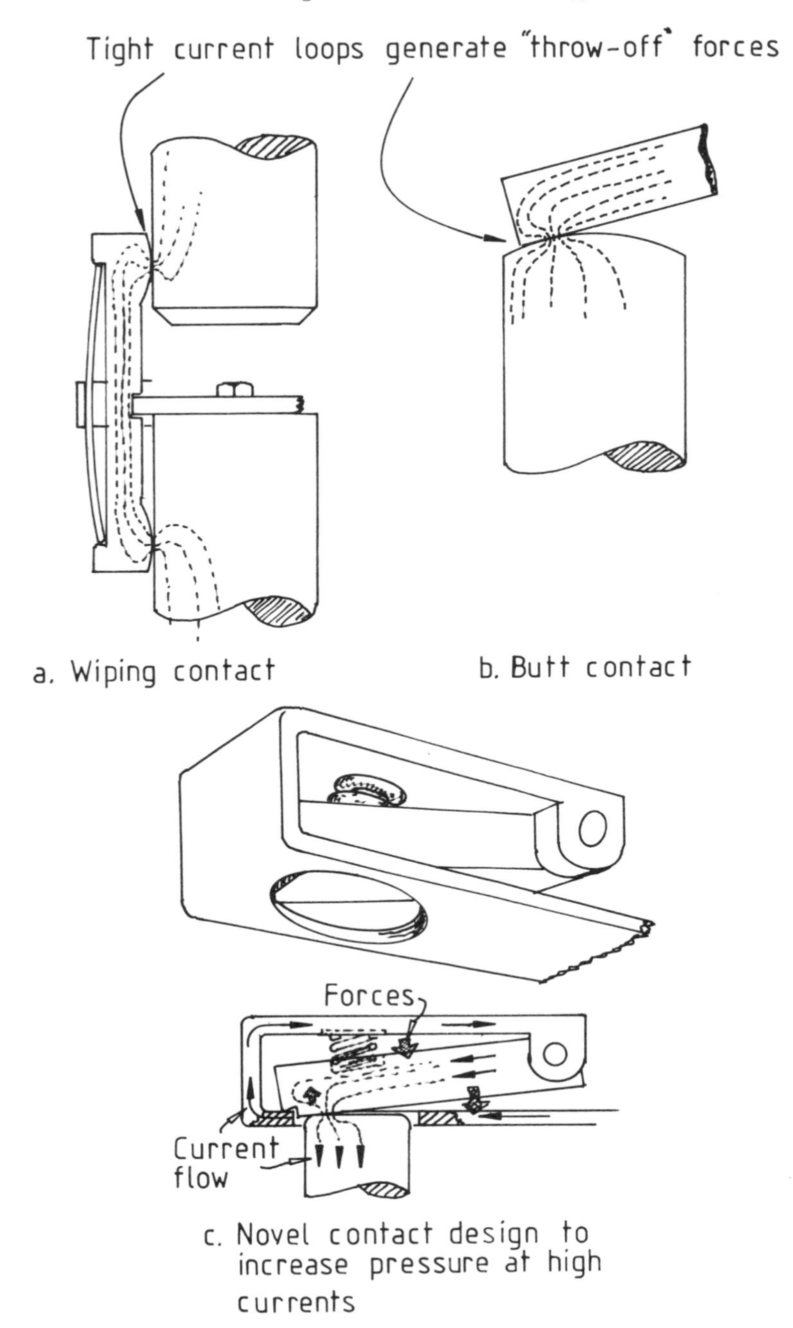

Fig. 9.4 Illustrating pinch effect and a way of offsetting it.

frequency of a system of parallel bars, and gives an indication of the sort of techniques that can be used to facilitate calculation in this field.

APPENDIX: NOMOGRAM FOR THE DETERMINATION OF THE FUNDAMENTAL NATURAL VIBRATION FREQUENCY OF LATERALLY DEFLECTED BEAMS

The basic frequency condition is:

$$f = \frac{K_1}{L^2}\sqrt{\left(\frac{EIg}{Ad}\right)}$$

where E = Young's modulus – lb/in^2 × 10^6
I = second moment of area – in^4
d = density – lb/in^3
A = cross-sectional area – in^2
g = gravitational acceleration = 386 in/s^2
L = length between supports – in
K_1 = constant depending on the end conditions of the beam

Putting $K_M = \sqrt{\left(\frac{Eg}{d}\right)}$ as the material constant and $K_S = \sqrt{(I/A)}$ as the section constant, we can tabulate useful values as follows.

Characteristic	*Aluminium*	*Copper*	*Brass*	*Steel*	*Porcelain*	*Epoxy resin*
			Material			
Density, d	0.098	0.322	0.304	0.28	0.09	0.065
Young's modulus, E	9.5	18	12	30	10	3
K_M – in/s × 10^5	1.93	1.47	1.235	2.03	2.07	1.33

In the following table of values for K_S, y = depth of a rectangular beam or the inside diameter of a tubular beam and Y = outside diameter of a tubular beam or of a solid cylindrical beam.

	Type of beam		
	Rectangular section	*Rod*	*Tube*
K_S	$\frac{y}{2/3}$	$\frac{Y}{4}$	$\sqrt{\left(\frac{Y^2 + y^2}{4}\right)}$

K_1 has the following values:

End A condition	*End B condition*	K_1	
Hinged	Hinged	1.571	Simply supported beam
Built-in	Built-in	3.56	Encastré beam
Built-in	Free	0.57	Cantilever
Built-in	Hinged	2.46	Propped cantilever

To use the nomogram, Fig. 9.7:

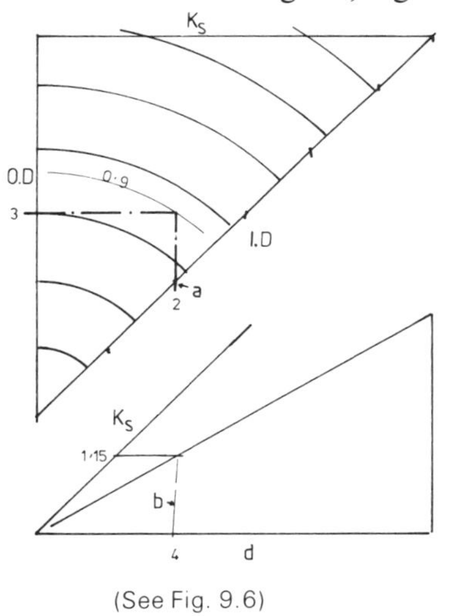

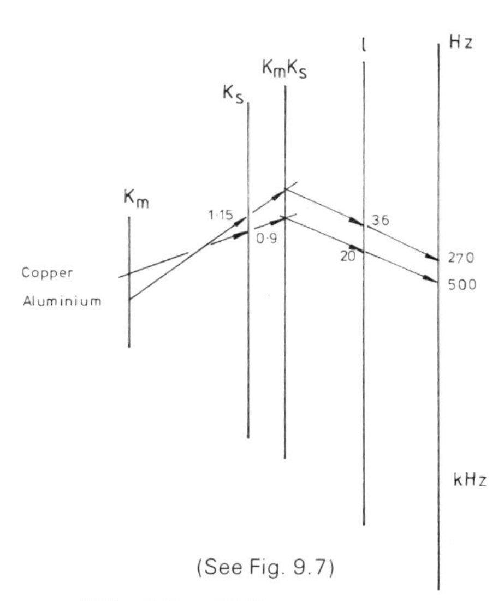

Fig. 9.5 Diagram showing the use of Figs 9.6 and 9.7.

1. Find K_M and K_S either by calculation or by the use of the charts – Fig. 9.6. The use of these charts is shown in Fig. 9.5. The upper chart relates to cylindrical conductors, or beams, and is used to find the value of the section constant. For solid bars, look for the outside diameter up the left-hand side of the diagram, and read off the value of K_S alongside it. As an example, the section constant for a 3 in diameter solid bar is 0.75. If the bar is tubular, look for the inside diameter along the diagonal scale on the right of the chart. Follow a line vertically until it meets a line drawn horizontally from the outside diameter scale. The curved line where they intersect gives the section constant. An example is shown in Fig. 9.5, which can be followed in Fig. 9.6. A tube 3 in outside and 2 in inside diameter, at a in Fig. 9.5, leads to the result K_S=0.9. If the bar is rectangular, the lower chart is used, the depth taken from the bottom scale and traced up until it meets the sloping line. Tracing across horizontally to the left-hand scale then gives the value of K_S. The example b is for a bar of 4 in depth and has a section constant of 1.15.

2. In Fig. 9.7, draw a line between the chosen values of K_M and K_S and note where it intersects the $K_M K_S$ line. The actual value is usually immaterial.

3. Then draw a line from this intersection point through the required value of 1 to intersect the frequency line.

On the right-hand side of Fig. 9.5, the two examples already mentioned are drawn in. The case of the tubular bar is now assumed to be of 20 in span and to be made of copper. The point on the first scale marked 'copper' is joined to the point 0.9 on scale K_S, and projected to the next scale. From the intersection point, a line is drawn through the length of 20 in and projected to the frequency scale where the fundamental vibration frequency is found as 500 Hz. A similar procedure b for the rectangular bar, assumed to be made of aluminium and 36 in long, gives the result 270 Hz.

The frequency thus obtained is for a simply supported beam. The frequencies for other conditions can be obtained by multiplying this result by the following constants:

Encastré beam	2.265
Cantilever	0.365
Propped cantilever	1.565

Applying these factors for the examples given as an illustration shows how the natural frequency for a busbar, for example, is changed by the form of fixing at the support. If a flexible fixing is used, then a frequency of 270 Hz was calculated for the aluminium busbar, and if this support was rigid enough to be considered encastré, then that frequency is increased to just over 600 Hz.

In practice the truth lies somewhere in between, since the joint will not be mathematically true to either definition. The most important use of these calculations is to ensure that the natural frequency is not near the supply frequency so that a resonant condition is created, which would magnify the electromagnetic forces.

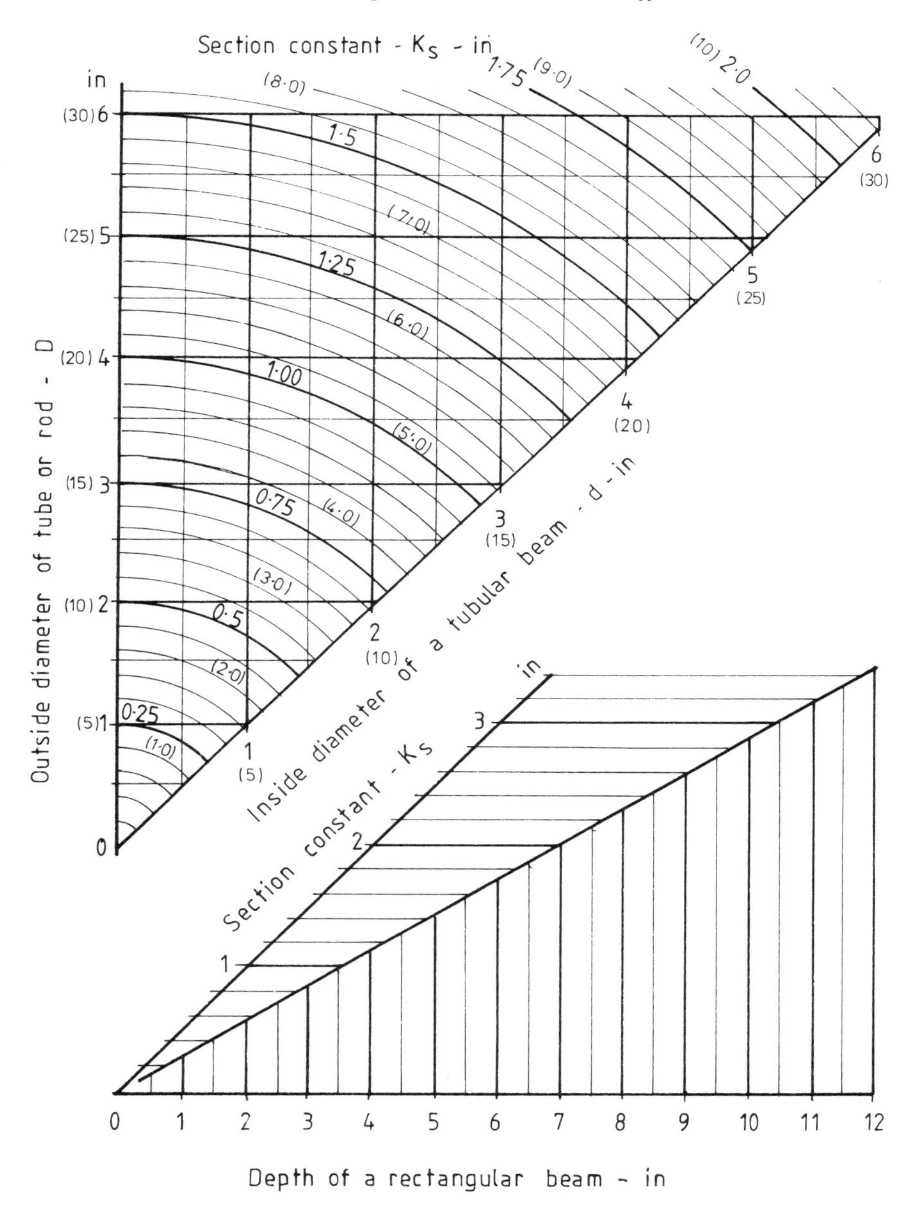

Fig. 9.6 Chart to find constants for nomogram in Fig. 9.7.

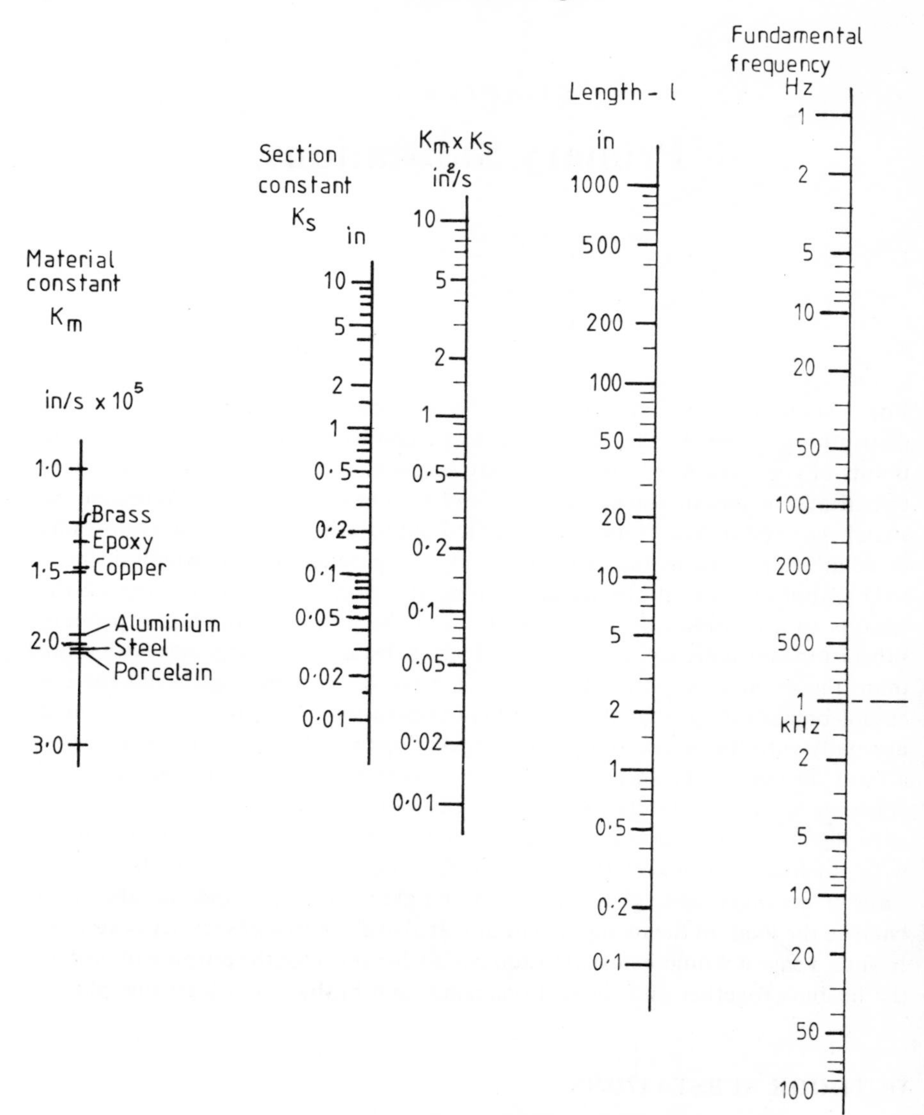

Fig. 9.7 Nomogram to determine the natural vibration frequency of busbars.

CHAPTER 10

Primary Substations

For switching devices to do their job of controlling and protecting the distribution network, they first have to be connected into the system. These points of connection then provide ready access to the network for the ancillary operations of measurement, earthing and testing of the system. Whether the switchgear and its accessories are mounted outdoors or indoors, the philosophy of connection remains the same; it is only the practical details which differ.

Distribution substations usually follow the principle of connecting all the circuits to a common busbar through a switching device, which in a primary substation is usually a circuit-breaker. If the substation is a large one with more than one point of supply, this busbar is often divided through a bus-section circuit-breaker to give greater flexibility of operation. This procedure, operated normally with the bus-section circuit-breaker open, reduces the area affected by a fault on one of the incoming supplies and then allows the reconnection of consumers once the faulty incomer has been disconnected.

In the largest and most important substations the use of two busbars, usually with off-load transfer of the outgoing feeders, but sometimes with on-load transfer arrangements, gives even greater flexibility of use and the ability to balance the load on the incoming circuits. It also gives greater security of supply. In such a case it would be usual to add circuit-breakers for the purpose of linking the busbars together and for sectionalising each busbar in at least one place.

OUTDOOR SUBSTATIONS

Where space is not a problem and the distribution network is mostly connected by overhead lines, the substation is often arranged in a fenced-off compound, using equipment designed for use outdoors.

The busbars are usually rigid, and frequently tubular, although connection by stranded overhead line is not unknown.

For the purpose of creating an obvious disconnection of a feeder or supply transformer, each circuit is connected through a disconnector, and this frequently incorporates a separate earthing switch. Figure 10.1 is an example of a reasonably large outdoor 33 kV substation, taking incoming supplies from the transmission system and supplying a large industrial user at 33 kV, an interconnection at 33 kV to another similar substation and feeds into the local

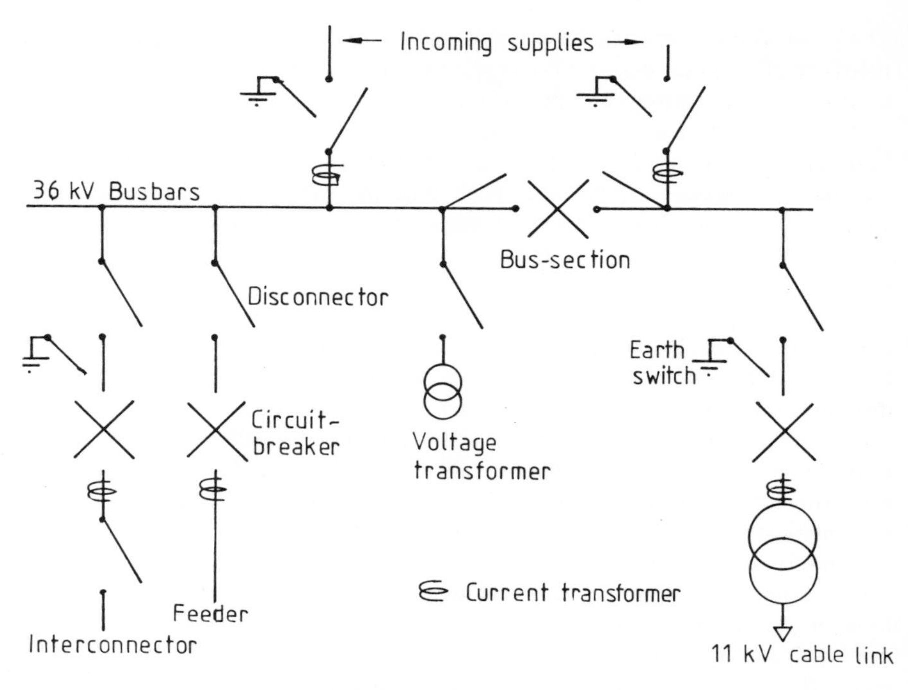

Fig. 10.1 A 36 kV substation.

Fig. 10.2 A 36 kV substation. (*Courtesy Eastern Electricity Board.*)

11 kV distribution network. Figure 10.2 is a photograph of an actual substation of this type. The usual description applied to equipment with outdoor bushings used in such substations is 'open-terminal'.

All the components of such a substation are mounted on earthed metal structures, commonly made of galvanised steel, and the height is governed by national standards, e.g. BS 162, such that the earthy end of all insulating supports is at the limit of reach of the average person, say 2.2 m. This is to prevent any operator or maintenance worker unwittingly encroaching on the earth clearance of any conductor.

Disconnectors usually use air as the insulating medium for the gap, and the support insulators for the contacts generally also serve as the connection points for the conductors to the circuit-breakers, busbars or lines. Whether or not disconnectors are fitted on the circuit side of the circuit-breaker depends on the possibility of a back-feed to that circuit-breaker, due to further interconnection in the system beyond. In the simplest case where all the feeders are dead-ended, only the incoming circuit-breakers need to be isolated on both sides, and possibly not then, if there is only one in-feed.

Some types of circuit-breaker have room internally for the fitting of current transformers. This used to be the case with bulk oil, open-terminal circuit-breakers, and has advantages over the alternative of providing separately mounted, and therefore separately insulated current transformers. In those cases where the protection system requires current transformers on both sides of the circuit-breaker this advantage becomes even more pronounced.

Voltage transformers have to be separately mounted and therefore separately insulated. Combined current transformers and voltage transformers are also available and this economises on insulators.

Outdoor circuit-breakers incorporating vacuum interrupters need additional protection as the insulation of the vacuum interrupter is only suitable for use in indoor substations. Figures 10.3 and 10.4 show a modern open-terminal vacuum circuit-breaker employing cast aluminium housings for the interrupters. These housings are sealed and filled with a suitable insulation medium. Both SF_6 and dry air are used as filling media, depending on the voltage rating of the equipment. In both cases a pressure slightly in excess of atmospheric is used so that possible leakage can be detected.

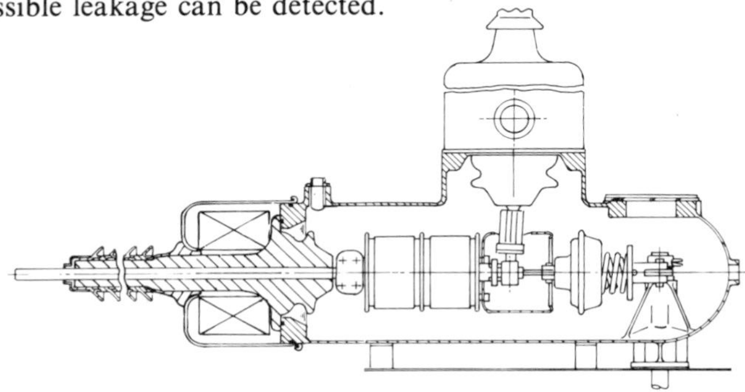

Fig. 10.3 Sectional drawing of GEC type OX circuit-breaker. (*Courtesy GEC Distribution Switchgear Ltd.*)

Fig. 10.4 The GEC type OX circuit-breaker. (*Courtesy GEC Distribution Switchgear Ltd.*)

Fig. 10.5 An outdoor disconnector. (*Courtesy Joslyn Manufacturing and Supply Co.*)

The bushings on these circuit-breakers are interesting as they do not employ the usual porcelain weather sheds, but are made from a cast polyurethane insulated conductor over which are nested a series of interlocking EPDM weather sheds. (EPDM is Ethylene Propylene Dough Moulding compound.) Modern production techniques can now also produce these EPDM weather sheds in continuous lengths which can be cut to the length needed for the different voltages. It is a well proven technique used to insulate cable tails when the cable is brought up a pole for connection to an overhead line.

By slipping off the sheds the way is open for the removal of the current transformers, should it be required, for example, to change the ratio. So this new

Fig. 10.6 An outdoor current transformer. (*Courtesy GEC Instrument Transformer Division.*)

Fig. 10.7 An outdoor voltage transformer. (*Courtesy GEC Instrument Transformer Division*).

Fig. 10.8 A cable termination to the overhead system. (*Courtesy Pirelli General.*)

design retains the useful and advantageous facility for fitting current transformers on the circuit-breaker terminals.

Figures 10.5 – 10.9 illustrate other items of plant used in outdoor substations.

The operation of an outdoor substation usually requires the provision on the site of a small control room in which panels are mounted to house all the operating equipment and protective relays, etc. In a distribution substation these would be fairly simple and all the items relating to one circuit would be on one panel, as they would be on an indoor metal-enclosed switchboard. The disconnectors and earthing switches are usually manually operated and the

Fig. 10.9 An earthing switch fitted to a disconnector. (*Courtesy South Wales Switchgear.*)

circuit-breaker usually has an additional control switch in the kiosk which is attached to the circuit-breaker to house the mechanism. Another accessory fitted to this kiosk is a small heater so that the operating mechanism and its associated wiring and release coils are kept dry.

The techniques employed to gain access to circuits for earthing and testing are the same in principle whether the substation is indoor or outdoor, the differences being only in detail.

Interlocks are provided so that the disconnectors associated with a circuit-breaker cannot be operated unless the circuit-breaker is open. It is not practicable for this interlock to be mechanical, so special key type locks are used, connected to the operating mechanism. In its simplest form it is impossible to remove the key from the circuit-breaker when it is closed. If the circuit-breaker is opened and the key removed, it is impossible to close the circuit-breaker until the key is replaced and turned to the operating position. Once the key is free, it can be inserted in the corresponding lock on the relevant disconnector, which can then be opened or closed. An opening operation will possibly release another key which will allow the earthing switch to be closed. Closing the earth switch will then trap the key, making it impossible to reclose the disconnector, and that in turn makes it impossible to reclose the circuit-breaker.

It can be seen that this system is capable of infinite variation and it is complemented by key exchange boxes which require several keys to be inserted to release one – or more – other keys. This may be necessary where several feeds are possible to the point where an earth is to be applied (Fig. 10.10).

Once the interlocked earth has been applied, portable, clip-on earths with cable connection may be added for additional security, and similar clip-on test connections can be used for test purposes.

After the operational facilities required have been provided and switchgear with the correct electrical switching characteristics has been installed, the most

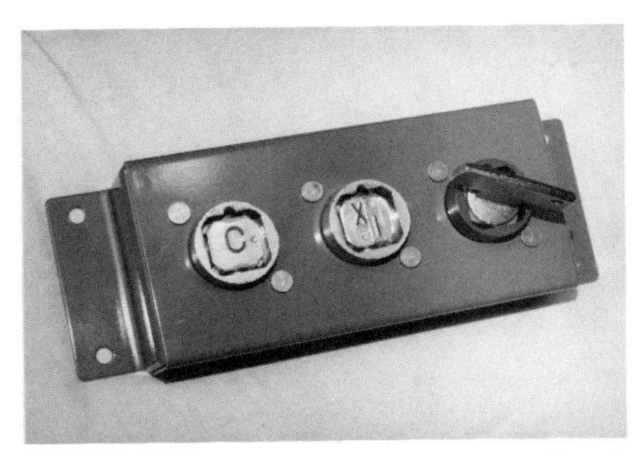

Fig. 10.10 A key interlock and exchange box. (*Courtesy Castell Safety International Ltd.*)

important feature of the outdoor substation is its ability to stand up to the climatic conditions for considerable periods of time without attention. This requirement can be divided into the effect of the weather on the mechanical parts such as the operating mechanisms, metal housings and supports on the one hand and the insulation components on the other.

Taking the mechanical components first, it is common practice to zinc coat steel structural components by hot galvanising, or by zinc spraying if the galvanising process cannot be applied, e.g. for reasons of possible heat distortion. The mechanisms are usually protected inside sealed compartments and heaters are often fitted to ward off condensation. Bearing in mind the imperative need for circuit-breaker mechanisms to trip correctly when asked, it is important to make all bearings resistant to corrosion. In considering the insulation parts, this comprises insulation between the live plant terminals and earth, whether this is in the form of support insulators for busbars or disconnectors, or the bushings fitted to circuit-breakers and instrument transformers. The electrical stresses to which these insulators are subjected derive from the long term stressing at the service voltage over many years, augmented by overvoltages created by electrical storms or by switching operations. The question of overvoltages deriving from switching operations has been dealt with in Chapter 3.

Lightning creates voltage surges with very steeply rising fronts when it strikes on or near the overhead line conductors. The lightning surge value is independent of the system rated voltage, so is a more severe condition on a distribution voltage line than it would be on a line at transmission voltage.

Equipments that can be subjected to overvoltages of atmospheric origin are said to be 'electrically exposed'. The impulse voltage tests referred to in Chapter 6 are a test of the ability of an insulator to withstand high voltages with fast rising wavefronts.

This is why switchgear and transformers, which are often connected to overhead lines, are always given an impulse rating; motors, which are invariably cable connected, are not necessarily given an impulse rating.

In substations where there are serious risks of atmospheric disturbances which could lead to severe stressing of the insulation, arcing horns or surge arrestors are often fitted. Surge arrestors are components with special characteristics, usually mounted in a porcelain housing. The major part of the contents is made up of special resistance material of non-linear characteristic such that the resistance reduces as the voltage increases in accordance with a power law, the index of which is in the region from 4 to 10 or more. Mathematically:

$$RV^N = K$$

where R = resistance, V = voltage, K = constant, N = index.

With the smaller values of N there is a gap, or several series-connected gaps, between the system and the non-linear resistance (NLR). Therefore, in the event of a high voltage surge these gaps flash over and the energy is dissipated in the NLR. As the voltage falls, the resistance quickly rises and the current falls to the point that the arcs across the gaps cannot be maintained. A surge arrestor of this type is illustrated in Fig. 10.11.

Modern NLRs made from zinc oxide have values of N of the order of 10 or greater. They can therefore have such a high resistance at system voltage that they do not require the series gap (Fig. 10.12).

Arcing horns are a simpler means of protection and are fitted to the line insulators, or the plant bushings. These take the form of metal rods attached to the two ends of the insulator, and the gap between the ends is set so that it will flash over at a voltage well above the line voltage but less than a value which could damage the plant. Therefore it only comes into use when lightning surges with peak values close to the tested impulse level are reached.

Table 1.3 indicates that quite wide margins exist for the setting of the arcing horn gaps. As will be realised, the flashing over of the arcing horns initiates a fault on the system which needs a circuit-breaker to clear it. Surge arrestors are normally self-clearing.

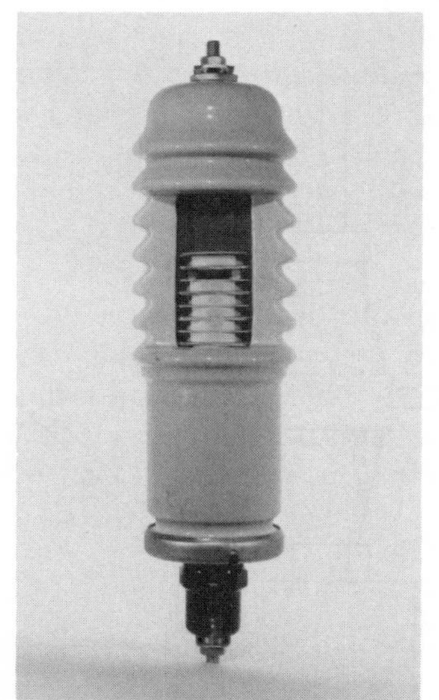

Fig. 10.11 Surge arrestor with series gaps. (*Courtesy Joslyn Manufacturing and Supply Co.*)

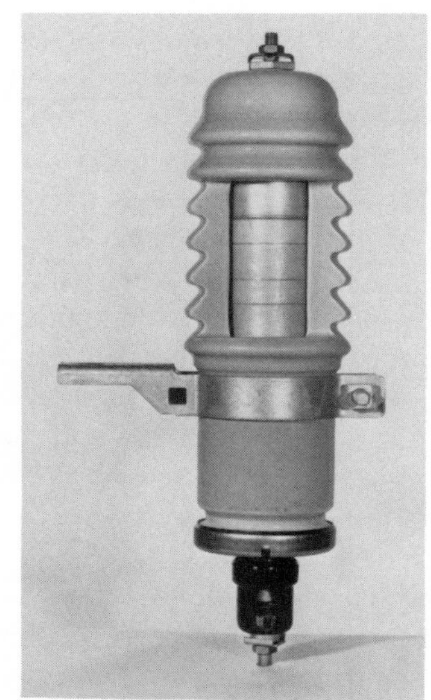

Fig. 10.12 Gapless surge arrestor. (*Courtesy Joslyn Manufacturing and Supply Co.*)

INDOOR SUBSTATIONS

Many and varied have been the types of indoor substation built over the years during which metal-enclosed switchgear has been available. Until quite recently in some countries the user built his own switchboard, using components purchased from a variety of sources. The interior of the building is divided up into a number of cells corresponding to the circuits that are to be connected. Busbars are run along the back wall above the cell divisions, or along the ceiling, mounted on porcelain stand-off insulators. Floor-standing circuit-breakers, with their mechanisms accessible from the front of the cell, are connected to the busbars through disconnectors, also mounted on the back wall or on the roof, whichever is the most convenient. The other side of the circuit-breaker is connected through a current transformer to the cable terminals (Fig 10.13).

All the components for this type of installation are procured by the user, probably from several suppliers, which presupposes a knowledgeable user, contractor or, perhaps, consultant.

While the introduction of 'factory assembled' switchgear took place many decades ago in the UK, it is only in the post war years that it has been in general use in some European countries and South America.

However, the subject to be dealt with here is the factory built metal-enclosed

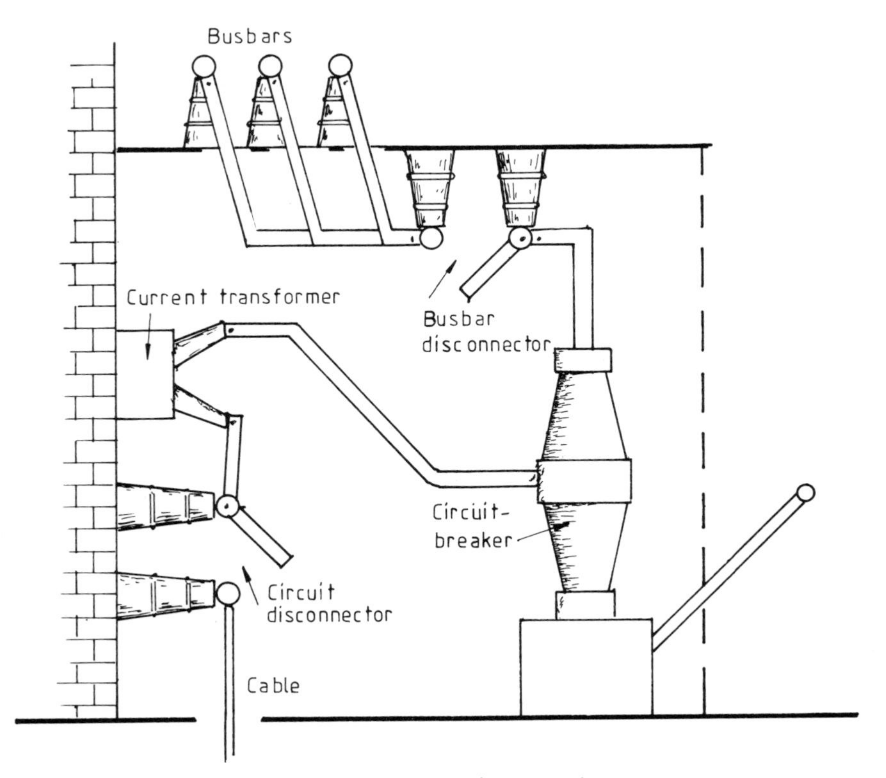

Fig. 10.13 Cellular type indoor switchgear.

switchgear equipment which is delivered to the substation virtually complete. It only needs the busbars linking together, the main cables connecting and the small wiring completing, to be ready for commissioning.

Today, such equipments are defined, and certain criteria established, in International and National standards. IEC 298 is the International standard (which is virtually repeated verbatim in BS 5227). In the USA, NEMA Standard SG5 contains constructional detail for this type of equipment.

Before looking further at the ways in which the needs of indoor 'factory-assembled' switchgear have been fulfilled, it would be useful to examine what these needs are. First of all, the object of the enclosure of each factory-made cell is to safeguard the operator against electric shock. Therefore the principle of construction must be that any part of the equipment normally accessible to the operator in carrying out his duties must be earthed metal.

Within the enclosure certain components have to be mounted and connected together in such a way that the operational requirements can be met. These operational requirements are:

1. *Control*. That is, switching the circuit on and off under all possible conditions of use.
2. *Earthing*. The facility to apply an earth connection readily to the circuit connections, and also (more rarely and perhaps, therefore, less readily) to the busbars.
3. *Testing*. The provision of convenient facilities for making test connections to the circuit, and (again less often) to the busbars.
4. *Maintenance*. Safe access to all those components which require inspection, servicing and, ultimately, overhaul.

The hardware required to do this in a metal-enclosed switchboard comprises the following:

1. *Busbars*. A three-phase set of conductors connecting a series of cubicles together. Because a fault on the busbars will shut down several consumers, it is common for the busbars to be in a separate, metal-enclosed compartment, so that they remain unaffected by any problems elsewhere in the cubicle.
2. *Busbar connections and disconnectors*. Conductors which link the busbar into each cubicle. If the busbars are in a separate chamber, this involves some form of insulation. Some means of disconnecting the switching device is essential so that the busbars can be left energised at all times, whatever operations are being performed in any one cubicle.
3. *Switching device*. This may be of any kind, but in a primary substation it is most likely to be a circuit-breaker.
4. *Circuit disconnector*. The need for this is dependent on the type and importance of the connected circuit.
5. *Current transformers* (CTs). Almost always on the circuit side of the switching device, facilities are required for the connection of some CTs. Depending on the duty and importance of the connected circuit, up to four or five CTs may be required on each phase.
6. *Voltage transformer* (VT). On some panels of a switchboard there is usually

the need to provide a voltage reference. The incoming supply panel almost certainly requires one. It is normally connected in between the CTs and the cable terminations.

7. *Cable terminations.* Facilities need to be provided to support and connect to the circuit conductors one or more cables per phase, depending on the current rating. These cables are almost always contained within the enclosure, frequently in a separate compartment.

To avoid any confusion between the terms metal-clad and metal-enclosed, definitions based on those in BS 5227 are reproduced here.

Metal-enclosed switchgear. Switchgear assemblies with an external metal enclosure intended to be earthed and complete except for external connections.

There are three sub-divisions:

1. Metal-clad switchgear.
2. Compartmented switchgear (with one or more non-metallic partitions).
3. Cubicle switchgear.

Metal-clad switchgear. Metal-enclosed switchgear in which components are arranged in separate compartments with metal partitions intended to be earthed. Note that:

1. This term applies to metal-enclosed switchgear with metal partitions providing the required degree of protection (see Table 10.1) and having separate compartments at least for the following components:

(a) Each main switching device.

(b) Components connected to one side of a switching device, e.g. a feeder circuit.

(c) Components connected to the other side of the switching device, e.g. busbars. Where more than one set of busbars is provided, each set shall be in a separate compartment.

2. Metal-enclosed switchgear meeting all the requirements of (1) may use an insulating shutter barrier as part of the shutter arrangement, the combination of which provides the required degree of protection, and satisfies the test requirements for partitions and shutters made of insulating material.

Compartmented switchgear (with non-metallic partitions). Metal-enclosed switchgear in which components are arranged in separate compartments as for metal-clad switchgear, but with one or more non-metallic partitions providing the required degree of protection.

Cubicle switchgear. Metal-enclosed switchgear, other than metal-clad and compartmented switchgear. This term applies to switchgear having a metal enclosure and having either:

(a) A number of compartments less than that required for metal-clad or compartmented switchgear.

(b) Partitions having a degree of protection less than that required by the specification.

(c) No partitions.

For completeness the following definition for insulation-enclosed switchgear is included, based on IEC Recommendation 466.

Insulation-enclosed switchgear. Switchgear assemblies with an external insulation enclosure and complete except for external connections. (Extra insulation tests, including measurement of leakage current, are prescribed in the above-mentioned Recommendation for this class of switchgear.)

MEETING THE REQUIREMENTS

The switchgear designer has to concentrate more on the risks of things going wrong and the consequences of failure, than on the normal day-to-day functions of the equipment. Switchgear joins the ranks of nuclear power stations and airliner design in that the most important element is that the equipment shall not break down. Discussing now in order the elements just reviewed, the following factors are important in examining the possibilities on offer.

Table 10.1 Degrees of protection (based on BS 5490:1977)

The 'degree of protection' definitions and tests for compliance are contained in BS 5490:1977, and are specified by the two letters IP, followed by two 'characteristic numerals' defined below.

First numeral	*Description*	*Second numeral*	*Description*
0	Non-protected	0	Non-protected
1	Protected against solid objects greater than 50 mm	1	Protected against dripping water
2	Protected against solid objects greater than 12 mm	2	Protected against dripping water when tilted up to 15°
3	Protected against solid objects greater than 2.5 mm	3	Protected against spraying water at an angle up to 60°
4	Protected against solid objects greater than 1.0 mm	4	Protected against splashing water
5	Dust-protected	5	Protected against water jets
6	Dust-tight	6	Protected against heavy seas
		7	Protected against the effects of immersion
		8	Protected against submersion

BUSBARS

In the main, these will be made of copper. From time to time the relative prices of copper and aluminium on the metal markets have made aluminium look an attractive alternative to copper. On balance, however, the potential problems that might arise from the use of aluminium, or more specifically, from the points where connection may have to be made between aluminium and copper, have tended to prevent its use in any substantial way. Low voltage switchgear, where the busbar system can be quite complex when circuit-breakers are stacked one above the other as well as side-by-side, sometimes have a busbar system of welded aluminium bars, particularly in the USA.

If the copper/aluminium joints are carefully made, with scrupulous cleanliness and proper attention paid to oxide removal, followed by measures to ensure that moisture cannot enter the joint, then long periods of trouble-free service can ensue.

With the supreme need for safety and reliability in switchgear, most medium voltage switchgear manufacturers play safe and employ copper.

There are two ways of creating the busbar links between panels, depending on the view taken of the need to segregate the busbar chambers between cubicles. One method is to cut a large hole in the side sheet of the panel so that busbars with fixing centres equal to the panel width can be bolted between the busbar connections in each unit. The joint from one busbar to the next is made at this common point. This thus creates a common busbar chamber running from the end of the switchboard to the bus-section (or the other end if there is no bus-section). If the busbars are uninsulated, then, in the event of an electrical breakdown between the busbars for any reason, there is the risk that the resultant arc will be driven by the electromagnetic forces it creates along the busbars, away from the point of in-feed. The resultant damage could then be widespread. However, if the busbars are insulated, then the effects of the fault are contained at the initial site and the open busbar chamber minimises the consequences as it provides escape for the pressure shock wave generated by the initial arc. This tends to be the British solution (Fig. 10.14).

The alternative solution is to fit insulated barriers at the junction between panels, which form the busbar supports in the form of rather specialised three-phase – or sometimes single-phase – bushings to which the connections in the cubicle are bolted. This confines the busbar fault to the unit in which it originated, but does not provide pressure relief as described above. This approach is favoured by some continental manufacturers, and is used in America in a simplified form. It is essential that no air spaces are created near the busbars where they pass through the inter-unit barriers otherwise, in the event of condensation on the insulation surfaces, high local voltage stresses can be created in these air spaces with consequent electrical discharge and possible insulation damage.

The current rating of the busbars is greater than that of the feeder circuits, and it is usual for the current rating of the incoming supply units, the bus-section and the busbars throughout the switchboard to be the same. Technically there is not

Fig. 10.14 Busbar chamber with inter-unit busbars. (*Courtesy Yorkshire Switchgear Ltd.*)

always a need for this and in a large switchboard it is logical for the busbar rating to be reduced as more units are interposed between the feeder and the incomer. For reasons of standardisation and maximum flexibility of use, so-called 'tapered busbars' are rarely used nowadays.

BUSBAR CONNECTIONS AND DISCONNECTORS

The form that these take is dependent on the basic unit concept and the way in which the disconnection of the switching device is to be achieved.

Cubicles with fixed switching devices
When the switching device is either a switch-disconnector or a plain disconnector, then it is unusual to provide separate means of disconnection from the busbars or circuit. Partly this is due to the fact that such devices only require infrequent maintenance, are simple and thus rarely give trouble, and are generally used with circuits that are of minor importance.

In the early days of metal-enclosed switchgear containing circuit-breakers, designs existed in which the circuit-breaker was fixed and the isolation from the busbars was achieved by linking a disconnector between the busbars and the circuit-breaker. This created problems when maintenance had to be carried out and led to the introduction of the withdrawable circuit-breaker.

Since the introduction of circuit-breakers which are claimed to need no maintenance during their normal working life span, designs with fixed circuit-breakers have been re-introduced. Some of the first designs of vacuum circuit-breaker switchgear followed this principle and had three-position disconnectors, or selectors, fitted on the busbar side of the circuit-breaker so that circuit earthing could be achieved by setting the selector into the earth position and then closing the circuit-breaker. The voltage, normal current and short-time current ratings of the selector match those of the associated circuit-breaker. These cubicles were metal-enclosed as distinct from metal-clad in order to reduce the amount of insulation. The selectors were in the busbar compartment and the circuit-breaker was in a compartment which communicated with that containing the circuit components. Figure 10.15 illustrates the principles of the fixed circuit-breaker cubicle.

However, this arrangement was never popular with users, who bought it with reluctance since in spite of the claims concerning the freedom from maintenance, users felt uneasy about the consequences of a mechanical breakdown which would make it necessary to shut down the switchboard to obtain access to the offending components. Although some access to the circuit-breaker could be obtained with the busbars alive, by putting the selector in the earth position and applying earths to the cables, the ability to withdraw the circuit-breaker totally from the switchboard is preferred. This enables the user to replace the defective circuit-breaker, either with one not in use or with one from a less important circuit, while the damage is repaired.

Another factor is that working on a switchboard with the busbars and some of its circuits live requires stringent safety precautions and a permit-to-work system

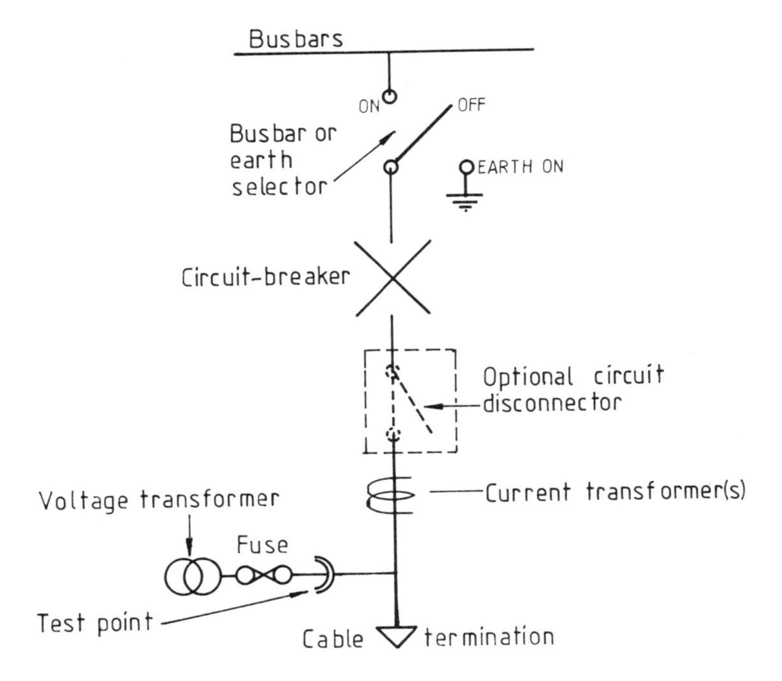

Fig. 10.15 Fixed circuit-breaker cubicles.

(see Chapter 15), whereas the withdrawn circuit-breaker is just a piece of mechanical equipment offering no electrical dangers, which can be taken to a workshop for attention.

As a result, British manufacturers eventually introduced new withdrawable designs of vacuum circuit-breaker switchgear and sales of the fixed type virtually ceased.

However, from about 1982, metal-enclosed switchgear design in Germany entered a new phase. Manufacturers of vacuum circuit-breakers have re-introduced the fixed circuit-breaker concept with disconnectors between the busbars and the circuit-breaker. A disconnector between the circuit-breaker and the circuit connections is rarely required. The only case where it may be desirable is on an incoming supply panel where it would be used to isolate the supply. The one difference between these new designs and the British designs of the early 1970s is that all the active components are enclosed in gas-filled compartments.

Of the five manufacturers now making such equipment, one uses dry air as the gas, and the others use SF_6. Three designs are totally phase segregated, with aluminium diecastings, steel fabrications and cast resin insulation with a thin steel cladding forming the individual chambers in the three cases. The main objective of this form of construction is to ensure that no part of the insulation can come into contact with the ambient air. The term 'climate-proof' has been coined to describe such equipment. Typical examples are shown in Figs. 10.16 – 10.21. These equipments suffer from the same drawbacks as the original British designs described earlier and also have the additional disadvantage of being inevitably more expensive. Their success in the market will depend on the users present acceptance of vacuum circuit-breaker reliability and his fear of insulation failure in hostile environments.

Metal-enclosed switchgear with withdrawable circuit-breakers

With the few exceptions referred to above, circuit-breakers are disconnected from the switchboard by withdrawal. This achieves a number of desirable objectives:

1. It is a relatively cheap method of disconnection from the busbars and the circuit.
2. It makes the circuit-breaker available for maintenance, away from its electrical environment.
3. It creates easy access to the circuit connections and the busbars for earthing and testing.

A common technique in this field is to provide what is colloquially called a 'spout' insulator. This is a hollow, cylindrical insulator at one end of which is a plug contact connected through the insulation into the busbar or circuit chamber. The circuit-breaker has a mating contact at the end of each bushing, which engages with the spout plug when the circuit-breaker is fully plugged in.

For the cubicle still to be metal-enclosed when the circuit-breaker is disengaged, it is necessary to fit metal doors, usually called 'shutters', which automatically close off the entrance to the spouts when the circuit-breaker is

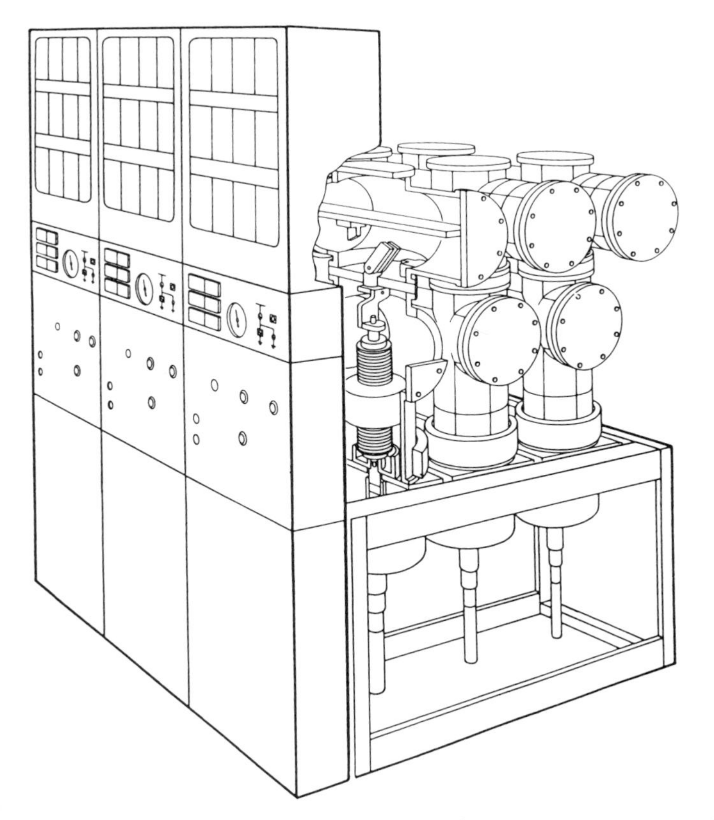

Fig. 10.16 The Siemens gas filled fixed v.c.b. unit. (*Courtesy Siemens Ltd.*)

Fig. 10.17 The Siemens gas insulated unit. (*Courtesy Siemens Ltd.*)

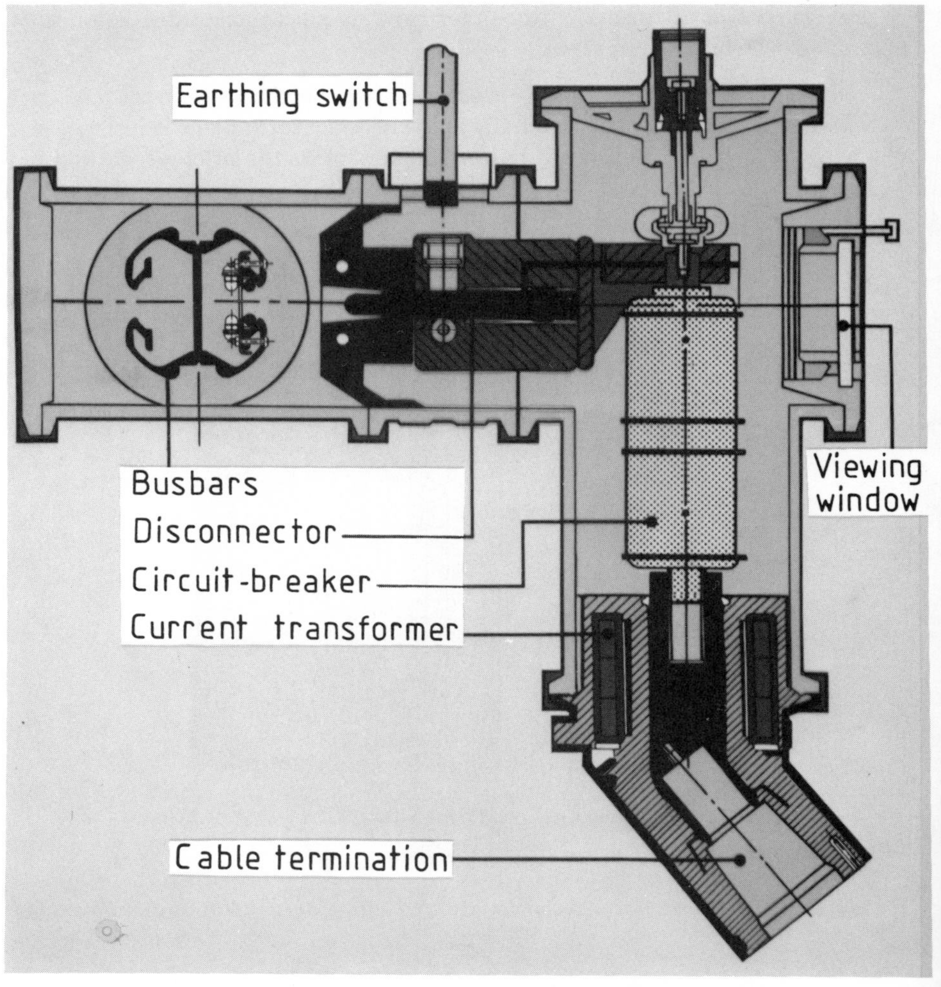

Fig. 10.18 The Brown Boveri fixed v.c.b. unit. (*Courtesy BBC Brown Boveri Ltd.*)

withdrawn, and which are automatically opened as it is re-introduced. The crucial aspects of design of such spout insulators are:

1. The even distribution of electrical stress over the internal surface of the spout between the live plug and the earth plane, whether the circuit-breaker is in or out and whether the shutters are open or closed.
2. To maintain the integrity of the insulation under conditions of heat and humidity, with possible condensation if high humidity and wide-ranging temperatures coincide. The question of providing ventilation in such insulators, which will help to prevent condensation on the one hand but perhaps reduce electrical creepage and provide a communications channel between otherwise

segregated compartments on the other hand, is a permanent discussion point among switchgear designers.

Factors relevant to the above discussion are whether the circuit-breaker is withdrawn vertically or horizontally and where the earth plane is in relation to the spout insulator. Several variations are possible in the latter respect and Fig.

Fig. 10.19 The Brown Boveri sealed air insulated unit. (*Courtesy BBC Brown Boveri Ltd.*)

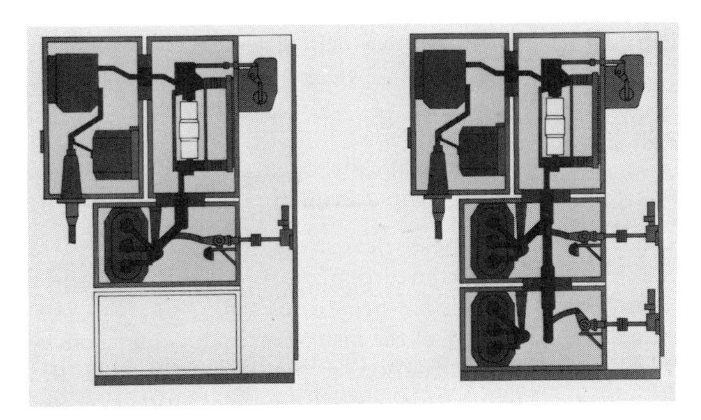

Fig. 10.20 The Calor Emag gas filled fixed v.c.b unit. (*Courtesy Calor-Emag Elektrizitäts-Aktiengesellschaft.*)

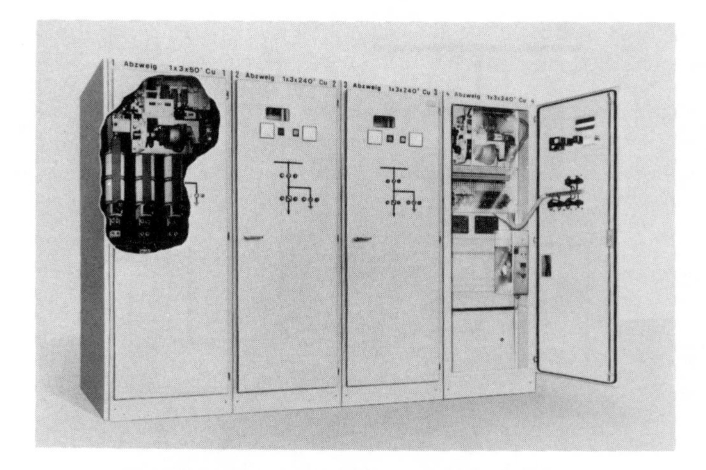

Fig. 10.21 The Calor Emag gas insulated unit. (*Courtesy Calor-Emag Elektrizitäts-Aktiengesellschaft*.)

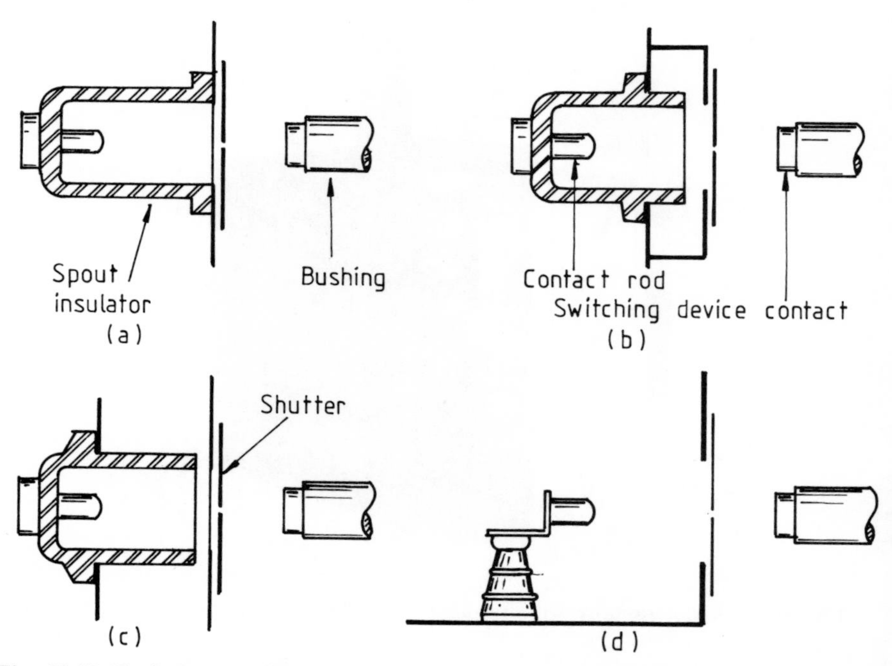

Fig. 10.22 Typical mounting of spout insulators: (a) on division sheet; (b) through division sheet; (c) spaced from division sheet; (d) no spout.

10.22 illustrates some of them. In Fig. 10.22(a), the spout is mounted like a bell on the division sheet so that the creepage distance is only the inside insulation length. Figure 10.22(b) shows a modification of this, which is easy to achieve with the use of cast resin materials, and which increases the creepage length by

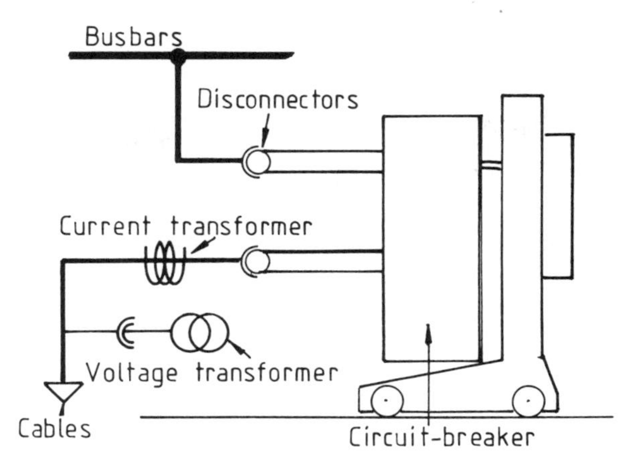

Fig. 10.23 A cubicle with a horizontally withdrawable circuit-breaker.

Fig. 10.24 A horizontally withdrawable circuit-breaker equipment. (*Courtesy Brush Switchgear Ltd.*)

letting the spout project through the division sheet. Now care has to be taken not to reduce the creepage and clearances in the busbar or circuit chamber to dangerous levels as the earth plane is moved towards the connection point.

A further variation is shown in Fig.10.22(c), where the spout is taken away

Fig. 10.25 Typical cubicle with a vertically withdrawable oil circuit-breaker. (*Courtesy GEC Distribution Switchgear Ltd.*)

Fig. 10.26 Carriage with a horizontally withdrawable circuit-breaker arranged for integral earthing by transfer. (*Courtesy Yorkshire Switchgear Ltd.*)

from the division sheet and an air gap is added to the insulation creepage distance. In this case, special arrangements have to be made to ensure that the metal-clad definition is not impaired as a communication channel between the circuit-breaker compartment and the busbar or circuit chamber may be introduced.

If a simple metal-enclosed design is acceptable, the simplest form is to do away with the spout and just provide a hole in the division sheet, covered as usual by shutters, through which the circuit-breaker bushings project and engage with contacts connected to the busbar or circuit conductor supports, as illustrated in Fig 10.22(d).

THE SWITCHING DEVICE

Although devices based on switches, as opposed to circuit-breakers, are not common in primary substations, they are occasionally used to control local supplies, e.g. to a small distribution transformer providing low voltage supplies for the substation. Another application where the voltage transformer rating is of such magnitude that it is too heavy to be conveniently mounted on the switchgear cubicle is to have an extra panel on the switchboard to house the voltage transformer, together with its fuse protection and disconnection features. This is frequently done in the USA where voltage transformer power ratings tend to be higher than in most other parts of the world.

So the majority of switching devices are circuit-breakers and almost all of them are withdrawable for reasons already mentioned. Therefore the circuit-breaker has to be mounted on a carriage so that it can be moved into the cubicle or taken away for maintenance. This carriage has some mechanical arrangement of levers or a screw to move the truck, or carriage, at least the last part of the way into contact to overcome the frictional resistance created by the disconnecting contacts and the shutter-operating mechanism.

If the circuit-breaker bushings are horizontal then the one act of entering the carriage fully into the cubicle will place the circuit-breaker in the service location and it only remains to close it for power to be transmitted through the connected circuit. This type of equipment is known as horizontally withdrawable switchgear, shown diagrammatically in Fig. 10.23, with a typical example shown in Fig. 10.24.

Bulk-oil circuit-breakers in particular (Fig. 10.25) usually have bushings mounted vertically in a metal housing usually referred to as the 'top-plate'. This also carries all the mechanical parts and the insulated drive rods to the moving contact structure (see Fig. 4.11). The oil is contained in a tank on to which the top-plate is bolted. In order to engage the bulk-oil circuit-breaker in the service location it has to be pushed into the cubicle to a preset location and then raised, so a carriage for such a circuit-breaker must have motion in two dimensions. This facility is used when the oil tank has to be removed for maintenance or to provide a facility known as 'integral earthing' by circuit-breaker transfer which is described in more detail in Chapter 15.

A two-dimensional carriage can also be provided for horizontal drawout switchgear but the lifting feature is then an extra facility not normally required for day to day operation (Fig. 10.26).

Other features that need to be fitted to the circuit-breaker carriage and inside the cubicle are associated with the location of the carriage and the automatic operation of the shutters. If we exclude the earthing by circuit-breaker transfer until we deal with the special topic of earthing, then both shutters open automatically as the circuit-breaker is put into the service location. It is quite common for some locking feature to be provided so that entry of the circuit-breaker can be prevented if there is some operational reason for this. It is also generally possible to lock the circuit-breaker into the service location, either by a padlock in the racking mechanism, or a special bolt which can then be padlocked. Figures 10.27 and 10.28 illustrate some of these locking features.

Fig. 10.27 Shutters and locking features. (*Courtesy GEC Distribution Switchgear Ltd.*)

Fig. 10.28 Circuit-breaker racking gear with locking features. (*Courtesy GEC Distribution Switchgear Ltd.*)

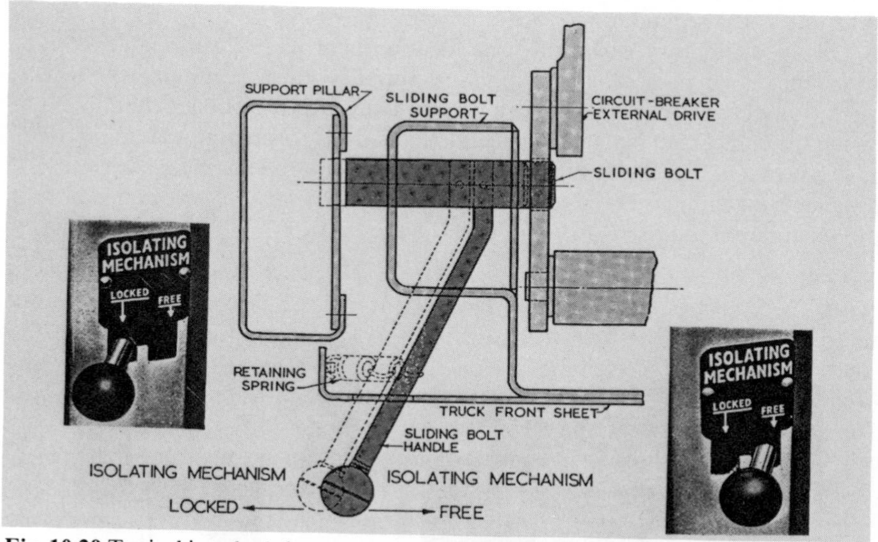

Fig. 10.29 Typical interlock for preventing isolation of a closed circuit-breaker. (*Courtesy GEC Distribution Switchgear Ltd.*)

An essential interlocking feature for all switchgear with withdrawable circuit-breakers is that of preventing the circuit-breaker from being operated unless it is first correctly located, and in reverse, to prevent the circuit-breaker from being withdrawn except when it is open (Fig. 10.29).

Sometimes position switches are fitted into the cubicle to give remote indication of the location of the circuit-breaker, in addition to the mechanical indicators fitted to the cubicle. Always there are auxiliary switches operated by the circuit-breaker which are connected into the protection and indication systems for many purposes.

As the circuit-breaker also needs auxiliary power supplies for open and close releases, some form of secondary plugs and sockets are required. This term is used to differentiate from the primary plugs and sockets used to connect the circuit-breaker into the cubicle. These supplies also drive electric motors providing power to rewind springs or to give assistance with the racking gear. As it is useful not to have these circuits connected until the circuit-breaker is in its service location, one method of engaging these contacts is to have mating, self-aligning blocks of contacts, one part on the carriage and the other part in the cubicle so that both the primary and the secondary circuits are completed by the one act of racking the circuit-breaker into the service location (Fig. 10.30). Alternatively, there may be a separate plug which is manually inserted into a socket before the circuit-breaker is located in the cubicle. In this case some form of interlock is desirable to ensure that the secondary circuits are made before the circuit-breaker can be placed in the service location. A typical method of achieving this is by a mechanical device which prevents the use of the racking gear unless the secondary circuits are made.

This is satisfactory when only one location is provided, although even then some means of providing the auxiliary supplies must be found so that the operation of the circuit-breaker can be checked whilst it is outside the cubicle, or inside the cubicle but isolated. Typically this is done through an extension cable, as illustrated in Fig. 10.31. When there are also alternative earthing locations, the designer's ingenuity is taxed to the utmost to provide safe, reliable connection of the secondary circuits required in each of the locations.

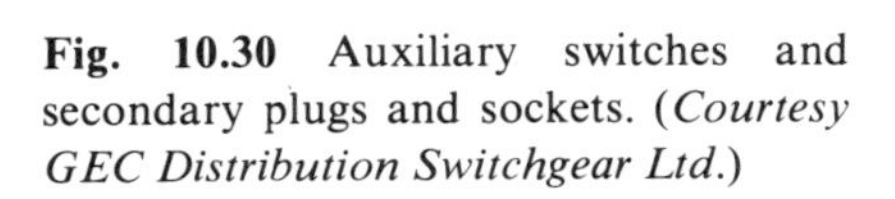

Fig. 10.30 Auxiliary switches and secondary plugs and sockets. (*Courtesy GEC Distribution Switchgear Ltd.*)

Fig. 10.31 Plug and socket extension cables. (*Courtesy Yorkshire Switchgear Ltd.*)

INSTRUMENT TRANSFORMERS

Current transformers are necessary to provide a suitable low voltage current proportional to the load. The characteristics of current transformers are considered in relation to the duty they have to perform in Chapter 13. It is sufficient to note here that they can take two forms – those with a straight conductor as a primary 'winding' and those that have more than one turn as a primary winding. The former are referred to as *bar primary current transformers*, and the latter as *wound primary current transformers* (Figs. 10.32 and 10.33). If the core and secondary windings are a separate component slipped

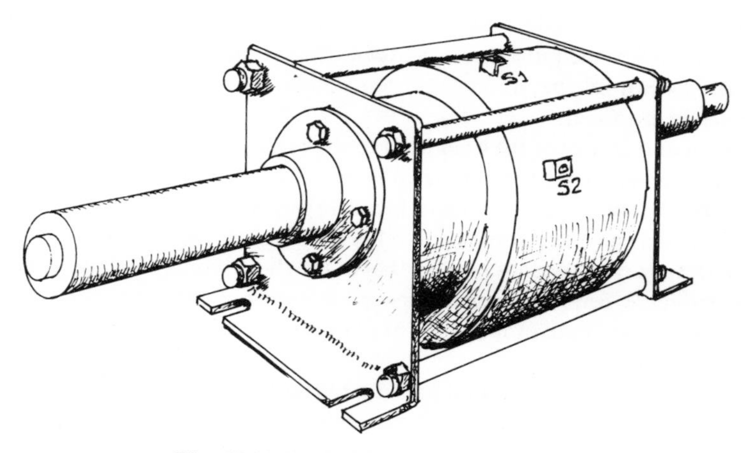

Fig. 10.32 Bar primary current transformer.

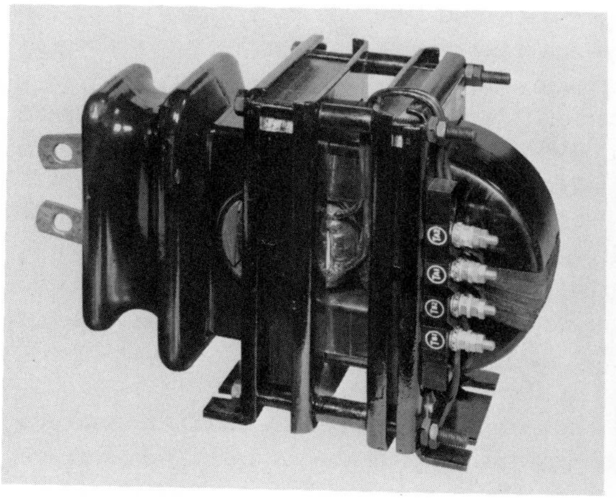

Fig. 10.33 Wound primary current transformer. (*Courtesy GEC Distribution Switchgear Ltd.*)

Fig. 10.34 Ring CT on a circuit bushing. (*Courtesy GEC Distribution Switchgear Ltd.*)

over a fully insulated primary bushing they are referred to as *ring type current transformers* (Fig. 10.34).

Bar primary current transformers may carry the insulation on the primary conductor, which then incorporates an earth electrode under the outer surface, or the secondary windings and iron circuit may carry the main insulation and accept bare conductors as the primary. This latter technique is common in the American market. Continental Europe tends to favour the wound primary current transformer, usually cast in a block of resin with primary and secondary terminals. Except for the very lowest ratios, say below 100 A primary current, the UK market prefers the bar primary current transformer.

It will be appreciated from this that the current transformer primary rating is not necessarily the same as the current rating of the circuit, in fact it is more often less, and sometimes considerably less. This is because the current transformer is more accurate when the ratio nearly matches the actual load current, which may be only a fraction of the standard rating of the panel in which the current transformer is mounted.

The design and layout of the switchgear cubicle plays a part in determining the form of the current transformer. If it is more convenient to include a long bushing between the circuit spouts and the cable compartment, than to provide space for a wound primary design, then the bar primary type will be used even at quite low ratios. If the converse is the case, then wound primary current transformers have been used up to quite high ratios with only a single turn primary, usually laminated for ease of manufacture. An 800 A primary rating is quite possible for such a CT.

A problem with low ratio wound primary current transformers arises from the electromagnetic forces generated at high fault currents. Damage to the multi-turn primary windings and their insulation can occur at excessive short-circuit currents. It is also very difficult and expensive to provide short-time current durations of more than 0.5 s or 1.0 s and this may be a limitation on the application of the protection system, if the switchgear otherwise might have a 3 s short-time rating, which is common in the UK.

Voltage transformers also have to be fitted if a voltage reference is required. They are necessary if distance or directional protection is used, if a voltmeter or a wattmeter is fitted or if under-voltage releases are needed. Often it is sufficient to fit one voltage transformer to each section of busbar, in which case it is usually located on the incoming unit. If there is more than one incoming unit they would all be fitted with voltage transformers. To avoid parallel circuits, automatic switching of the voltage reference to those instruments and meters that require it is usually provided through the auxiliary switches on the incoming circuit-breakers and the bus section (Fig. 10.35).

In many parts of the world it is necessary to provide protection in the event of a winding failure in the voltage transformer and this generally takes the form of a fuse. A consequence of this is that the voltage transformer is mounted in a disconnectable way so that access to the fuses for changing can readily be achieved. This may be by withdrawal from some voltage transformer spouts, like those for the circuit-breaker and fitted with lockable shutters, or by translation

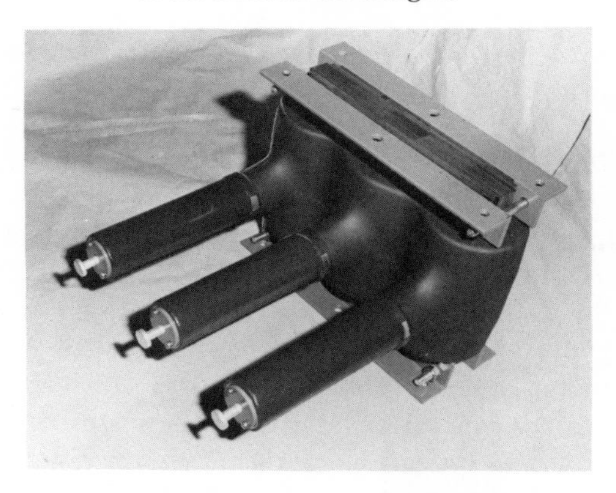

Fig. 10.35 Voltage transformer.. (*Courtesy GEC Distribution Switchgear Ltd.*)

and/or rotation in its enclosure. In the latter case some means has to be provided to isolate the secondary windings in case a back feed energises the voltage transformer, and before access can be gained to change fuses an earth must be applied to the primary winding.

CABLE TERMINATIONS

This is an area which is extensive enough to warrant a chapter of its own. Many and varied are the arrangements, sizes and types of cable for which termination facilities have to be provided, and national differences are also quite pronounced, as explained in Chapter 14.

VARIATIONS IN METAL-ENCLOSED SWITCHGEAR DESIGN

Metal-enclosed switchboard panels can be arranged in as many ways as there are switchgear designers, and fashions have changed over the years, partly as a result of the introduction of new materials, such as casting resins. The most common varieties in the world are panels with fixed switches or withdrawable circuit-breakers.

Cubicles containing switches are usually simple, as instrument transformers are rarely fitted and, as has already been stated, isolating means are only fitted if the switching device is a vacuum interrupter. This is because operators still feel uneasy about the effectiveness of the small gap in the vacuum interrupter, particularly when there is no easy way of monitoring the level of vacuum that exists. So in that case it is usual to arrange a disconnector in series with the vacuum interrupter, to isolate it from the busbars before earthing and testing operations are performed on the associated conductors, as seen in Fig. 10.15.

As an additional safeguard against accidental contact with live busbars, some continental switch-disconnectors have a facility for the operator to insert an insulating screen in the disconnector air gap which shuts off the busbars and

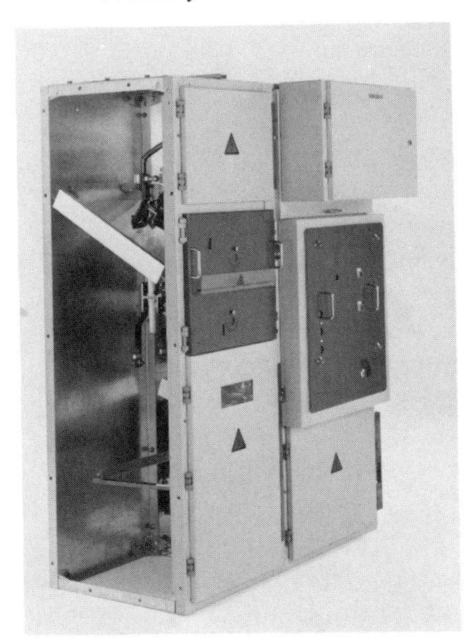

Fig. 10.36 A typical continental switch cubicle with an insulating screen in the switch contact gap. (*Courtesy Siemens Ltd.*)

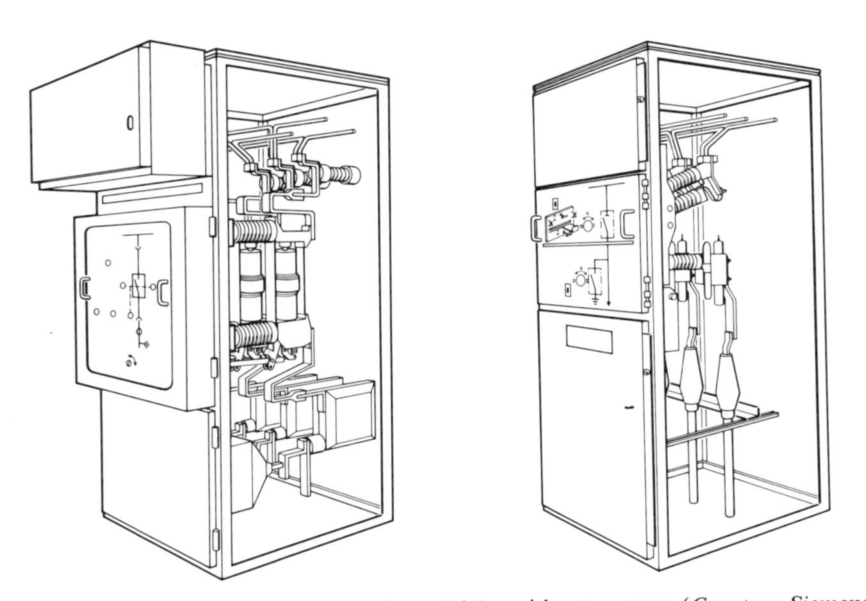

Fig. 10.37 Metal-enclosed circuit-breaker cubicles without screens. (*Courtesy Siemens Ltd.*)

switch contacts connected to them. Figure 10.36 is a typical unit of this type.

The construction of cubicles housing circuit-breakers is more complex and more subject to variation than those containing switches. The simplest metal-enclosed units are factory made copies of the early locally built cells. Behind an earthed metal door is a horizontally withdrawable truck which engages with isolating contacts supported on insulators, and connections (normally bare copper) are made from these contacts to the busbars on the one hand and to instrument transformers and cables on the other (Fig. 10.37). The cables usually rise from a cable basement in continental substations, so they enter the cubicle through its base and terminate either on the current transformer terminals or on support insulators adjacent to them. The cable tails are insulated by plastics sleeves and the terminals protected by plastics shrouds.

However, this type of equipment is now rare, and the more usual arrangement is the metal-clad construction which has metal partitions dividing the interior into separate compartments for the busbars, switching device and the circuit components, as shown in Fig. 10.38. The partition between area 2 and area 4 is not always present, although it is increasingly being provided, particularly in the UK, as users are becoming increasingly concerned about the spread of damage within the equipment in the event of a fault in the cable compartment. Area 4 is the most usual site for breakdown to occur in metal-enclosed switchgear. Figures 10.39 – 10.41 show some typical equipments.

In some parts of the world the partition between the circuit-breaker compartment and the busbar and circuit compartments is made of insulating material with fire-resisting properties, and the shutters are of the same material. This is no longer metal-clad but enables the size of the cubicle to be reduced, though it introduces the need to make extra tests to demonstrate that dangerous

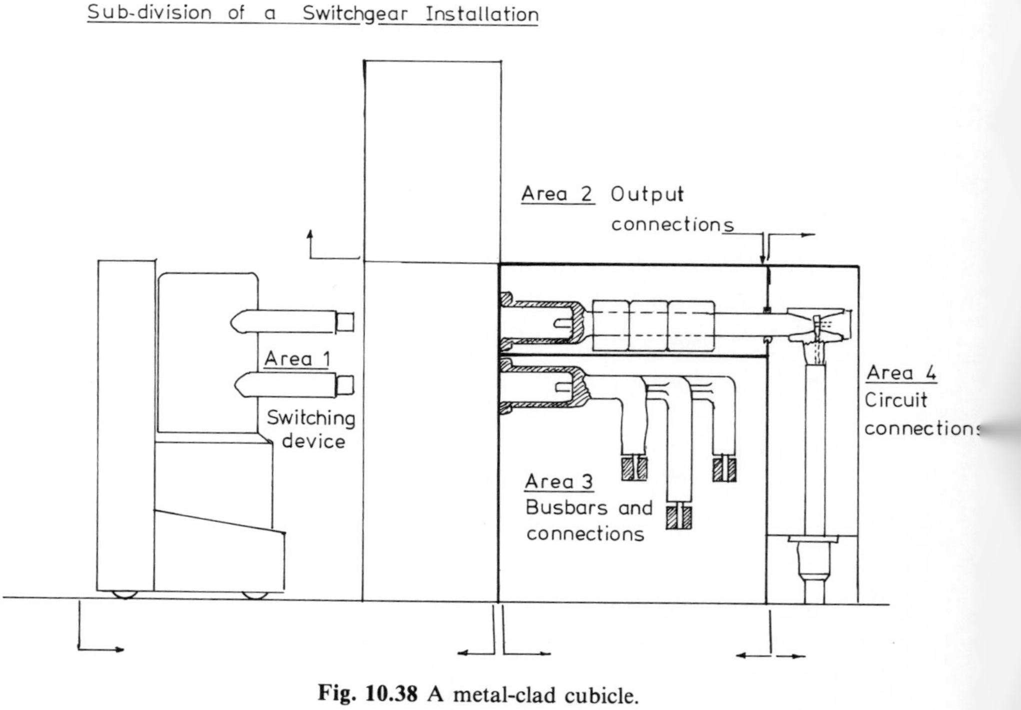

Fig. 10.38 A metal-clad cubicle.

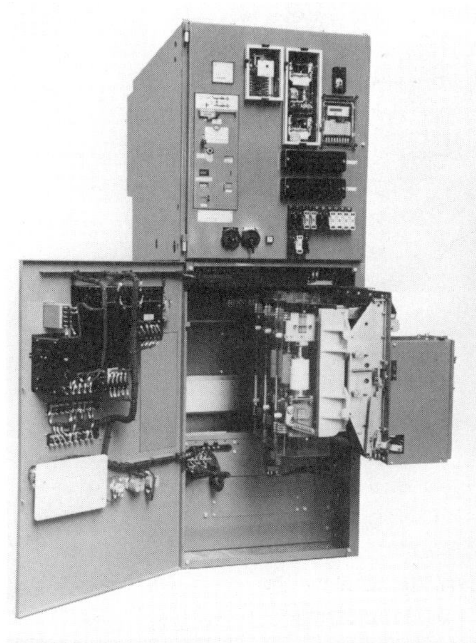

Fig. 10.39 Vacuum circuit-breaker and cubicle. (*Courtesy NEI-Reyrolle Distribution Switchgear.*)

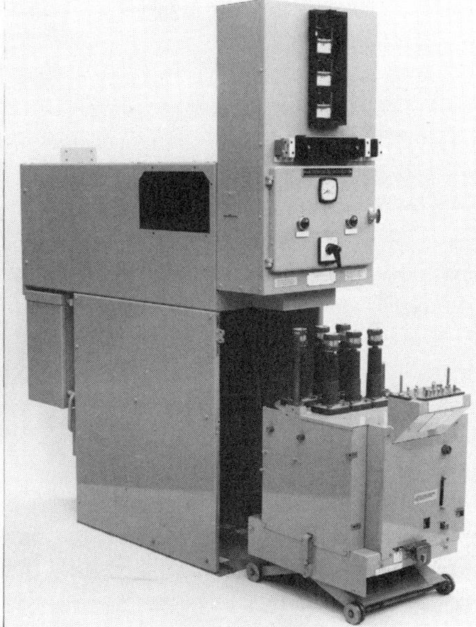

Fig. 10.40 12 kV rotating arc SF_6 circuit-breaker and cubicle. (*Courtesy South Wales Switchgear Ltd.*)

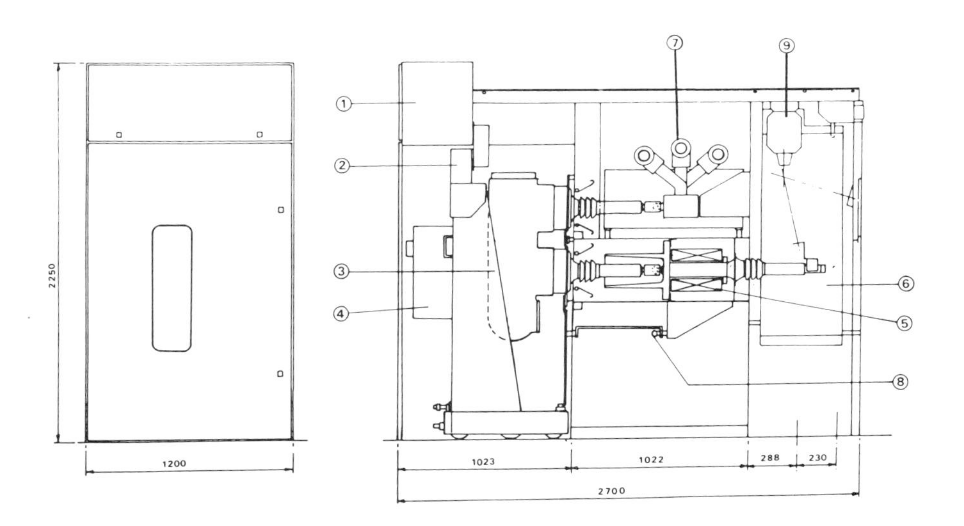

1) WIRING PANEL. 2) SECONDARY PLUGS AND SOCKETS. 3) CIRCUIT BREAKER.
4) OPERATING MECHANISM. 5) CURRENT TRANSFORMERS. 6) CABLE COMPARTMENT
7) BUSBARS. 8) CIRCUIT EARTHING PLUGS. 9) VOLTAGE TRANSFORMER.

Fig. 10.41 36 kV vacuum circuit-breaker and cubicle. (*Courtesy GEC Distribution Switchgear Ltd.*)

electric charges do not accumulate on these insulating partitions which may be harmful to the operator if he were to touch them when the circuit-breaker is withdrawn. These tests are prescribed in the international standard IEC 466, which covers 'insulation-enclosed switchgear'.

Early British designs of vacuum switchgear either utilised the same principle as the previous bulk oil circuit-breakers, so that the moving portions were interchangeable, or used fixed circuit-breakers with disconnectors on the busbar side. The former arrangement made it possible for owners of existing switchboards from the relevant manufacturer to exchange his old circuit-breaker for one using vacuum technology.

Other variations in the form of construction for metal-enclosed switchboards arise from the way in which facilities are provided to meet the user's requirements for carrying out earthing, testing and maintenance. These operating practices are described in detail in Chapter 15, but the constructional consequences are discussed here. The major effects derive from the number of circuit-breaker locations that have to be allowed for, whether a door has to be provided for the circuit-breaker compartment and what arrangements are to be made for the protection of a voltage transformer. If a cubicle door is specified, does it have to be closed when the circuit-breaker is disconnected? Some users prefer to be able to prevent unauthorised interference by always being able to lock the cubicle, while others like to be able to see the circuit-breaker location at a glance.

REDUCING THE CONSEQUENCES OF AN ARCING FAULT

Electrical breakdown between conductors, or between conductors and earthed metal, inside metal-enclosed switchgear is rare, but when it happens the consequent arcing fault can have serious consequences for personnel in the vicinity and causes extensive damage to the equipment. A few years ago, German and Scandinavian manufacturers developed designs of cubicles which would direct the consequences of an internal arcing fault away from those areas likely to be occupied by operating personnel. Test procedures were also developed so that criteria could be established to prove that this objective had been achieved. These tests are described in Chapter 18 and of necessity are performed at a high power test station, where the roof, the back wall and one side wall of a typical substation can be simulated by mounting the equipment for test in a corner of a test cell and suspending a false ceiling over it. The other two sides are open to the test enclosure.

The usual method of construction to achieve success in this test is to arrange for the top covers and/or rear covers of the cubicle to be easily torn from their fixings to release the pressure wave into the rear of the substation, and to construct those parts of the cubicle that might form the earth electrode for internal arcing faults of metal thick enough not to burn through in the prescribed time period. However, with fault currents in the tens of kiloamperes and durations in the order of tens or even hundreds of milliseconds, the fireball so

Fig. 10.42 Typical cubicle with 'arc-proof' features.

Fig. 10.43 Insulated conductor system. (*Courtesy Yorkshire Switchgear Ltd.*)

released is of such magnitude that the question should be asked, what the consequences still might be in a small, totally enclosed substation. Figure 10.42 illustrates a unit of this type.

British engineers tend to hold the view that rather than try to deal with the consequences of an arcing fault in this way, efforts should be concentrated on preventing the faults occurring in the first place. Many of the cubicles in which so much attention has been paid to strengthening doors and walls on the one hand, and enfeebling them on the other, have bare conductors inside, either as busbars or connections, or both. This fact must be a significant factor in the frequency of the incidents which gave rise to the requirement to introduce these arrangements.

In British switchgear most, if not all, of the internal conductors are insulated and means are provided to insulate the joints between conductors such as those that occur in the busbar chamber (Fig. 10.43). This avoids direct breakdown paths between conductors and to earth and the incidence of internal arcing faults in such equipment is so small as not to warrant the expense and complication of 'arc proofing'. However well intentioned preventive measures are, breakdown can never be totally eliminated so it is prudent to take precautions to limit the damage on those rare occasions when trouble does occur.

The most likely site of electrical breakdown is in the cable compartment, and one useful way of reducing damage here on switchgear of very high fault rating is to phase segregate the cable box. Ideally this requires the use of single-phase cables and effectively each is terminated in a separate box, constructed by putting metal partitions between the phases in the cable compartment. A phase-to-phase fault is then impossible, and as it is usual to earth the neutral through a resistance in systems of high fault level, the earth fault current is limited to a few kiloamperes, and the damage can be contained in the single-phase compartment if reasonably high speed earth fault protection is available.

Some compact designs of switchgear use insulating sheets to insulate metal partitions when these are close to live conductors, or to place between the phases so that they can be mounted closer together than purely air clearances would allow. It is desirable to ensure that the materials used for such barriers do not give off quantities of smoke or acidic vapours in the event of an adjacent arcing

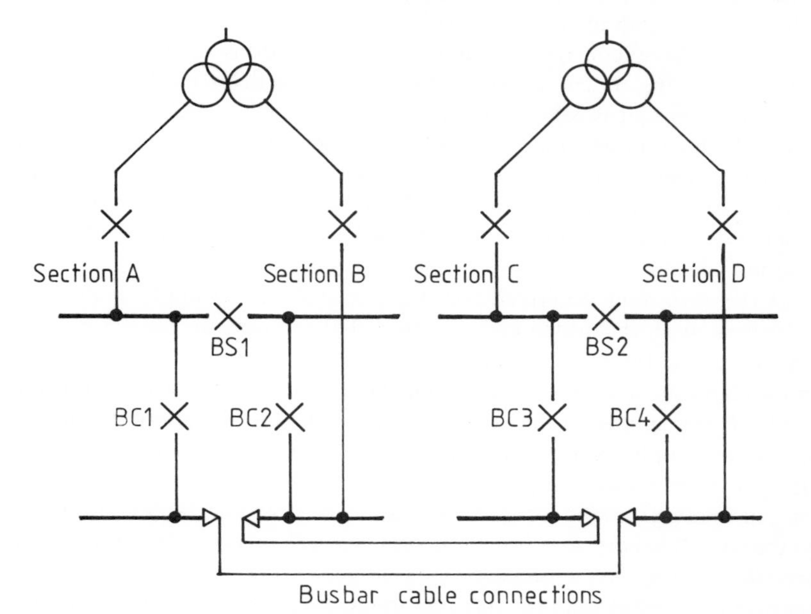

Fig. 10.44 A novel double-busbar system.

fault. The consequences of this can be extensive in terms of refurbishment of the switchboard, or even of the substation, even though the arcing damage is contained within the compartment where it originated.

DOUBLE-BUSBAR SWITCHGEAR

Another variation, although one of operational principle rather than construction, is the double-busbar switchboard. This is less used nowadays than hitherto. Originally the second busbar was provided in an important substation to make the consequences of a busbar fault less serious, as the circuits could be switched quite quickly to the alternative busbar. For this reason the busbars were usually called the 'main' and 'reserve' busbars. Other terms for the latter were 'emergency', or even 'hospital', busbar.

With increasing interconnection of distribution systems, so that alternative sources of supply can usually be arranged in the event of a breakdown, the original use of the double-busbar switchboard diminished. There are, however, some uses where, on a large switchboard with more than one incoming supply, the load can be balanced more evenly by making use of the two busbars.

An ingenious use of a double-busbar switchboard for quickly restoring connection in the event of a loss of an incoming supply is illustrated in Fig. 10.44. Two special transformers with double secondary windings are used for the supply, making effectively four incoming supplies. These are fed into four of the sections of busbar in two double-busbar switchboards linked as shown by cables. Normally the system runs with the bus-section circuit-breakers (BS1 and BS2) closed and the bus-coupler circuit-breakers (BC1 to BC4) open.

A fault on either winding of a transformer will necessitate the tripping of both secondary circuit-breakers, thereby depriving two sections of the busbars of supply. These sections are energised after a few seconds delay by the closing of the bus-coupler circuit-breakers on sections A and C (BC1 and BC3), irrespective of which transformer has been disconnected from the busbars. It will be seen that transfers of feeder loads between associated secondaries and between transformer units is reduced to a simple busbar selection operation and that under no condition are the transformer secondaries paralleled, thus restricting fault levels.

Most of the variations described for single-busbar panels apply to double-busbar versions, with the added complication that the ease of transfer plays a part in dictating the type of construction.

The most elaborate and expensive arrangements occur when on-load transfer is required, i.e. the supply to the consumer is switched to the alternate busbar without loss of supply. Since this implies at least a brief period of parallel connection of the two busbars, precautions need to be taken to ensure that this does not cause any problems of synchronisation or potential fault conditions. Usually the two busbars are first coupled through a circuit-breaker as a safeguard. With fixed circuit-breakers the changeover can be effected by using two disconnectors and closing the one relating to the reserve busbar before the other is opened.

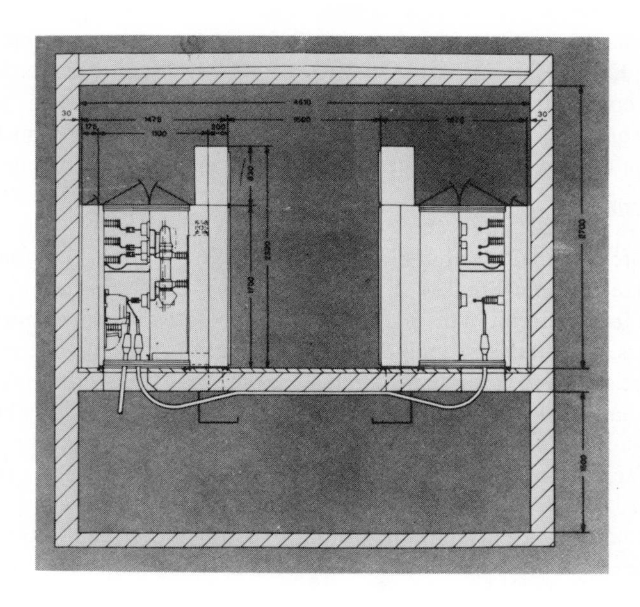

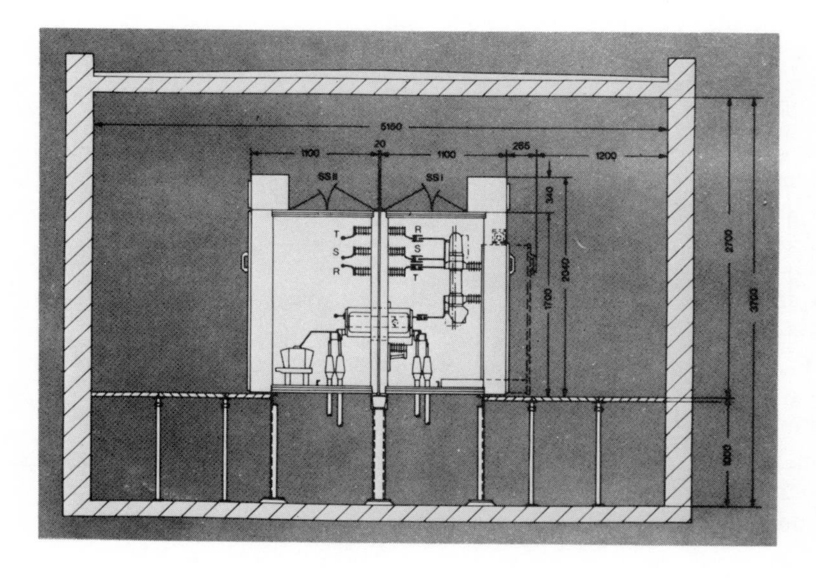

Fig. 10.45 Continental type of double-busbar layout. (*Courtesy Sachsenwerke Licht- und Kraft-Aktiengesellschaft.*)

When the circuit-breakers are withdrawable, on-load changeover requires there to be two circuit-breakers per circuit, or the extra space and expense of fitting the disconnectors just mentioned for a fixed circuit-breaker installation. A fairly common arrangement used on the Continent with horizontally withdrawable circuit-breakers makes use of two similar single-busbar switchboards. For ease of operation these are typically mounted facing one another on opposite sides of the substation, although for ease of circuit connection a back-to-back arrangement might have been preferred (Figs 10.45).

The more popular double-busbar switchboards employ off-load transfer. With a fixed circuit-breaker a selector is used to choose which busbar is connected to the circuit. This may have just two positions corresponding to the two busbars, or it may also include an earth position allowing the earthing of the circuit to be achieved through the circuit-breaker. The usual interlock is included to ensure that the selector cannot be operated when the circuit-breaker is closed.

With switchboards incorporating withdrawable circuit-breakers some form of transfer facility is required. An analogous arrangement to the double-switchboard described in connection with on-load transfer omits one of the circuit-breakers from each circuit. Busbar selection is then changed by opening a circuit-breaker, removing it from the cubicle, inserting it into the corresponding cubicle of the other switchboard and closing it again. This seems an expensive and time consuming procedure, but it avoids adding complication to a simple carriage which would arise if every circuit-breaker had to have movement in the vertical direction to allow for busbar selection.

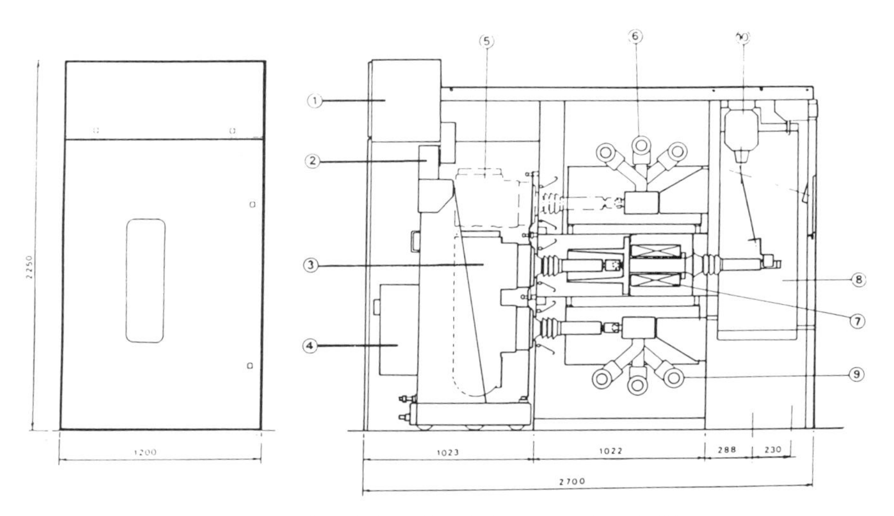

1) WIRING PANEL. 2) SECONDARY PLUGS & SOCKETS. 3) CIRCUIT BREAKER.
4) OPERATING MECHANISM. 5) BREAKER IN TOP BUSBAR SERVICE POSITION.
6) TOP BUSBARS 7) CURRENT TRANSFORMERS. 8) CABLE COMPARTMENT.
9) BOTTOM BUSBARS. 10) VOLTAGE TRANSFORMER.

Fig. 10.46 36 kV double-busbar switchgear with vacuum (SF_6 insulated) circuit-breaker. (*Courtesy GEC Distribution Switchgear Ltd.*)

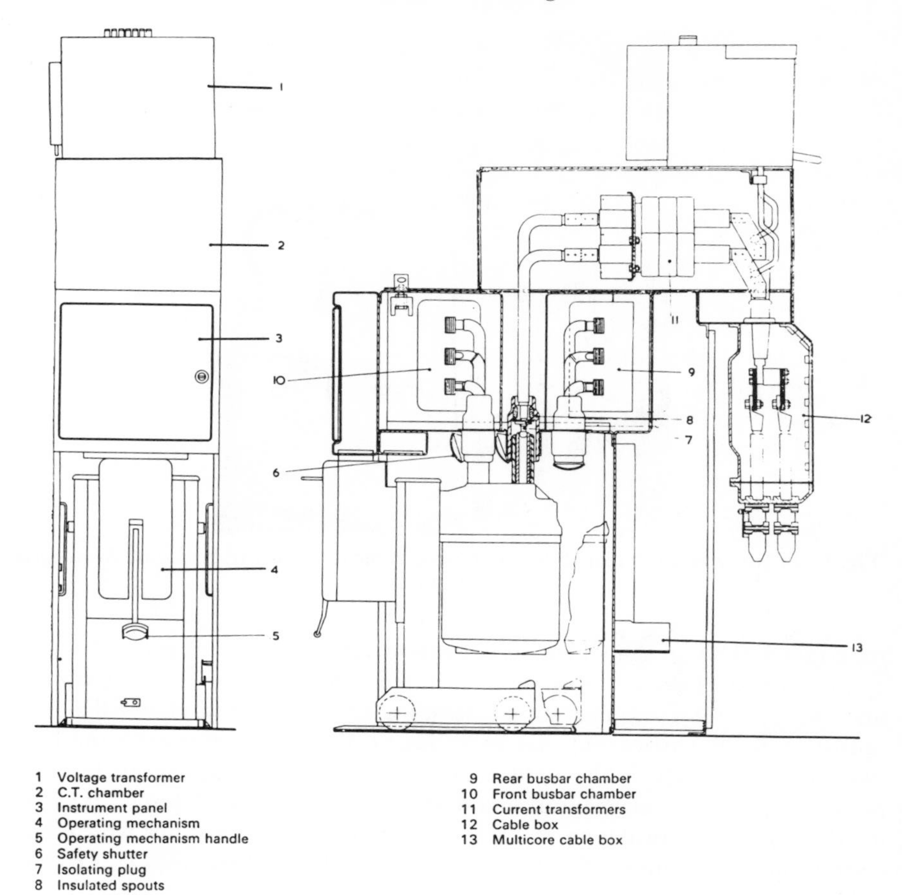

Fig. 10.47 Vertically isolated double-busbar cubicle. (*Courtesy GEC Distribution Switchgear Ltd.*)

Another solution is to fit a two-position selector between the busbar spouts and the two busbars. However, this is space consuming and needs a lot of insulation if the busbar chambers are to be kept fully segregated, which is usually a requirement for double-busbar switchgear.

With the UK type of vertically withdrawable circuit-breakers where motion in the vertical and horizontal direction are already provided for other reasons, busbar selection by circuit-breaker transfer can easily be provided. The same applies to horizontally withdrawable circuit-breakers if the second motion is already provided on the standard single-busbar carriage for another reason, such as integral earthing (Figs. 10.46, 10.47 and 10.48).

LOCKS AND INTERLOCKS

The type of lock usually employed to control the operation of the switchgear is

Fig. 10.48 Horizontally isolated double-busbar cubicle. (*Courtesy Yorkshire Switchgear Ltd.*)

the padlock. This prevents an unauthorised operation from being performed and ensures that operations are carried out in a correct sequence. The key type of interlock, whereby one operation releases a key which then allows the next operation to be carried out, has been described in connection with outdoor switchgear. This type is also used occasionally on indoor switchgear, but it is more usual to use mechanical interference type devices. Another type of interlock is the electrical type, where one operation actuates a switch which thus permits the next operation, but these are not generally liked because of the risk of failure.

The fundamental interlock that is required on all metal-enclosed switchgear has already been mentioned. This is the arrangement that makes it impossible to disconnect a switching device from the busbars and/or circuit while it is closed. Usually this interlock also ensures that the switching device is properly connected and in the correct location before it can be closed. It is preferred that this interlock be wholly mechanical as that will lead to greater integrity. There are also some minor points to watch when designing this interlock:

1. If the circuit-breaker is closed and attempts are made to operate the disconnecting arrangements, these attempts must not cause the circuit-breaker to open.

2. The feature that prevents the circuit-breaker from being closed unless it is properly located or connected should not allow the mechanism energy to be released if attempts are made to close it before it is properly connected. There are arrangements that introduce mechanical interference with the mechanism which effectively prevent it from closing but still permit attempted closure. The mechanism energy is then discharged against the mechanical device, which may cause damage. It also creates a condition from which it is difficult to recover.

Other locking and interlocking requirements mostly stem from earthing and testing requirements. Chapter 15 deals with this subject in detail in studying all the operational facilities, and in those countries which follow the philosophy of providing operator safety by mechanical prevention, a number of extra interlocks are required. One provision in those countries is that all earths to the system conductors shall be made through a fully rated, and tested, earthing switch of a design which has been proved to have a fault-making capacity, and a short-time rating, equal to the short-circuit rating of the switchgear. This affects the construction of the cubicle and the circuit-breaker carriage, depending on the method chosen by the manufacturer to fulfil this requirement. Possible ways include:

1. Providing locations within the cubicle, and transfer facilities on the circuit-breaker carriage so that the circuit-breaker itself can be used to apply the earth
2. Providing an add-on component that converts the circuit-breaker carriage into an earthing device, by extending one set of bushings and connecting the other three bushings to earth.
3. Providing a separate earthing carriage, which replaces the circuit-breaker.
4. Fitting an integral earthing switch to the cubicle.

The locking and interlocking philosophy that has to be followed as far as possible relates to the procedure laid down in the operating rules developed by the user organisations. In summary this means:

1. It must be impossible to have the circuit-breaker connected in the service location whilst earthing is taking place, or until the earth has been removed again.
2. Whilst the required earthing operation is being set up, it must be possible to padlock the shutters over the live connections in the closed position.
3. If the circuit-breaker is being used for the application of the earth, the electrical tripping must be made inoperative, but in such a way that the circuit-breaker cannot be restored to the service location without first restoring the tripping function. Mechanical tripping must always be possible.
4. Where the earthing is provided by circuit-breaker transfer, the earthing locations must have separate padlocking facilities, so that the operator must have specific authority, and the issue of a key, before earthing can be carried out, and different keys are required for circuit earthing and busbar earthing.

When the circuit-breaker needs to be operated in a different location than the normal service one, auxiliary supply leads may have to be provided for such items as spring winding motors, closing release coils, position and auxiliary switches, etc. Whatever arrangements are provided for this, it must not be possible to put the circuit-breaker back into service without restoring the correct connections for normal service.

DEGREES OF PROTECTION

One of the criteria applied to metal-enclosed switchgear is the degree to which it prevents penetration by foreign bodies, including human fingers and hands, and water.

There is an international standard on this subject, IEC 529, which is repeated in BS 5490. These establish decreasing levels of permitted penetration by foreign objects and also degrees of resistance to the entry of water. They also prescribe tests for proving both types of penetration against each of the different levels.

A code is used to indicate the degree of protection, the basic elements of which are the prefix letters IP (Index of Protection) and two digits. The first of these indicates the permitted size of foreign body that is allowed to enter the enclosure, and the second the degree of protection against water ingress. Table 10.1 records the levels of protection appropriate to each digit.

Typical levels of protection appropriate to indoor switchgear are IP 30 and IP 40. Degrees of protection applied to internal partitions, not normally accessible when the switchgear is in service, are reduced by one level. As an example the shutters on metal-enclosed switchgear could be IP 20. In some high security applications a modest degree of water resistance might also be specified for indoor switchgear, such as power station auxiliary switchgear, and that used on offshore oil installations.

Outdoor switchgear typically has a degree of protection between IP 33 and 44. The former is resistant to the entry of water falling at an angle of 60° and the latter to water from any direction.

The recently introduced fully sealed, gas-filled, fixed circuit-breaker equipments obviously have a very high degree of protection such as IP 68.

CHAPTER 11
Secondary Switchgear

That part of an electric utility's distribution system which leads directly to the transformers to provide the low voltage supplies to the consumers is the largest part of the network. Therefore it is that part of the system which benefits most from economy in the use of switchgear. Also, the consequences of a fault on this system are more limited in the number of consumers disconnected than would be the case with a fault in a primary substation. On the other hand, the simpler switchgear used on the secondary network means that it takes longer to find a fault and carry out the switching routines needed to reconnect as many consumers as possible pending repair of the fault.

In this chapter the medium voltage secondary networks are described. The various countries of the world have developed their own preferences for the operating voltages; in the UK for example, the most common voltages are 33 kV and 11 kV, although there are one or two areas which use 22 kV. In other parts of the world the use of voltages in the 24 kV region are more common. Secondary (low) voltages also vary, those of American origin being in the 120 V region and European based voltages closer to 230 V.

Optimisation of the planning of the secondary distribution network therefore balances the investment in switchgear against the cost and inconvenience of the loss of supply to a minimum number of consumers. This area of study is important today in the light of the technological changes which can affect many aspects of this optimisation.

In most rural areas of the world, and even in some urban areas in developing countries, much of the secondary distribution network uses overhead lines to distribute the power from the primary substations, which would themselves consist of outdoor, open-terminal switchgear as discussed in the preceding chapter. This is because the cost of an underground, cable connected system is very much more than an overhead network, even though the latter is obviously more vulnerable to climatic influences and mechanical damage. Different operating principles and different switchgear types have evolved for the overhead system as compared with the cable network.

OVERHEAD (AERIAL) NETWORKS

One fundamental difference between faults on overhead lines and those on cables is that a majority of the former are transient in nature. They are usually due to natural causes, predominantly lightning storms, but also due to bridging of conductors by large birds, falling tree branches or clashing of conductors arising from a combination of ice loading and strong winds. In most of these cases the fault current destroys the cause, or the cause is short-lived, so that as soon as the fault current has been interrupted by the primary circuit-breaker, the supply can be restored immediately without harm. Fault levels are usually low in these medium voltage systems, less than say 16 kA, and given high speed interruption, the damage will be slight.

It is therefore logical to arrange for the source circuit-breaker feeding an overhead line network to have an operating system such that it recloses immediately after tripping in the probability that the supply will be restored. This can usually be achieved with a 'dead time' of about 0.3 s, which means that the break in supply will be barely perceptible to the consumer.

For circuit-breakers situated within the aerial network and mounted on poles, this reclosure operation is made more sophisticated. The resulting device is called a *recloser*, since this is its major function. Nowadays it is fitted with an electronic, often digital, control system giving a wide range of characteristics. The load current and fault current ratings need only be quite modest as there is a respectable amount of impedance between the likely site of a fault and the power supply, and loads are usually light.

The first function of the control system is to grade the initial opening time of the recloser so that it is inversely proportional to the fault current,in other words, it reacts quickly to large faults. Then it is necessary to be able to adjust the time setting between the trip and the first reclosure (the dead time). The difference between the recloser operation and the primary circuit-breaker is that the latter is usually designed to give only one reclosure and then lock-out after a second trip, while the recloser is programmed to perform several reclosures at variable intervals.

Four operations is quite usual, and the first two might be virtually instantaneous, but the later ones would have a longer time interval whilst supply is disconnected. During these periods other apparatus might be brought into play, such as the automatic sectionaliser. Thus the control system has to be able to adjust the time intervals between successive reclosures, perhaps modify the inverse time characteristic for the later operations, and determine the number of opening operations permitted before lock-out. A power supply for the electronic circuits is likely to be difficult to provide from external sources at the usual locations of such equipment, so reliable long-life batteries are now used.

Another characteristic of the rural automatic multi-shot recloser is the provision of the power supply for the close and trip releases. It is now usual to take this from the overhead line to which the recloser is connected, in the form of a coil designed for the system voltage. Figures 11.1 and 11.2 show typical reclosers, one being in section so that the operating coil can be seen. Both these reclosers use rotating arc SF_6 techniques for current interruption. The philosophy of use as described above is much followed in the USA, and the usual standard to which these equipments are made and tested is the American Standard C37.60.

Fig. 11.1 Reyrolle type ESR SF_6 auto-recloser. (*Courtesy NEI-Reyrolle Distribution Switchgear.*)

Fig. 11.2 Sectional view of Brush type PMR SF_6 auto-recloser. (*Courtesy Brush Switchgear Ltd.*)

Fig. 11.3 Automatic sectionaliser. (*Courtesy Brush Fusegear Ltd.*)

Where the overhead line has no branches, a single auto-recloser at its feed point will protect it and eliminate transient faults without consumers being aware. However, where there are spur lines joined to the main feeder along its length, it is useful to be able to restore supply to the healthy parts of the system in the case of a permanent fault. This can be done if the spur lines are connected through automatic sectionalisers. These are disconnectors which can detect the passage of fault current and sense time, so that they can operate in conjunction with the auto-recloser and isolate a spur line during the dead time of the recloser.

A typical device is shown in Fig. 11.3, which has a small current transformer

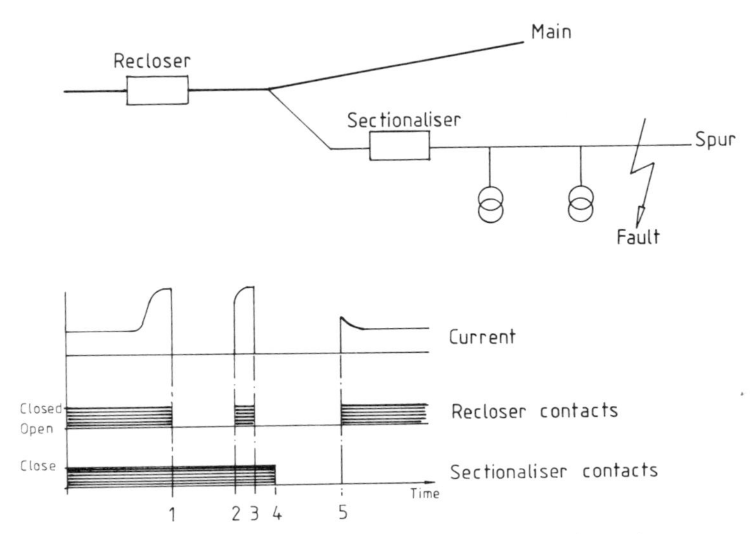

Fig. 11.4 Illustrating the use of sectionalisers with reclosers.

round the conductor which forms the contact rod of the sectionaliser and which is connected to a printed circuit. The passage of fault current above a certain value, e.g. three times the normal load current, starts the logic circuit. The first pulse is short, as the recloser trips and then recloses. If a second pulse of fault current is detected at the sectionaliser, this establishes that the fault is on the spur and that it is probably permanent. As soon as the recloser opens again, and the current at the sectionaliser has fallen to an insignificant level, the sectionaliser is opened. Then, when the recloser restores supply to the system, hopefully the remaining consumers will be re-connected.

Figure 11.4 illustrates this system. The diagram shows a main overhead feeder to which is connected a spur line to which two distribution transformers are connected. A fault is shown to have occurred on this spur line, just before point (1) on the current/time graph in the lower part of the illustration. At point (1) the recloser contacts open and interrupt the fault current. The sectionaliser anticipates a reclosure and therefore its contacts remain closed. At point (2) the recloser closes again and finds that the fault has persisted. This is detected by the automatic sectionaliser, which now prepares to open as soon as the current has fallen below the chosen threshold value. At point (3) the recloser opens again, the current disappears, and the sectionaliser opens at point (4). When the recloser closes yet again at point (5), it finds that the fault has gone and therefore remains closed.

Another protective device used in connection with an overhead line system is the expulsion fuse. This is a tube containing a fine wire, one end of which usually has a spring connected to it so that when an overload occurs on the line and the wire melts, the spring draws the ends apart and rapidly extends the arc. The pressure developed causes the gases generated from the lining of the tube to be discharged through the end of the tube and the arc extinguished. The energy

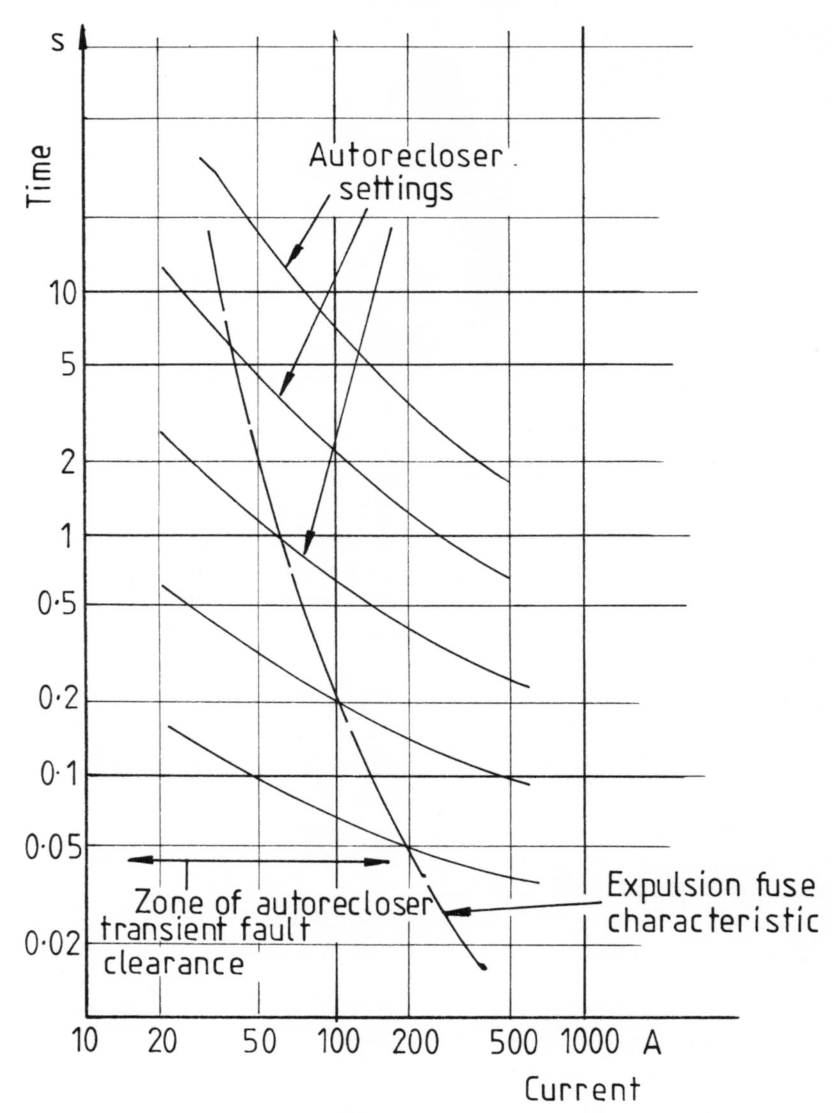

Fig. 11.5 Time/current curves for reclosers and expulsion fuses.

releases the fuse carrier so that the carrier swings open, thus putting an isolating gap in the line and giving an indication to the service engineer that the fuse has operated.

These fuses cannot deal with large currents, but serve a useful purpose in protecting spur lines without actuating the feeder circuit-breaker which could make the whole system dead. When used in connection with a recloser, the time–current characteristic of the fuse should match the characteristic of the circuit-breaker or recloser so that the recloser operates first on transient faults but the fuse gives back-up protection on the spur for permanent faults. This is

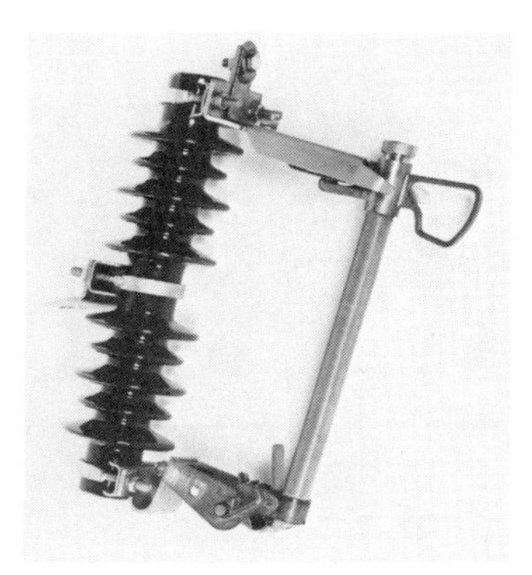

Fig. 11.6 An expulsion fuse assembly. (*Courtesy South Wales Switchgear Ltd.*)

illustrated in Fig. 11.5, which compares typical time-current characteristics for reclosers and fuses. Figure 11.6 shows a typical expulsion fuse assembly.

Off-load sectionalisers have been discussed above, but many aerial distribution networks in the world use load-break switches for their control. These devices, which usually have the necessary fault making capability, can be used as an economic adjunct to automatic reclosers to improve service on such networks.

In addition to the current sensing technique mentioned above for detecting a dead line, the sensing of voltage also has its uses. Figure 11.7 shows a ring connected network (or loop system, as designated in the USA) fed from two

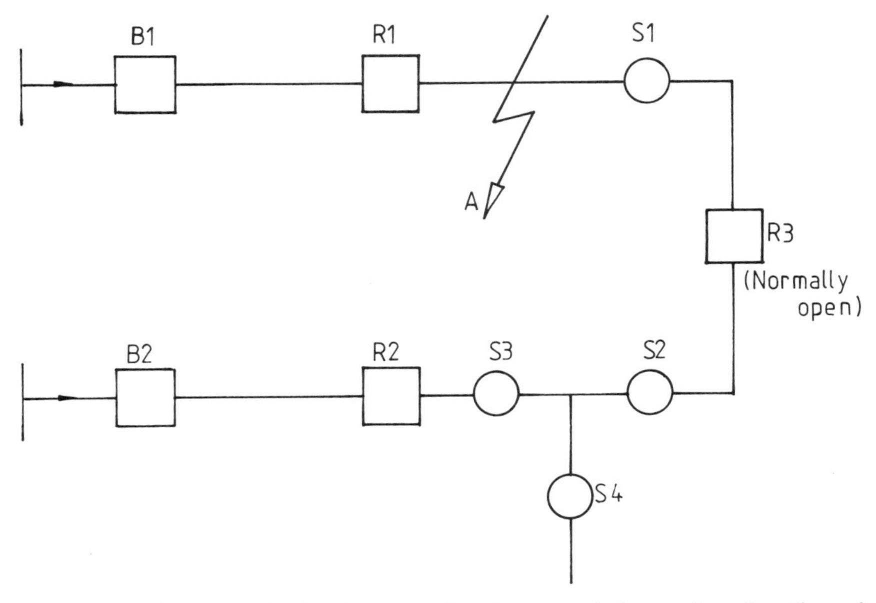

Fig. 11.7 Aerial network showing use of reclosers, switches and sectionalisers for protection.

Fig. 11.8 A vacuum switch, reclosing sectionaliser. (*Courtesy Joslyn Manufacturing and Supply Co.*)

Fig. 11.9 A pole-mounted load-break switch. (*Courtesy South Wales Switchgear Ltd.*)

primary substation circuit-breakers B1 and B2. R1, R2 and R3 are reclosers in the ring, and S1 and S2 are load-break switches connected between reclosers R1 and R3, and R2 and R3 respectively. R3 forms the normally open point in the ring, and the switches have automatic mechanisms and voltage sensing devices on both sides (towards R1 and R3, in the case of S1). The control system of the switches is assumed to be that they will always trip, with a variable delay, in the event of a loss of voltage on either connected line and will reclose after a preset interval. If a second loss of voltage occurs within a preset period after the switch recloses it will lock-out, and not reclose again. After that period it resets.

Consider now a fault on the line at point A. The recloser R1 operates and instantly recloses. S1 ignores this short loss of voltage and if the fault is transient, all is normal again. If the fault persists, R1 opens again and remains open long enough for S1 to open also. When R1 recloses fault current is re-established so it opens again. S1 remains open as the voltage disappears during the lock-out period. R3 could have the capability of recognising loss of voltage and be timed to close after a reasonably long delay. In the case described, when it recloses, supply is restored to all consumers except those connected to the faulty section.

If the permanent fault is between S1 and R3, the switch permits the second reclosure of R1 to restore supply up to S1 and when R3, having detected the voltage loss, eventually closes, it will immediately trip again when it senses the passage of fault current. This example can be extended in many ways, and in Fig. 11.7, a spur has been added controlled by an automatic sectionaliser S4 and a further automatic switch at S3. This will give protection to the three sections of line between R2 and R3, if the time settings of S2 and S3 are adjusted appropriately, and prevent faults on the spur from denying supply to consumers on the ring. Figures 11.8 and 11.9 illustrate typical switches.

One difference between medium voltage supplies in the USA and those in Europe is that American practice is to supply domestic consumers with single-phase medium voltage lines, the small single-phase distribution transformer being mounted close to the load. This obviously has an influence on the design of pole mounted switchgear used in connection with these systems. These differences extend into areas of the world which are influenced by American practices, such as the Philippines and parts of Latin America.

UNDERGROUND (CABLE) NETWORKS

Overhead electrical systems are cheaper to install than those connected by underground cable, but are more vulnerable to damage and climatic influences. In areas where consumer density is high, loads are correspondingly large and systems are complex, then the underground system has advantages. In urban environments too, there is increasing pressure from the public to remove the untidy lines for aesthetic reasons. During the last 10 years or so, the American utilities have had an intensive programme of undergrounding urban distribution systems.

In Europe generally, urban networks have been cable connected for a long time, and the common system in use is the ring main. All distribution at voltages above 1 kV is almost universally three-phase, and with the traditional use of paper insulated cables, these are three-phase also. The introduction of elastomeric cable insulation materials in recent years, particularly on the continent, has caused a swing towards single-phase cables, and this has some influence on connection practice and the design of equipment.

Figure 11.10 shows a ring main secondary distribution system, where each distribution transformer is connected to the ring through a protective device. The diagram is not representative, since several alternative arrangements are shown which would not co-exist in practice. At A the transformer is connected through a circuit-breaker with a switch, or disconnector, between the ring and the tapping point on both sides. The connection at B is made through a fuse-

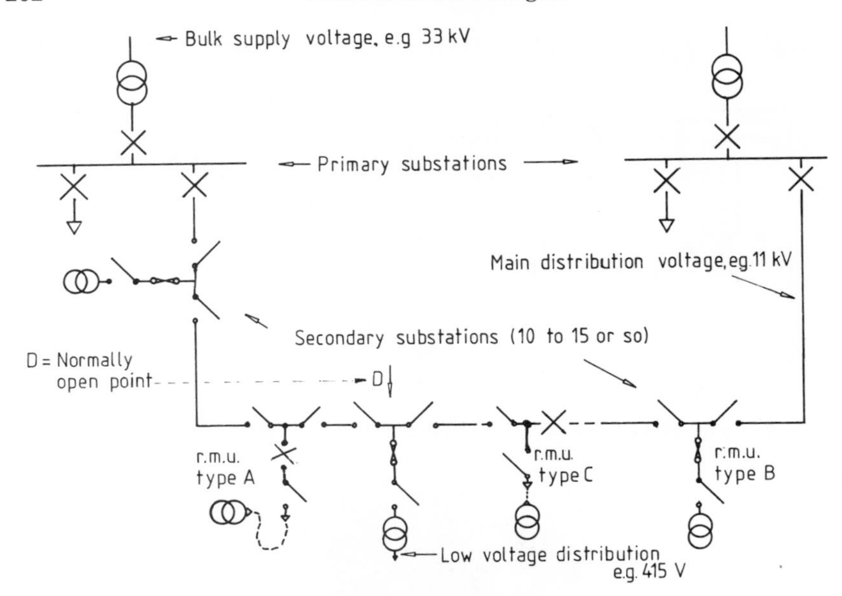

Fig. 11.10 UK ring main system.

switch, and the version at C has a circuit-breaker in the ring connection and a switch for the transformer.

In the days before the medium voltage load-break switch had been developed, and reliable medium voltage fuses did not exist, arrangement A was the traditional system. Arrangement C was a variant which recognised that faults on the ring cables were far more frequent than faults on the transformer and therefore connected the protective device into the ring. Operation of the circuit-breaker used feeder protection techniques, necessitating pilot wire connection between successive tapping points. This made the system expensive and prevented its wider adoption.

The leading technique today is that shown at B and this has led to the development of the one-piece ring main unit, or ring main equipment, combining two switch-disconnectors and a fuse-switch into a single equipment. The UK version has a nationally used Standard, produced by the UK supply authorities, ESI Standard 41.12. This document goes into considerable detail concerning the provision of earthing and testing provisions, which are discussed in more detail in the context of system operations in Chapter 15.

The switches have a load-breaking rating suitable for the relatively low currents involved. This used to be 400 A, but with the modern introduction of systematically related ratings this has tended to increase to 630 A. These switches comply with the relevant International and British Standards, IEC 405 and BS 5463 respectively. The fault-making earth switches which are incorporated in the UK equipment comply with the relevant Standards, IEC 129 and BS 5253. Continental equipments have simpler facilities for earthing and testing, although the wide proliferation of the UK based systems in the world has led to an

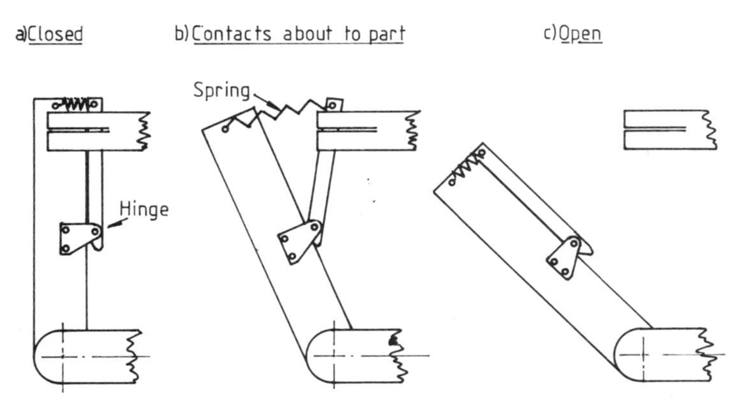

Fig. 11.11 Operation of a 'flicker blade' contact system.

Fig. 11.12 The 'Tyke' oil-filled ring main equipment. (*Courtesy Yorkshire Switchgear Ltd.*)

increased use of similar techniques elsewhere, and manufacturers have had to introduce similar facilities in order to remain competitive in the international market.

In the same way that arc interrupting techniques differ between the UK bulk oil circuit-breaker and the continental minimum oil circuit-breaker, the design of switch-disconnectors has also followed different paths in UK and Europe generally. In the UK the predominant type of switch is still that using oil in a metal-clad enclosure, with the oil providing the arc interrupting medium and the main insulation. On the Continent, the 'hard gas' principle is the chosen technique, and air has been the dominant insulation choice.

Early oil switches had direct manual mechanisms, the load switching being rendered somewhat independent of the operator's actions by the use of 'flicker blades' on the contacts. These were small contact blades hinged to the edge of the main contact blades with springs holding the edges in contact – see Fig. 11.11. When the switch opens the friction between the flicker blade and the contact overcomes the action of the spring until a mechanical stop prevents further separation of the two blades and the flicker blade is forced to follow. When the flicker blade comes out of contact the spring takes over again and the contact opening is sensibly independent of the operator.

However, it was in the fault making that the dangers of the direct manual mechanism were apparent. While it was possible to demonstrate that the switch could be closed against a fault on a high power test plant, the degree of physical effort applied could not be guaranteed to be repeated every time such a switch was closed in service, and some fatalities have occurred. For many years now spring driven mechanisms have been fitted to such equipments which ensure that the effort applied to the switch when closing is always sufficient for safe closing on fault. When used for opening, the same mechanism avoids the need for the flicker blade and thus simplifies the design of the contacts. Many thousands of

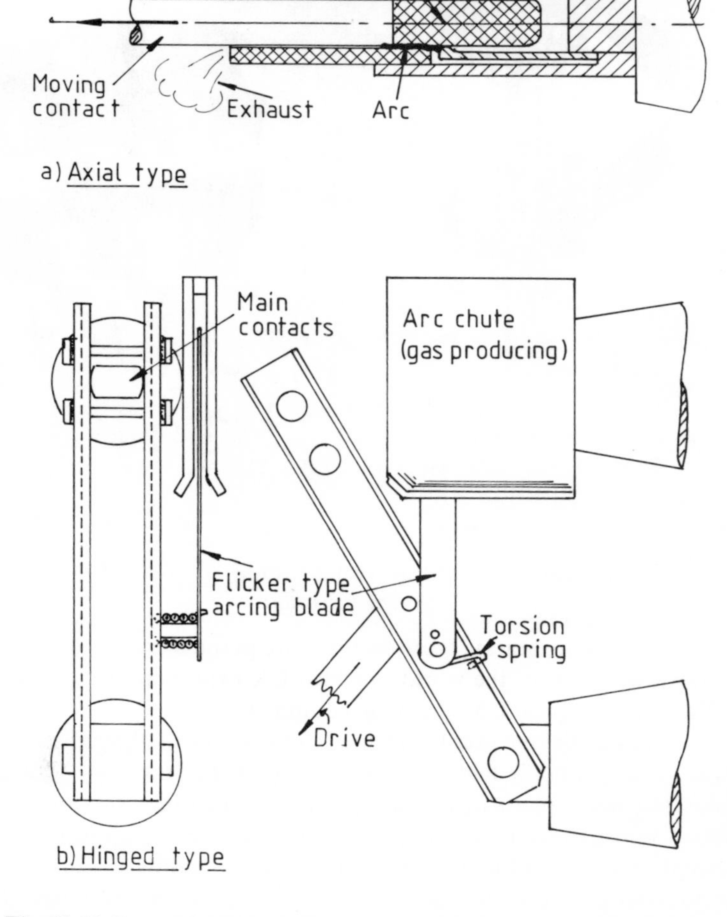

Fig. 11.13 Comparing axial and hinged types of 'hard gas' switches.

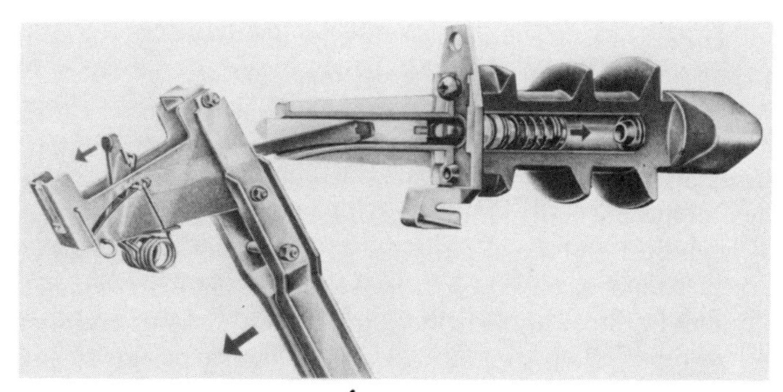

Fig. 11.14 Hard gas quenching chamber, swinging contact. (*Courtesy Sachsenwerk Licht- und Kraft-Aktiengesellscaft.*)

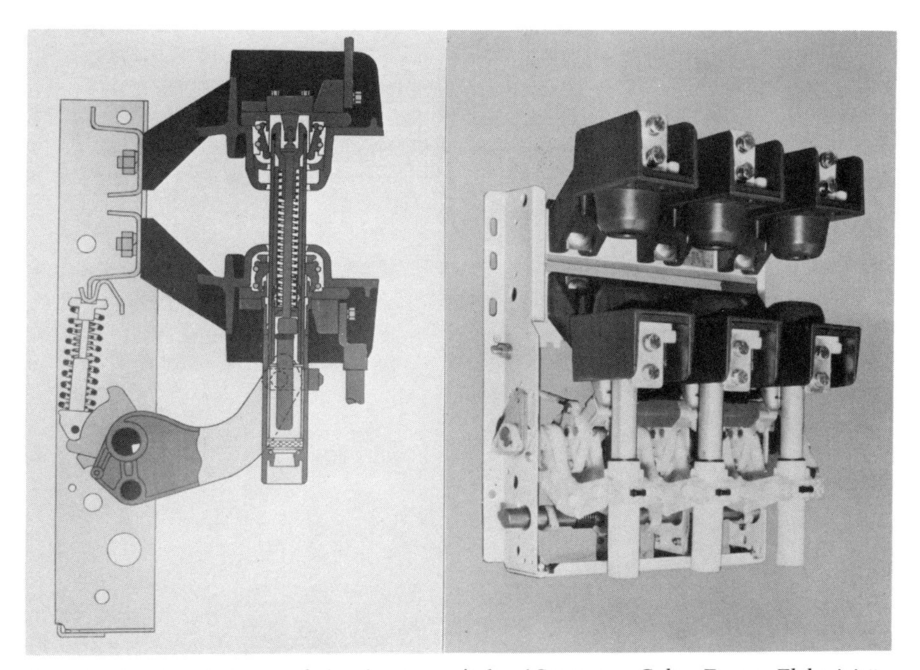

Fig. 11.15 Axial type of hard gas switch. (*Courtesy Calor-Emag Elektrizitäts-Aktiengesellschaft.*)

ring main units incorporating these switches are in satisfactory service in the UK and many other parts of the world to which UK exports find their way. Figure 11.12 illustrates a typical UK ring main equipment.

The hard gas technique uses the property of some insulating materials to emit quantities of gas, usually hydrogen, when subjected to the heat from an arc. If this emission takes place in a suitably shaped enclosure these gases can be directed to have a powerful cooling effect on the arc, which is quite adequate for the currents and power-factors involved in load switching.

The physical arrangements used divide into two, one with a hinged blade, to which an auxiliary flicker blade is attached, the auxiliary blade working in a small arc chute attached to the fixed contact. The main contacts open first and then the flicker blade parts and draws the arc inside the arc chute which has a very narrow space between its walls to ensure maximum gas producing effect. The alternative technique uses axial travel, again with main contacts which part before the arcing contacts. This also uses the flicker principle so that the opening action of the arcing contacts is under independent spring control. The arcing contact is inside a tube of gas producing material and usually has a probe of similar material attached to its end so that the arc is drawn in an annular space between two gas producing walls. Figure 11.13 shows these two arrangements diagrammatically, and Figs. 11.14 and 11.15 show actual arrangements by different European manufacturers.

European switchgear using these techniques takes two forms. The oldest, but still current, mounts the switches in metal enclosures with busbars in the form of a small switchboard (Fig. 11.16). Fuses can be mounted in series with any of the switches, for the protection of the attached loads. The other form makes much use of cast resins for the enclosure of the conductors and generally uses axial type interrupters, with the object of keeping the size to a minimum. These 'compact' equipments can either be extensible, and built up into switchboards, or prefabricated as ring main equipments with three switches, one of which is arranged to have fuses connected in series.

A very compact type of ring main equipment, first introduced some 15 years ago and which has achieved much popularity because of its small size and low price, is shown in Fig. 11.17. As a contribution to its design, it employs manual single-pole switching and this imposes some limitations on its acceptability, for instance in the UK where three-phase switching at medium voltage is standard. The opening operation uses the axial hard gas technique and it is rated at 400 A.

Fig. 11.16 Extensible type European ring main equipment. (*Courtesy Fritz Driescher.*)

Fig. 11.17 Hazemeyer type 'Magnefix' compact ring main equipment. (*Courtesy Holec Nederland BV.*)

Closing on fault is achieved by the operating handle which contains a precharged spring which drives the moving contacts into the closed position when correctly located on the equipment and released. Tubes at the side of the unit adjacent to the transformer switch contain the protective fuses.

Irrespective of the technology employed, most of these equipments have interlocks to prevent maloperation, for instance to prevent the engagement of earth and test connections unless the relevant switch has been opened. These interlocks have been taken to the highest degree in the UK standard design as described in Chapter 15.

Where the USA has turned to underground distribution, a similar arrangement of open loop supply is used and similar switching arrangements and fuse protection is provided. Since the switch equipment is usually mounted on a concrete pad by the road side, the term 'padmount equipment' is used to describe the arrangement. Whilst the arrangement of 600 A loop switches and the numbers of 200 A taps is subject to considerable variation, the most common arrangement is shown in Fig. 11.18. The majority of these units employ air switches using hard gas interrupters, mounted in a metal enclosure but of a 'live front' layout that allows access to live conductors when the doors of the loop side are opened for operation. It is very much a case of overhead practice brought to ground level (Fig. 11.19).

All operations are made using insulated handling poles, colloquially called 'hot sticks'. Air insulated fuses are used and the 200 A tap circuits are made using plug-in right-angle cable connectors, of the type described in more detail in Chapter 14.

These can be unplugged once the fuse is withdrawn, using the hot stick, or the live-break version can be employed which has a load break, hard gas switch incorporated internally, and can thus be operated on-load.

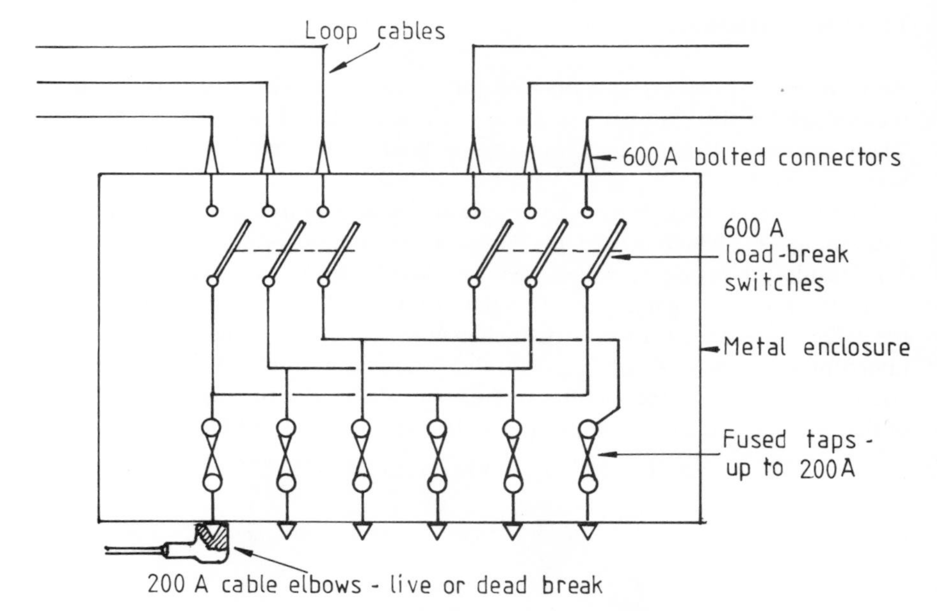

Fig. 11.18 An American padmount unit.

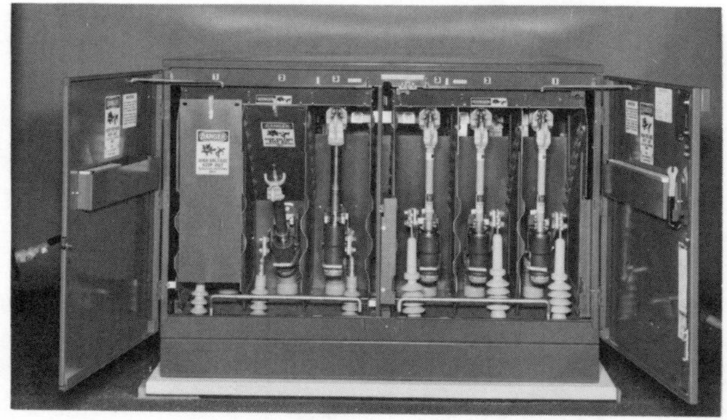

Fig. 11.19 An American air-insulated padmount unit. (*Courtesy S & C Electric Company*.)

Other versions of the padmount unit are oil filled and use oil switches or vacuum switches for the loop circuits. The fuses are also oil immersed and this allows for higher continuous current ratings than for fuses mounted in air, the full 200 A tap rating being matched on the oil filled units. Because there is no switch as such on the tap side of these equipments, full range fuses, capable of dealing with overloads as well as faults, are required. To obtain this requirement has needed the connection in tandem of a general purpose, current limiting fuse and an expulsion fuse. Recently a combined fuse in a common barrel has been produced.

The fuses used in Europe are for fault protection, the switch being capable of switching overloads, a feature which is part of the test procedure for these switching devices, although there is now on the market a British fuse with a full range characteristic.

FUTURE TRENDS

As with circuit-breakers, there is a recent strong trend in Europe to use modern technology for the load switching and fault making, and for the same reasons. These are either to reduce the amount of maintenance required or to improve safety, or both.

There are now many European designs of switches and ring main equipments containing SF_6, which is used as the insulation and arc extinguishing medium. This results in compact, sealed metal-enclosed equipments, which are being supplied in increasing numbers. This pattern of development is also apparent in the UK. There are also a much smaller number of designs based on vacuum interrupters, usually using SF_6 gas for insulation in order to reduce the size. The comparative rarity of the vacuum version is probably due to the relative expense of that medium for such a simple application (Figs. 11.20 and 11.21).

Fig. 11.20 SF_6 compact ring main unit arrangement. (*Courtesy Norwegian Brown Boveri Ltd.*)

Fig. 11.21 A vacuum switch ring main equipment. (*Courtesy GEC Distribution Switchgear Ltd.*)

The introduction of elastomeric cables and the desire to terminate these in air-filled chambers has led to a new industry in cable terminations, described in Chapter 14. These have had an impact on the design of secondary switch equipment, and all new designs at present use the proprietary systems that are appearing.

The other area that is undergoing reappraisal with the introduction of these new switchgear units is the application of fuses for distribution transformer protection. In the UK, it has been the normal practice to include the fuses inside the oil tank, where they have been well protected from environmental influences and were quite accessible on the relatively rare occasions when they needed to be replaced. A survey carried out by the UK Electricity Boards has shown that the need for fuse replacement arising from distribution transformer faults is extremely low, the theoretical life of any one fuse running into hundreds of years!

The other need for this fuse is to protect against faults on the low voltage connections on the transformer side of the fused feeders. The conclusion drawn from this study is that the fuses can be dispensed with if a system can be found that will detect transformer or LV faults at the feeding substation. This is a difficult problem, since the currents involved differ very little from likely load currents in the ring cables. At the time of writing, research is continuing.

Fuse protection is thus likely to continue in the medium term, but the fuse mounting and accessibility for changing has to be rethought for the new generations of fuse-switches. Sealing of the switch in gas makes it immune from

Fig. 11.22 Brush 'Falcon' SF_6 ring main equipment. (*Courtesy Brush Switchgear Ltd.*)

Fig. 11.23 Visitrace detecting and transmitting CTs. (*Courtesy the Electricity Council.*)

Fig. 11.24 Visitrace receiving equipment and display. (*Courtesy the Electricity Council.*)

environmental influences, but poses virtually insoluble problems if ease of access is to be the chief criterion. So the fuses are mounted in air, which is seen as a retrograde step by some users, particularly those used to having them immersed in oil. Some users who have them in air now would prefer a better medium, an attitude that is common in the dustier and more humid parts of the USA. So an acceptable solution to the elimination of the fuse whilst retaining protection for the branch circuit is very desirable.

As the fault levels are quite low, a breaking current in the range 10–12 kA being normal, present developments in simple SF_6 switching technology are producing some very cost effective circuit-breakers that could put the clock back to the days when the circuit-breaker ring main unit was common. Sophisticated electronics can now be used for high speed detection and operation, and there is already at least one such equipment on the market (Fig. 11.22).

The reduction of the inconvenience to consumers that arises from the opening of the primary circuit-breaker due to a cable fault is another area where much development is going on. At present, in many networks it is necessary to visit each secondary substation in turn to see whether the fault passage indicator has operated. This is a device fitted round the cable which operates to give an indication if a fault current passes through the cable. Arrival at a substation which is not showing an indication indicates the length of faulty cable and steps can then be taken to open the ring switches at each end of that length and to close up the normally open point, thus restoring supply to all consumers. This can obviously take an appreciable time, and a first step towards improvement is a system of indication at the primary substation of the location of the fault. Then the engineer knows immediately where he has to go to restore supplies.

The system increasing in popularity for this purpose in the UK was developed by the UK Electricity Council's research laboratories. At each substation two current transformers are fitted round each three-phase ring cable. One of these is the detector and the other is a transmitter, which transmits a high frequency burst of energy along the dead cable after the operation of the source circuit-breaker. Each substation transmits its signal a predetermined period after the disappearance of the fault current and these signals are decoded at the primary substation and displayed as signal lights on a panel. Each light can obviously be identified with a particular substation, due to the time code. No signals will be received from those substations past the faulty cable, so the last lamp to light indicates that the faulty cable is the one extending away from that substation. This indication process takes literally only a few seconds, and immediate action can then be taken to restore supply. Figure 11.23 shows the CTs and Fig. 11.24 the receiving equipment.

The next step for even further improvement is to have remotely controlled switches at the substations, so that the engineer does not have to visit the sites before restoring supply. This is obviously possible, but at a price, and electricity supply is a financial business like any other and subject to the same laws of cost/benefit analysis. A further step, once remote operation is possible, also follows logically,and that is the use of automated supply restoration procedures, using a simple microprocessor. Once more the cost rises, and the return is likely to be even more marginal, unless the supplies are valuable and the primary substation is unattended.

Secondary substations cover a wide range of equipment, much of which is installed outdoors, or in simple, probably unheated, prefabricated buildings. In the UK the connection of the ring main equipment to the distribution transformer(s) is often by direct mounting on to the transformer tank, the fuse-switch bushings projecting into the oil for the direct connection of the HV winding. In most other parts of the world cable connection is more usual, particularly in the USA where each tap circuit is a single-phase supply to a transformer, often many metres away.

Where the medium voltage switchgear is mounted outdoors, an appropriate degree of protection is specified, although this does not have to be dust-tight provided that the water penetration element is of the right order. Figures of IP33 can be adequate for well designed equipment. Construction that is robust and vandal-proof is an increasing requirement, as obviously it is dangerous as well as inconvenient if inadvertent operation is possible, or even operation by an ill-intentioned member of the public!

CHAPTER 12

Low Voltage Switchgear

The types of low voltage switchgear normally used in distribution systems have basic similarities to the switchgear used at medium voltages. The circuit-breakers are usually of the air break type, and there are switches, disconnectors and fused devices as defined in Chapter 1. Fuse-switches play a larger part in LV distribution than in medium voltage distribution, although this is changing with the greater use of the moulded case circuit-breaker.

LOW VOLTAGE CIRCUIT-BREAKERS

Dealing first with the circuit-breaker, this is fundamentally of two types, the fully rated type which is rated and tested similarly to its medium voltage equivalent, and the more recently introduced type usually referred to as a moulded case circuit-breaker (mccb), from its method of construction. This has a simpler rating and testing structure and is intended to be maintenance free and to be used in less important applications. It was first developed by the American Westinghouse Company as an improvement over the fuse-switch, but now occupies an important portion of the LV switchgear market. The construction of the LV air circuit-breaker is simpler than the medium voltage equivalent as only a short arc and elementary arc chute are needed to achieve the objective of increasing the arc voltage above the level of the supply voltage. Fault levels can be high at these voltages, however, and the electromagnetic forces are then high because of the small sizes involved. The rapid introduction of a high arc voltage has the effect of reducing the short-circuit current and the concept of a prospective current rating, as used for current-limiting fuses, is commonly applied in this field. If the LV air circuit-breaker is fitted with contacts that can open by breaking the latching arrangements in the event of a substantial fault current so that there is no need to wait for a trip impulse, then a significant degree of current limiting can be achieved. Figure 12.1 illustrates such a current-limiting air circuit-breaker.

There is a need to introduce some new definitions of terms that are used to describe the way in which low voltage circuit-breakers are made and used:

Rated operational voltage. The value of voltage to which the making and breaking capacities and the short-circuit performance categories are referred.

Rated insulation voltage. The voltage which designates the circuit-breaker and to which the dielectric tests, clearances and creepage distances are referred.

The rated operational voltage must never exceed the rated insulation voltage.

Rated thermal current. The maximum current that the circuit-breaker, equipped with any over-current releases, can carry in eight hour duty, without the temperature rise of any part exceeding the permitted limits.

Rated uninterrupted current. A value of current which the circuit-breaker can carry in uninterrupted duty.

Eight-hour duty. Duty in which the main contacts of a circuit-breaker remain closed while carrying a steady current long enough to reach thermal equilibrium but for not more than eight hours without interruption.

Uninterrupted duty. Duty in which the main contacts of a circuit-breaker remain closed whilst carrying a steady current without interruption for periods of more than eight hours (weeks, months or even years).

These current ratings are an attempt to recognise the situation whereby the contacts can become coated with dirt or oxide and the resistance then increase with consequent increase in temperature rise. If the circuit-breaker is operated the arcing followed by the wiping action of the contacts during re-closure will restore clean surfaces again. Obviously the rated uninterrupted current will be lower than the eight-hour rating when contact faces that can deteriorate in the manner described are used, or the circuit-breaker intended for continuous duty is made with silver faced contacts. It is obviously important that a potential user of LV circuit-breakers understands the significance of these rating definitions.

There is another important factor in the current rating of LV circuit-breakers that the switchboard manufacturer and eventual user must understand. It is quite common for these circuit-breakers to be temperature rise tested in free air,

Fig. 12.1 A low voltage current-limiting circuit-breaker. (*Courtesy GEC Installation Equipment Ltd.*)

considered as being under indoor conditions reasonably free from draughts and external radiation. This is because the circuit-breaker manufacturer often sells it as a component to be built into a switchboard by someone else, and the eventual condition of use is not known when the test is made. When the circuit-breaker is mounted in a switchboard the cooling conditions are often quite different from a free air situation, and suitable de-rating factors may need to be applied. In fact it is necessary to carry out thermal tests in the actual conditions of use to be really sure that it will be adequately cooled and ventilated for the required rating. The LV switchboard maker often mounts the circuit-breakers in a modular fashion in vertical columns as well as in horizontal rows to achieve a compact layout, and if uninterrupted duty is required, a full test in that arrangement is necessary.

The voltage ratings recognise the fact that most manufacturers make one frame size with the maximum required insulation voltage to meet the desired market place, and then can vary the short-circuit duty to suit the actual system voltage required in any one contract. Thus one design of equipment can have one rated insulation voltage and two or three rated short-circuit breaking capacities, each associated with a different rated operational voltage.

In a similar way to that described above, it is important for the switchboard manufacturer to know how the short-circuit tests have been made since these too are often made with the circuit-breaker virtually in free air, perhaps with screens to represent the possible location of the steelwork of an enclosure. Since it is an air circuit-breaker there is an emission of hot and ionised gas from its arc chute, and it is important to make sure that this does not cause any reduction in the insulation level inside an enclosure.

These matters are now covered in the standards which relate to factory built assemblies. The British Standard for these is BS 5486 and the standard for the LV circuit-breaker is IEC 157, which is reproduced as BS 4752.

The short-circuit duty is in two categories designated as P1 and P2. The difference is brought out in Table 12.1, reproduced from BS 4752.

For reliability and stability in a distribution system, with minimum outage time in the event of fault conditions, a category P2 circuit-breaker is the only one that should be specified, since it will interrupt its rated short-circuit current three times and still be capable of its normal service duty. Although the P1 circuit-breaker can interrupt its short-circuit current twice, it cannot then be restored to service with any confidence that it will continue to operate satisfactorily under normal conditions. For guaranteed reliability, the circuit-breaker would need to be replaced.

This difference in test requirement, and in particular the indeterminate after-test condition, has been a subject for controversy ever since it was introduced. The international standards committee has been working on a revised document since 1970, and this will be issued for international voting either in 1985 or 1986. The proposed changes are extensive and the P1 and P2 categories disappear. A similar concept remains in that it is still recognised that an LV circuit-breaker will interrupt safely a fault current that leaves the circuit-breaker in a condition where its characteristics, in respect of fault breaking, normal current carrying or protection operation, may fall short of its rating. This breaking capacity becomes its 'ultimate short-circuit performance' and it is also rated at a lower level of interrupting capability such that it is fit for further service with its performance unimpaired. It is suggested that this lower level of rating be declared by the manufacturer as 25%, 50% or 75% of its ultimate rating. This 'service short-circuit breaking capacity' is the value that should be used when selecting an LV circuit-breaker for application in a network.

LV circuit-breakers also have graded power-factor requirements as the short-circuit ratings are increased. This allows for the fact, as mentioned in Chapter 3, that in LV systems the resistance of the circuit plays a greater part in the limitation of the fault current, therefore at low values of maximum fault level the power-factor will be higher. The lowest short-circuit rating envisaged in BS 4752 is 1.5 kA and the standard power-factor at that level is 0.95. At the other end of the scale, a power-factor of 0.2 is standardised for a breaking current of 50 kA. This compares with the value of 0.15 for all ratings with medium voltage circuit-breakers.

Table 12.1 The two categories of short-circuit duty

Short-circuit performance category	*Rated operating sequence for short-circuit making and breaking tests*	*Condition after short-circuit tests*
P1	O – t – CO	Required to be capable of performing reduced service
P2	O – t – CO – t – CO	Required to be capable of performing normal service

O represents a breaking operation
CO represents a making operation followed, after the appropriate opening time, by a breaking operation
t represents a specified time interval (usually 3 min)

The temperature rise limits for LV circuit-breakers vary slightly from those specified for medium voltage equipment, particularly for silver faced contacts, where the only limitation imposed is that there must be no damage to neighbouring parts. In practice, due to the comparatively small size of LV equipment, the limitation of 70° C rise imposed on the terminals or the limits of

Fig. 12.2 Fully rated low voltage air circuit-breaker. (*Courtesy GEC Installation Equipment Ltd.*)

Fig. 12.3 Typical LV switchboard using circuit-breakers as Fig. 12.2 (*Courtesy GEC Installation Equipment Ltd.*)

Fig. 12.4 A moulded case circuit-breaker. (*Courtesy Ottermill Switchgear Ltd.*)

Fig. 12.5 Typical LV switchboard comprising moulded case circuit-breakers. (*Courtesy Ottermill Switchgear Ltd.*)

15° C for metal or 25° C for insulation material handles will establish a limit for the contacts also.

Operating mechanisms are usually either direct-manual for relatively small ratings, or independent-manual spring-operated for the heavier ratings. It is common for these circuit-breakers to incorporate overcurrent releases with inverse time characteristics as part of the design. These can be of a thermal type with instantaneous magnetic release at higher fault currents, and nowadays there are increasing numbers of designs using electronic devices to give a wide range of current and time characteristics. Figures 12.2–12.5 show typical circuit-breakers of the standard fully rated or moulded case types, with and without enclosures.

There exists also a range of disconnectors, switches and combinations of switches and fuses suitable for use in conjunction with the circuit-breakers and these other switching devices are covered by BS 5419. This Standard follows the philosophy of BS 4752 as discussed above in relation to ratings.

LV FUSEBOARD AND FUSE-PILLAR

The most basic and unsophisticated form of LV distribution switchboard is possibly the fuseboard illustrated in its common 'open' form in Fig. 12.6. Virtually the whole of the UK, and much of the rest of the world, relies on this type of switchgear for the distribution of the LV supplies from the distribution transformer.

If used in an indoor substation they will be of the open type as in Fig. 12.6, and if mounted outdoors they take the form of a feeder pillar in a single-

Fig. 12.6 An open type fuseboard. (*Courtesy EMMCO Ltd.*)

compartment weatherproof enclosure. The British arrangement comprises:

1. A disconnector operated by an insulated pole, which isolates the feed from the LV winding of the transformer.
2. Fuse ways incorporating fuse-links to BS 88, Part 5, in insulated handles. These are manually switched by pushing in or pulling out these fuses from the fixed contacts of the fuse pillar. The fuse contacts are tightened by a thumbscrew and wedges.

Each three-phase set of fuses is usually mounted in a vertical column. Busbars run across horizontally behind the fuse pillars, and cable connection is made at the bottom. The cable connections are either brought to the bottom, or the tails are brought up to the lower fuse terminals. As with most LV switchgear there is a neutral connection as well as the three phases. An earth bar is included for the connection of the cable sheaths. These fuse pillars have traditionally included little protection for the operator in the form of screens or shrouds.

Continental designs use a different form of fuse with spring contacts and conforming to the German DIN Standard 43620. Some of the more modern designs also incorporate shrouding for the live parts to a greater or lesser degree.

With increasing fault levels and consciousness of the need for the safety of personnel, there are a number of weaknesses with the traditional design, as used in the UK:

1. The large areas of uninsulated and accessible conductors present an electric shock hazard to anyone who has to work on the installation, and possible short circuits due to foreign bodies, particularly in tropical environments.
2. Switching by manipulating the fuse holders is dependent on the firmness of the operator. Should a fuse be closed on to a faulted cable the electromagnetic forces can be very severe in view of the tight loops and a prospective current that can be as high as 50 kA. This has given rise to the cumbersome practice of first making the circuit with a special form of spring-operated switch, before inserting the fuse.
3. When cabling has to be connected with other circuits live, temporary screens and insulated tools have to be used for fear of accidentally causing a fault on an adjacent circuit.

It has to be said, however, that in spite of these apparent dangers, there have been very few accidents on these equipments, which is a tribute to the skill and training of the operators. The fact that the electrical danger is very obvious must also play its part in the safety record. This comment leads to the conclusion that if an improvement in design is made, which removes that clearly evident danger, it must go all the way to making the equipment truly safe under all circumstances. A part-insulated pillar could paradoxically prove to be more dangerous than the bare version.

MODERN FUSE-PILLARS

There are new designs now entering the British market place, some of which could be called 'shielded', where there is no access to live conductors as long as

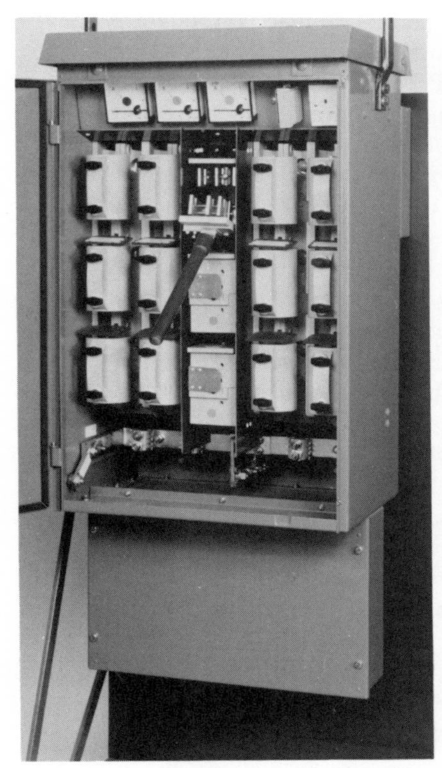

Fig. 12.7 Shielded type fuse-pillar. (*Courtesy GEC-Henley Ltd.*)

Fig. 12.8 Fully protected type of fuse-pillar (type SAIF). (*Courtesy EMMCO Ltd.*)

Fig. 12.9 Fuse unit from the SAIF system. (*Courtesy EMMCO Ltd.*)

Fig. 12.10 Fuse operating mechanism in position on a SAIF fuse-pillar. (*Courtesy EMMCO Ltd.*)

the fuses are in position, but need the operator to place shrouds over the live terminals if a fuse-link is withdrawn. Another version is not only fully protected, in that it has a degree of protection of IP 20 (see Table 10.1), but it also uses an independent spring operating mechanism to withdraw and replace the fuse-links so that there is no risk to the operator when he is performing those operations. Because of the comparative infrequency of these opening and closing operations, the mechanism is made detachable so that one per cubicle, or even one per engineer, only need be provided. Figure 12.7 shows an example of the shielded type and Figs. 12.8–12.10 demonstrate the features of the fully protected type.

Low voltage switchboards are of many types, and are to be found controlling and protecting supply systems in every conceivable type of installation. Single-phase switching is common in the less sophisticated and less important circuits, where fuses are used, such as in the fuse-pillars just described. There is no attempt made at these voltages, as there is at medium voltages in some parts of the world, to arrange for a fuse which blows on a fault to have the means to cause the other two phases to open.

CHAPTER 13

Instrument Transformers and Protection

One of the main reasons for using switchgear in any electricity system is to provide protection for that system in the event of a fault. Provision of highly efficient circuit-breakers that can remove those fault conditions in a few cycles is of no advantage unless the fault condition can be detected and the necessary opening signal generated and sent to the relevant trip coil. Thus the provision of an effective system of protection forms a large part of any switchgear installation that includes circuit-breakers.

The wide range of this subject, particularly in the transmission field where complex systems involve devices that can recognise the distance to the fault, the direction of the current flow and relative phase angles, etc., makes it a major study in its own right. However, it is such an important part of the switchgear scene that this book would not be complete without at least an overview of the basic principles of the types of protection applied to distribution systems.

Since protection is a separate science and has its own worldwide industry, it is important that there be a recognised standard interface between the signals derived from the network and the devices that interpret them and generate the signals to initiate action. The range of voltage and current ratings which are used in distribution systems require equipment to reduce them to these standard interface signals. This is the purpose of the instrument transformer. Generally, these are electromagnetic devices which transform the system voltages and currents to an agreed standard voltage and current. The relays and other protective devices are standardised to operate from these agreed outputs from the instrument transformers. The standard secondary voltage from the voltage transformer (VT) is usually 110 V, and the secondary of the current transformer (CT) is either 1 A or 5 A. The latter is more common in distribution systems where the distance from the CT to the relays is short and the voltage drop not important.

On low voltage circuit-breakers there is not the same need for instrument transformers, since the voltage levels are such that meters can be directly connected. Where normal current ratings are small, the conductors can be formed into coils which can be used to operate circuit-breaker releases directly. Current grading can be introduced by altering air gaps in magnetic circuits, and time delays can be introduced by the use of dashpots or mechanical escapements. Direct-acting trips of this type were used until quite recently on continental minimum-oil circuit-breakers, using insulated rods for the drive to the trip mechanism. Undoubtedly many of these are still in service.

Also used on low voltage circuit-breakers are thermal trips that depend on heating a bimetal strip which bends and releases the mechanism. There is a school of thought that favours the use of thermal devices to protect against overloads, because the effect of the overload is itself thermal. An electromagnetic trip device is often added to the thermal overload trip to deal with short circuits, where an instantaneous reaction is required.

Low voltage circuit-breakers of heavy rating usually have small CTs and nowadays probably use electronic devices to control the settings and decide when to trip the circuit-breaker. Even on moulded case circuit-breakers, originally introduced as a welcome alternative to fuses, very sophisticated protection arrangements can now be fitted.

CURRENT TRANSFORMERS

The parameters that define a CT are: primary current, secondary current, load impedance (burden, usually expressed in VA) and accuracy.

The primary is of very low impedance and has as few turns as are consistent with producing enough voltage to drive the rated secondary current through the load. CT burdens are standardised and are commonly 15 VA or 10 VA. This is decided by the impedance of the devices that are connected in the secondary circuit, which may comprise one or more relays and an ammeter. The flux density in the iron core is much lower than in a power transformer as it is important to keep losses to a minimum. Figure 13.1 is a vector diagram for a CT and shows how the phase angle error and the ratio error arise from the excitation current. With a burden of moderate inductance, the excitation current is nearly in phase with the secondary current, resulting in a small phase angle error. The ratio error is often compensated by a slight adjustment in the number of secondary turns.

Unless the CT has a high burden, or a very high accuracy, CTs with primary ratings above about 100 A have a single turn primary, usually in the form of a

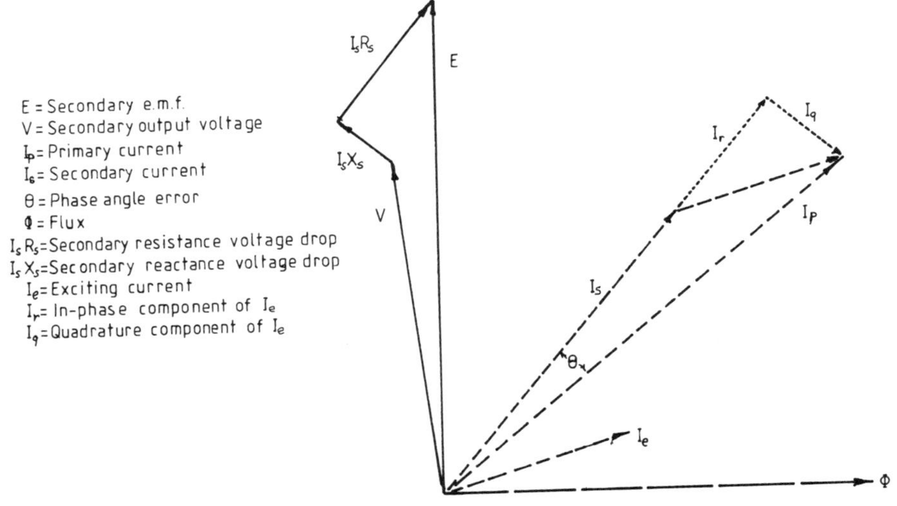

Fig. 13.1 Current transformer vector diagram.

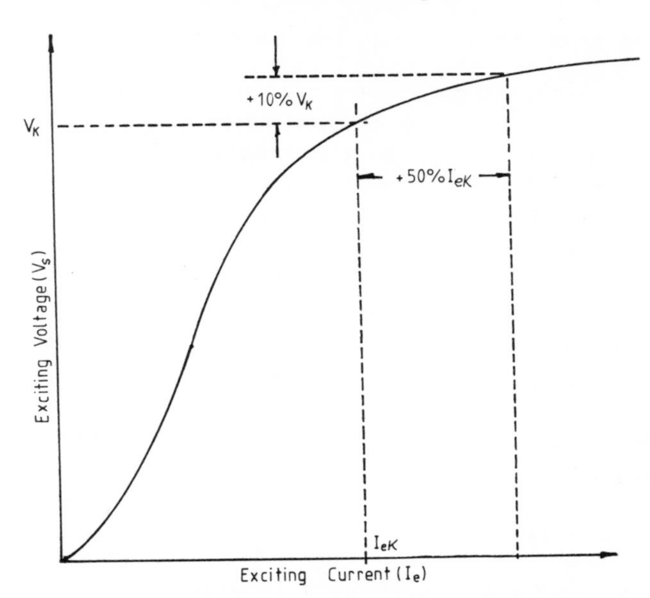

Fig. 13.2 Current transformer magnetising curve.

straight bar. There are two types of bar-primary CT, one which incorporates the primary conductor and its primary insulation, and the other, more often called a ring CT, which comprises just a core and secondary winding with enough insulation for the secondary voltage. The ring CT is then placed over an existing bushing or conductor in the equipment.

CTs used for measuring current, or as a current input to a wattmeter or watthourmeter, have the greatest accuracy requirements, particularly if connected to a tariff meter on which electricity charges are based. They only retain that accuracy up to, or slightly beyond, the normal load current.

On the other hand, CTs used for supplying protective circuits need to retain reasonable accuracy well beyond the normal current rating. When a CT is used for earth fault protection as well as general protection it is more useful to have an indication of the maximum EMF likely to be available from the CT. This is often expressed in terms of the 'knee-point' of the excitation curve obtained by

Table 13.1 Limits of error for measuring current transformers, accuracy Classes 0.1 to 1 (from BS 3938)

	% ratio error at percentage of rated current as shown			*Phase angle error at percent of rated current*, min		
Class	10–19	20–99	100–120	10–19	20–99	100–120
0.1	0.25	0.2	0.1	10	8	5
0.2	0.5	0.35	0.2	20	15	10
0.5	1	0.75	0.5	60	45	30
1	2	1.5	1	120	90	60

Table 13.2 Limits of error for measuring current transformers, accuracy Classes 3 and 5

Class	*% ratio error at percentage rated current as shown* 50	120
3	3	3
5	5	5

measuring the secondary current when a voltage is applied to it. Such a curve is shown in Fig. 13.2 and the actual 'knee-point' is defined as that point where an increase of 10% in the voltage produces a 50% increase in current. Current transformers defined in this way are designated Class X in the relevant British and International Standards. Tables 13.1 and 13.2 list classes of accuracy for measuring CTs, and Table 13.3 lists accuracy classes for protective CTs, according to BS 3938.

Table 13.3 Limits of error for protective current transformers, classes 5P and 10P (based on BS 3938:1973)

Class	*Ratio error at rated primary current*, %	*Phase angle error at rated current*, min	*Composite error at rated accuracy limit current*, %
5P	±1	±60	5
10P	±3	not specified	10

Standard accuracy limit factors are 5, 10, 15, 20 and 30.

The ratings of measurement CTs are expressed as rated burden and class, e.g. 15 VA, Class 1.

The ratings of protective CTs are expressed as rated burden, class and accuracy limit factor (the ratio of maximum rated current to normal rated current at which the composite error is not exceeded), e.g. 10 VA, Class 10P20.

Cold-rolled, grain-oriented transformer steel is used for the majority of protective CTs, and for those CTs which require very low losses to give a high accuracy, low-loss material such as Mumetal is used for the core. Cores for ring or bar primary CTs are usually wound from a long strip of steel, but wound primary (multi-turn) CTs often have built-up cores from stampings. Photographs of typical CTs are shown in Figs. 10.32–10.34.

VOLTAGE TRANSFORMERS

Under normal conditions the system voltage does not vary much, and maintaining accuracy of ratio and phase angle over a wide range is not such a problem as with a current transformer. However, in the event of a fault, the primary voltage of a VT used for protection purposes may fall to quite a low value, and its ratio must still be reasonably accurate. Tables 13.4 and 13.5 give the accuracy classes for VTs in accordance with BS 3941.

Table 13.4 Limits of error for measuring voltage transformers (BS 3941)

	0.8 to 1.2 times rated voltage *0.25 to 1.0 times rated burden at p.f. 0.8 lagging*	
Class	*Voltage error*, %	*Phase error*, min
0.1	±0.1	±5
0.2	±0.2	±10
0.5	±0.5	±20
1.0	±1.0	±40
3.0	±2.0	not specified

Table 13.5 Additional limits for protective voltage transformers (BS 3941)

	0.25 to 1.0 times rated burden at p.f. 0.8 lagging *0.05 to* V_f *times rated primary voltage*	
Class	*Voltage error*, %	*Phase error*, min
3P	±3	±120
6P	±5	±250

V_f = voltage factor from Table 13.6.

In the event of an earth fault the voltage across the primary winding of the VT could rise to 1.73 times its usual voltage and if the system neutral is earthed through a resonant coil, this condition could persist for some hours. This has to be recognised for VTs intended for use where these conditions could arise, and the British Standard includes voltage factors for VTs according to their intended use, as in Table 13.6.

PROTECTION

Fault currents arise in different ways in a distribution system, as explained in Chapter 2 and illustrated in Fig. 2.9. The most common is the earth fault between one phase and earth and it is usually important to remove this as quickly as possible, before it becomes a more serious fault involving the other phases. Often the current in an earth fault is limited by the system neutral being earthed through a resistance, and the extension to a phase-to-phase fault then leads to a

Table 13.6 Voltage factors and permissible duration of maximum voltage, V_f(BS 3941)

Voltage factor V_f	*Duration*	*Earthing conditions* *VT primary windings*	*System*
1.2	not limited	non-earthed	effectively or non-effectively earthed
1.5	30 s	earthed	effectively earthed
1.9	30 s	earthed	non-effectively earthed
1.9	8 h	earthed	resonant earthed

significant increase in the fault current. Because of the earth fault current limitation in this way it is often not possible to detect it by its magnitude, as this might well be less than the load current. A simple way of detecting an earth fault on a three-phase system is to measure the out-of-balance current caused by part of the current in the faulted phase taking an alternative path.

The connection of a CT in each phase and then paralleling the secondaries as shown in Fig. 13.3(a) gives an output, when an earth fault unbalances the phase currents. In a similar way the 'core-balance' CT shown in Fig. 13.3(b) only gives an output when the three-phase currents are unbalanced by the existence of an earth fault. Such an arrangement is also used by passing the CT secondary currents through a small 'core-balance' CT. The output current may be used to directly trip a circuit-breaker, or it may drive a relay which then exercises some control over the minimum current that is allowed to trip the circuit-breaker, or introduces a time delay if it is felt desirable.

In the event of a substantial overcurrent, it is even more important to remove the fault quickly, but care must also be taken to select the best point of switching for the most effective result in terms of isolating the fault and leaving the maximum number of consumers connected. In a radial system, that is a system

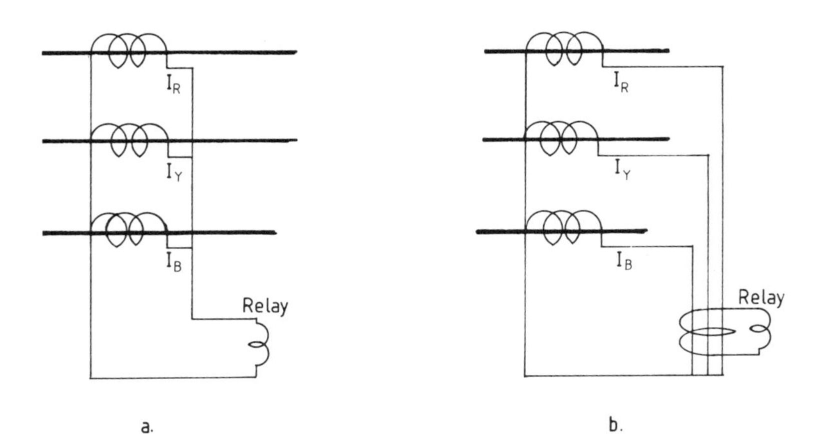

Fig. 13.3 Earth fault current detection.

of connection in which each feeder is only fed from one end, a fairly simple arrangement of time grading is acceptable. Figure 13.4 shows such a system and at each circuit-breaker a time delay is suggested. The interval between the relay operating times is 0.4 s and this means that if there is a failure of the protection at points 1 and 2, it is at least 1.2 s before the circuit-breaker at point 3 is tripped.

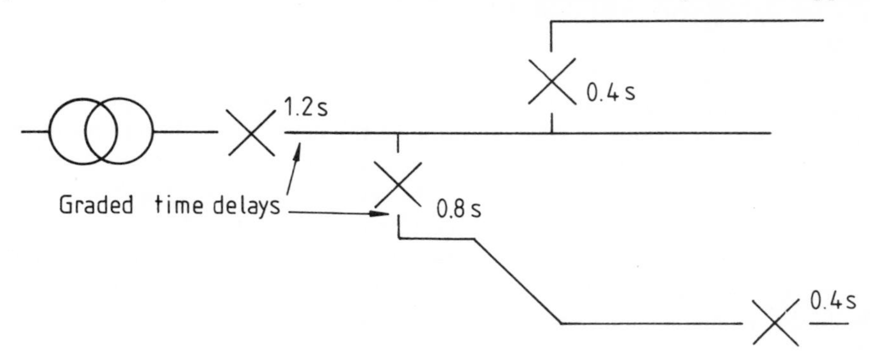

Fig. 13.4 Use of time graded discrimination.

It is also desirable to remove short circuits very quickly, and at the same time allow interruption of smaller faults to be slower. Thus there is a case for both time grading and current grading in the protective system. An arrangement that is capable of both types of grading in a somewhat rough and ready way, but has the virtue of not requiring an auxiliary supply to trip the circuit-breaker, is the use of fuses connected across the CT outputs, or in parallel with the trip coils, as in Fig. 13.5. These perform in a similar way to the direct-acting trips referred to earlier in this chapter. The circuit-breaker releases can be fitted with dashpots and armatures with adjustable air gaps also. Depending on the size of the fuse element and the magnitude of the secondary current, the fuse blows after a

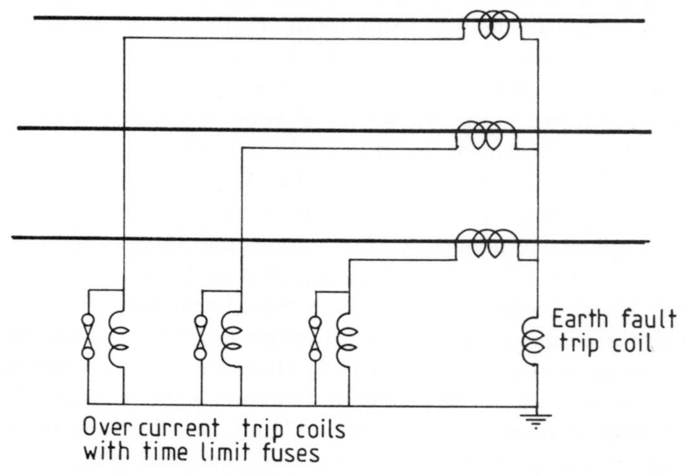

Fig. 13.5 Use of time limit fuses for time and current grading.

period of time and the secondary current is then diverted through the trip coil and the circuit-breaker opens. The accuracy of timing obtained with these devices is not sufficient for discrimination in a large network, but has often been used in less important installations, particularly if there is no ready source of auxiliary power. The trip coil magnetic circuit has to be designed for use with a.c., of course, when this system is used.

With the increasing complexity of distribution networks and the greater availability of auxiliary supplies, the use of time limit fuses, as they are called, has diminished, and relay protection is now the normal system. Relays give the user much greater control over the protection of the network, both in terms of the current and the time characteristics. As its name implies, the relay is interposed between the sensor device, the CT, and the actuator device, the trip coil to relay the information to the circuit-breaker that it should open, and when. This type of protection has been the subject of considerable change in technology in recent years, just as have the switching devices used with them. Three distinct stages can be noted in the development of the modern relay:

1. The electro-mechanical analogue relay.
2. The electronic analogue relay.
3. The electronic digital microprocessor relay.

OVERCURRENT RELAYS

The most common overcurrent relay used in the distribution system is the induction disc type of inverse time/current characteristic. This has a magnetic circuit in which there is a short gap and a coil fitted round the limb opposite the gap. The pole pieces of the gap are split and a short-circuited turn or copper shading ring is fitted round one part of the split pole piece, giving a phase displacement to part of the flux. An aluminium disc rotates between jewelled bearings with its surface in the air gap. As the current in the coil increases, the disc is subjected to a rotational force which is counterbalanced by springs. Contacts are connected to the disc so that as soon as it has completed a certain angle of rotation an auxiliary circuit is made and action is initiated.

By arranging for the coil to be tapped so that the magnetomotive force can be varied, and also for the spring resistance to be changed, the operating current can be altered. It will also be understood that moving the contact system so that the angle of rotation needed before the circuit is closed allows the operating time to be adjusted. The characteristic is also made more linear by adding a permanent magnet which produces a further flux which links with the disc and produces a braking action proportional to velocity. By altering all the factors mentioned a family of characteristics can be produced that control the pick-up current, and the current/time relationship of the relay. Figure 13.6 illustrates the principles described above and gives a typical time–current relationship for such a relay; a photograph of a typical example appears in Fig. 13.7.

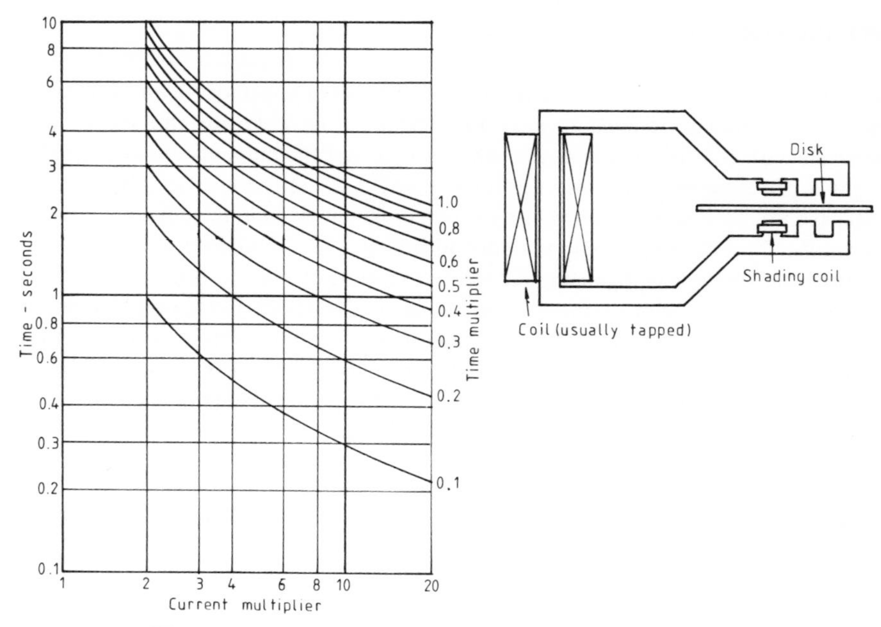

FIg. 13.6 IDMT induction relay time/current characteristics.

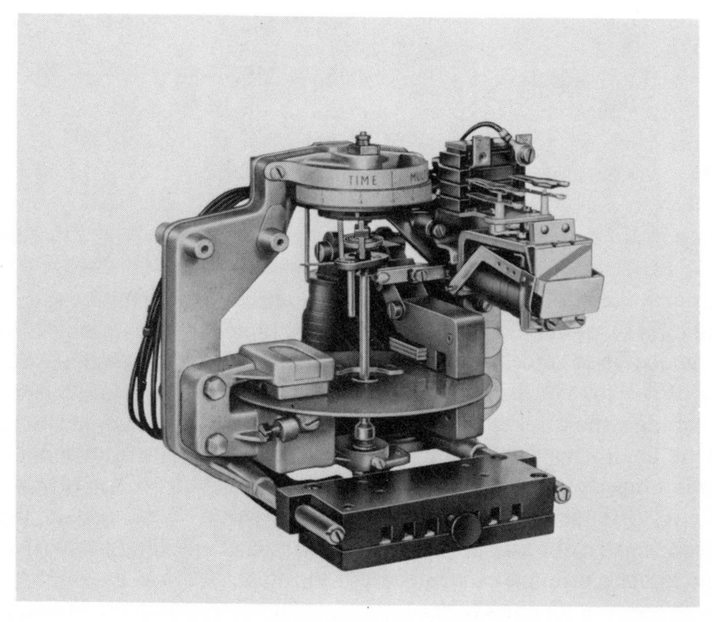

Fig. 13.7 Typical induction relay. (*Courtesy GEC Measurements Ltd.*)

UNIT PROTECTION

A different principle is often used to protect feeders supplied from primary substations, known as *unit protection*. The current entering the feeder is compared with that leaving it at the other end, and if the currents are not equal there is a problem requiring attention somewhere along the line or cable. This comparison can be achieved, either by connecting the secondary currents from CTs mounted on the switchgear at each end in series, as shown in Fig. 13.8(a), or by arranging the voltage from the secondary windings to be in opposition, as in Fig. 13.8(b). In the circulating current system, the residual current flowing through the relay is zero if the feeder is healthy. The sensitivity of the system depends on the manner in which the signals are passed between the substations, the relative characteristics of the CTs and any leakage currents due to system capacitance, etc.

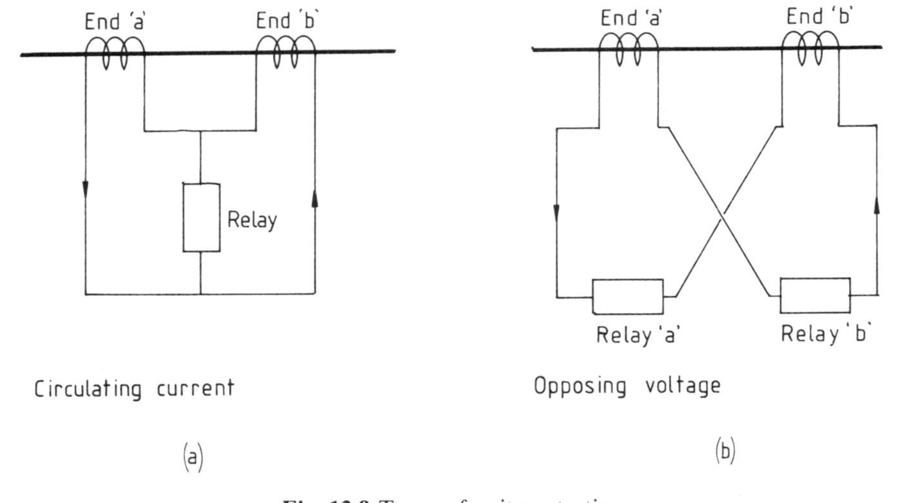

Fig. 13.8 Types of unit protection.

A very important consideration for unit protection systems is the relationship of the secondary current to the primary current in each of the CTs being compared, particularly as the primary current increases towards the maximum fault rating of the system. This must be maintained within defined limits, otherwise the out-of-balance (or spill) current can generate a trip signal when the fault is outside the zone of the protection system. In other words the spill current must remain below the threshold value for all through faults up to the full rating of the system. Problems in this area are introduced in the event of the fault current containing an appreciable d.c. component, as this is not transmitted through the CT. This is not too important if the CTs have very similar characteristics, so that the secondary currents still balance even though they are not reproducing the primary current very faithfully.

The voltage balance schemes form the basis of the well known proprietary unit protection systems, of which those carrying the trade names Translay and

Solkor are popular in the UK and many overseas markets to which UK manufacturers export switchgear.

All the unit protection schemes have a common requirement in that signals have to be exchanged between substations, often many miles apart. There are a number of possibilities, and those in general use are specially laid pilot circuits of light gauge conductors, or the rental of similar circuits from the telephone company. The choice depends on cost and availability. There is no space here to study the possible problems affecting the pilot circuits in the way of leakage currents, interference from power conductors, lightning surges if the pilots are not buried, etc. Specialist books exist where the subject can be pursued by those interested. As the voltage balance system has no current passing through the CT windings, an air gap is required in the iron circuits to prevent saturation of the cores leading to non-linearity. The flux has to be kept below the knee-point for the current range up to the rated fault level. In the proprietary systems this air gap is included in the relay.

So far the discussion of the operating principles of unit protection has dealt with it as if only single-phase currents were involved. In practice the three-phase situation has to be catered for, and although this could be dealt with as three single-phase systems and a neutral conductor, the cost of the pilot circuits would be prohibitive. So a summation CT is used to reduce the signal to single-phase as shown in Fig. 13.9. The three CTs at one end of the feeder are star connected and the four output terminals are connected to a specially wound CT with a single-phase output connected to the pilots. In this way the output is dependent on the distribution of the currents in the three phases of the primary circuit and it is usual to proportion the turns between the input points so that a greater output is obtained for earth faults than for phase faults.

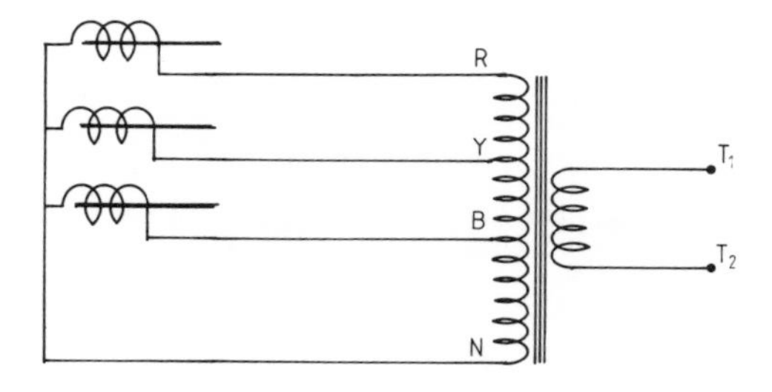

Fig. 13.9 Circuit using a summation current transformer.

The summation transformer also performs two other useful functions:

1. It isolates the pilot circuit from the C T secondary windings.
2. By having a higher number of turns on the secondary winding, the apparent impedance of the protective gear as seen by the CTs can be greatly reduced.

The errors deriving from the out-of-balance conditions that increase as the through fault currents increase can be offset, and the available sensitivity consequently greatly increased, by applying bias to the system. This is done by fitting restraining coils in the relay through which the secondary current passes.

Therefore, as the fault current increases, and the unwanted spill current tends to cause relay operation, the restraint forces also increase. By suitable design of the bias, an adequate sensitivity can be combined with good stability under through fault conditions.

BUSBAR PROTECTION

Mention must be made of the special case of protecting the busbar system of the switchgear. Because of the major shut-down that could arise in the event of serious damage to the busbars of a switchboard, it is important to ensure that such faults are quickly removed. Normal overcurrent protection often includes busbar faults within its protective zone, but the time setting at a primary board may be too long to achieve the desired objective of high speed disconnection. It is therefore common practice to fit separate busbar protection, particularly on large and important switchboards. The most common type in use is frame-earth protection, sometimes called 'frame leakage' protection.

The principle is simple, in that the metal enclosure of the switchboard is nominally insulated from earth and linked together to a frame-earth bar. The true earth bar that is connected to the station earth is run along the switchboard on insulators, and at some point a CT is connected between the frame-earth bar and the station-earth bar. Then in the event of a fault from any conductor to earth a current is detected by that frame-earth CT, the secondary of which is connected to an instantaneous relay. In a switchboard with divided busbars and a bus section switch or circuit-breaker, it is common practice to insulate the bus

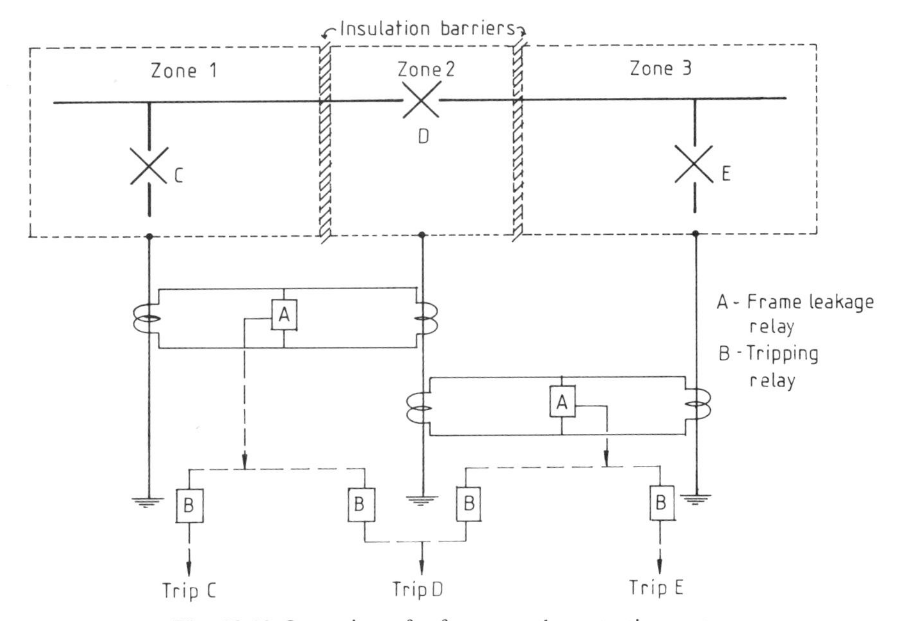

Fig. 13.10 Operation of a frame earth protective system.

section from both sections of the switchboard and apply CTs to the three zones so created. Figure 13.10 illustrates such a system and indicates how the relays operate the relevant circuit-breakers to eliminate the busbar fault whilst maintaining as much supply as is possible.

ELECTRONIC RELAYS

The relays described in connection with the protective systems discussed so far are all of the electro-mechanical analogue type. Their characteristics in terms of the relationship between input currents and output times are controlled by electromagnetic circuits, induction motors in the form of discs, and mechanical elements such as springs. Thus the relay characteristic shapes are similar and discrimination over a range of fault currents can be achieved without too much difficulty. Discrimination means the matching of relay characteristics and settings so that, in the event of a fault, the circuit-breaker nearest to the fault location trips first and if the fault persists, the next one further away from the fault then operates and so on.

Figure 13.11 illustrates this graphically. The curves show possible time/current settings of the over-current relays mounted at the points A, B and C of the system diagram. By setting the relay at A to curve A, relay B to curve B and relay C to curve C, the kind of discrimination described above is obtained. Now, however, if we consider a case in which the device at location F is a fuse-switch, it might have a time/current characteristic like curve F. This gives correct discrimination with location B for currents above I_F(A), but for currents below that level, the circuit-breaker at position B operates before the fuse-switch at F, thereby disconnecting a larger area than is necessary to just remove the faulted

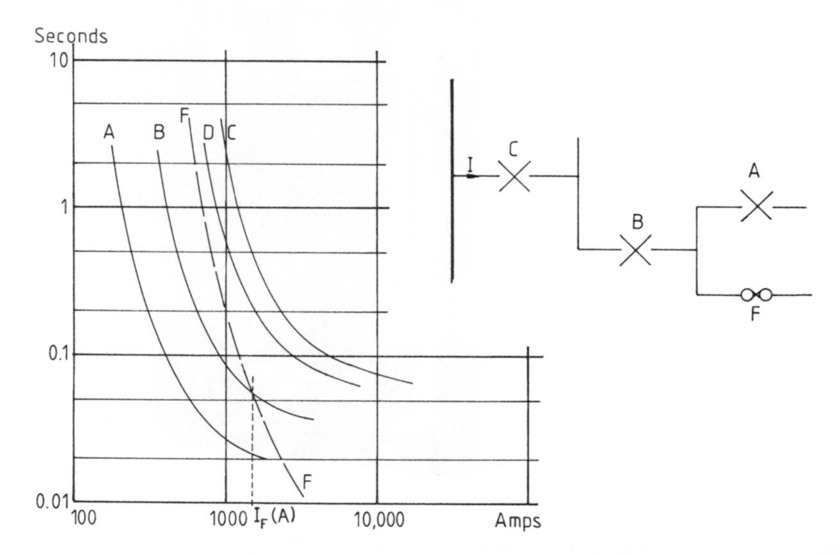

Fig. 13.11 Illustrating discrimination problems.

section. To avoid this, relay B has to be set to curve D, thus increasing the delay time for the back-up protection at B for faults beyond location A.

This example indicates the problems protection engineers have in matching the characteristics in a distribution system. The introduction of static relays incorporating the new technology of operation, with characteristics matching those of the established relays sufficiently well over the full range of time and current, requires complex circuits. This is one reason why static relays, using electronic logic and purely electrical delay systems, are expensive. Another reason for the high cost is the need to protect the somewhat delicate electronic devices against possible voltage disturbances being transmitted by the system.

This is easier to control in the next generation of relays which convert the current signals into digital values and then manipulate them as numbers. This makes it possible to match existing electro-mechanical characteristics quite readily, and also to provide smaller tolerances and a wider range of characteristics than with either the electro-mechanical or the electronic analogue types of relays. Typical of the new generation of relay is the MIDOS system produced by GEC Measurements Ltd. in the UK. Similar systems are made by the principal protective gear manufacturers in Europe and the USA.

These relays are much smaller in size and require much less power to operate than the traditional types. This ought to lead to a reduction in the size and burden of the driving CT, but for two important reasons this has yet to happen.

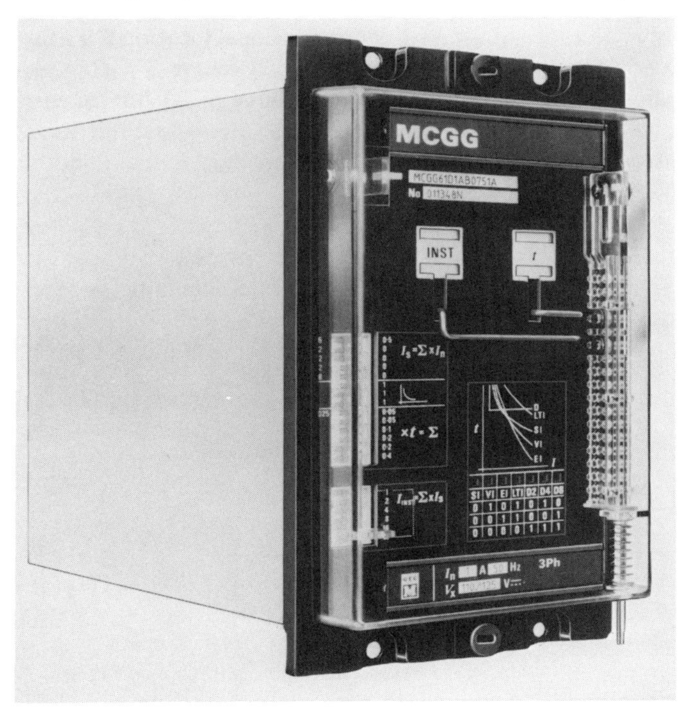

Fig. 13.12 Typical digital microprocessor relay – MIDOS system. (*Courtesy GEC Measurements Ltd.*)

The first is that there is no new agreed standard low-level international interface, so CTs are still made with 1 A and 5 A secondaries, and the second is that the first stage of the microprocessor relay is an attenuator to reduce the danger of voltage disturbances reaching the micro-chips. Figure 13.12 shows some microprocessor, digital relays with over-current and earth fault characteristics.

AUXILIARY POWER

Since all these relays require auxiliary power to energise the circuit-breaker trip coils, it is important that such a supply is always available, otherwise the protection becomes inoperative and the whole purpose of having the switchgear is lost. This means a supply with stored energy, usually a secondary battery. Therefore the tripping supplies are designed to work on d.c. and a range of standard voltages are specified in the international standards. Electronic relays also need a permanent supply for their circuits. The difference between the tripping supply and the electronic component supply is that the former is only required after the relay has operated, but the latter is a permanent, standing drain on the battery. This is not a problem when an a.c. power source is available where it can be arranged for the battery to be on permanent float charge.

It is possible to consider using the CT as a source of power for keeping the battery charged. It is also possible to consider charging up a capacitor from the CT output so that it can be discharged through the trip coil when the relay operates. If the relay has an inverse time characteristic then it can be arranged for enough energy to be available during the fault period whilst the relay is operating to trip the circuit-breaker when the relay contacts close.

One final point to make in connection with electronic relays is that they generally need a supplementary instantaneous relay with heavier duty contacts to make on to the trip coil circuit. The final element of the electronic relay is usually a thyristor or a reed switch. The trip coil current is always switched off by auxiliary contacts on the circuit-breaker, timed to open after the circuit-breaker has started to open. This ensures that in the event of any 'stickiness' in the circuit-breaker there is no danger of the trip impulse being removed before it has done its work.

METERING

Although this chapter is mostly about protection, these auxiliary circuits usually include instrument transformers to supply meters or instruments for measuring current, voltage or power. The CTs used to drive meters need a high accuracy, but only have to maintain this over a limited current range, i.e. up to a modest overload, not fault levels. Even so, in order to obtain this accuracy it is often necessary to use magnetic materials with specially low losses.

Ammeters and voltmeters, generically referred to as instruments, a term that includes such equipment as frequency meters, synchronising indicators and power-factor meters, do not always require the same degree of accuracy. If the idea is largely to indicate that the circuit is energised and the load is of an approximate value, the voltmeter or ammeter may be connected to a protection CT. Otherwise separate CTs are needed. Obviously, a VT of adequate accuracy is also required if power is to be measured, and these often have only two-phase windings if the two-wattmeter method of three-phase power measurement is used.

In view of the protective function of switchgear and its associated protective gear as described in this chapter, it is usual to provide back-up protection to guard against the possibility of the protection itself failing to operate. Typically a feeder may primarily be protected by a unit scheme and then have an inverse time over-current and earth fault relay set to operate if the primary protection fails. As the CTs may be involved in the primary system failure, the over-current system has its own set of CTs. So it is quite common to fit three sets of CTs to a feeder circuit, one set for unit protection of the feeder, another set for the over-current back-up and the third set for metering purposes. This last set may only consist of two CTs in the outer phases.

This chapter has concentrated on introducing the basic principles of the most common types of protection fitted to distribution systems. As will be realised, there are many factors affecting the functioning of the protective system, of which lack of space prevents inclusion here. There are also other systems of protection which detect such things as the direction of current flow or the distance of the fault from the circuit-breaker, and these are occasionally used in distribution systems, particularly in 36 kV networks.

CHAPTER 14

Cable Termination Systems

Historically, cables in medium voltage plant, such as switchgear and transformers, have been insulated with oil impregnated paper and terminated in metal boxes filled with a fluid insulant, such as bitumen compound. If such a joint is well made it satisfies the principal objectives of protecting the cable tails where the paper insulation is exposed and insulating the metal components where the cable is terminated.

With the introduction of oil-based plastics materials, a lot of research and development work was put into finding a better way of insulating cables than wrapping them in oily paper. The principle of extruding a plastics sleeve over the conductor had a lot of attractions, but the high dielectric stress involved in cable insulation made it difficult to achieve a reliable and economic application of this apparently simple technique. Today these problems have been overcome and elastomeric materials are increasingly used to insulate cables, but many supply engineers still remain unconvinced that they are better than the time-honoured paper.

With the accelerating trend away from switchgear designs using oil or compound as the main insulation or the arc extinguishing media, attention has been turned towards terminating cables without a mass filling medium. Early attempts in this direction using tapes and sleeves of insulating materials were aimed at replacing the compound in the same terminal box, and this proved to be a serious limitation.

The introduction of new materials with characteristics helpful to the application, such as the 'heat-shrinkable' plastics sleevings, has led to the present position where cables with either paper or elastomeric insulation can be successfully terminated with comparative ease in a cable terminating chamber of suitable dimensions.

One factor which complicates the task of fully insulating the cable termination is that the majority of connections between the cable and the apparatus bushing form a right-angle or elbow joint. The successful insulation of this elbow joint is a difficult and/or expensive operation depending on the medium used, and can depend significantly on the skill of the jointer.

The problems that arise in the cable termination zone stem from the substantial difference between the electrical stresses that exist within the uniform cylindrical electrode system of the cable, and those which arise with the termination hardware and the application of layers of supplementary insulation.

The introduction of semiconducting plastics materials has helped to alleviate stress concentration at the electrodes and thus ensure that the joint area is free from discharge. The use of a moulded, heat-shrinkable 'boot' which eradicates air spaces within the elbow joint also contributes to the effective jointing process.

With current thinking on simplified secondary distribution systems, however, more thought is being given to ways in which disconnection of the cable can be readily achieved should either the cable or the plant develop a fault.

SEPARABLE CABLE CONNECTORS

Another influence on this area of plant connection came from the USA, when aesthetic considerations encouraged the electricity supply utilities to put urban distribution systems underground. The problem of cable termination was then coupled with the desire to have a cable connected system with the same facilities and ease of operation as an overhead system, which led to the 'plug-in' cable termination, using a right-angle connector. The desire to handle the connector with the cable possibly live, using the hot-stick with which the American linesman had become adept, led to connector designs with a semiconducting outer sheath, which is connected to the earthed cable sheath, and an inner live sheath to shield the conductor and connection details. Figure 14.1 shows a section through a typical screened connector of the heavy duty bolted type originally designed for the connection of the ring, or 'loop' cables to the padmount units and rated at 600 A.

Where these connectors are coupled to the padmount units, a simple compartment enclosed by a hinged metal hood is used. The application is facilitated by the predominance in the USA of single-phase medium voltage distribution systems and the corresponding widespread use of single-phase

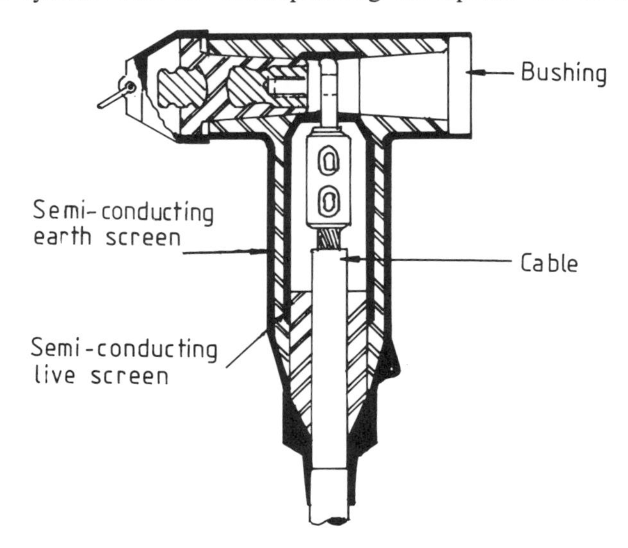

Fig. 14.1 Sectional drawing of a typical American pattern screened cable connector.

elastomeric cables. Following the initial development of a plug-in 200 A cable connecting elbow for the 'tap' side of the padmount unit, a similar bolted type for the 'loop' cables was developed.

In Europe, the existing use of single-phase cables with elastomeric insulation, often coming up from a cable basement straight into the switchgear cubicles, has led to widespread use of simple, in-line, shrouded joints in distribution switchgear. Today, there is available a range of screened elbow and in-line connectors, some with semiconducting, synthetic rubber screens and (more recently) some with an additional metal sheath.

These screened cable termination arrangements have been developed over the years into a comprehensive system, with features which simplify the day-to-day operation of a cable connected distribution system. These features include:

1. Connection plugs which enable additional cables to be connected into the first connector.
2. Earthing and testing conductors which can be attached to the cable end after removal of the connector closure cap.
3. A closure cap of sufficient capacitance to drive a voltage indicator so that it is possible to check whether the cable is live or dead.
4. Means for 'parking' a cable after it has been removed from the plant, either on an earthed or insulated stand.

So that connectors of different manufacture can be used interchangeably on any plant, it was necessary to standardise the interface. This led to the original publication of the American Standard ANSI 386, which sets down the essential dimensions for the taper at the end of the plant bushing. The connectors have a matching internal taper and are made from a fairly flexible material with a slight interference fit, so that when they are bolted or clamped up tight, a waterproof seal is made between the insulating surfaces. A light coating of silicone grease on the taper facilitates assembly.

The above mentioned standard gives details of tapers suitable for 200 A and 600 A connectors and covers the voltage range up to 38 kV. These dimensions

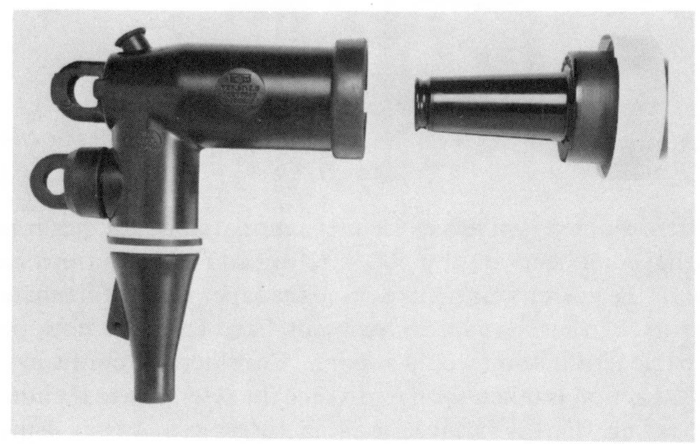

Fig. 14.2 A 200 A screened live-break connector. (*Courtesy of RTE Corporation, USA*.)

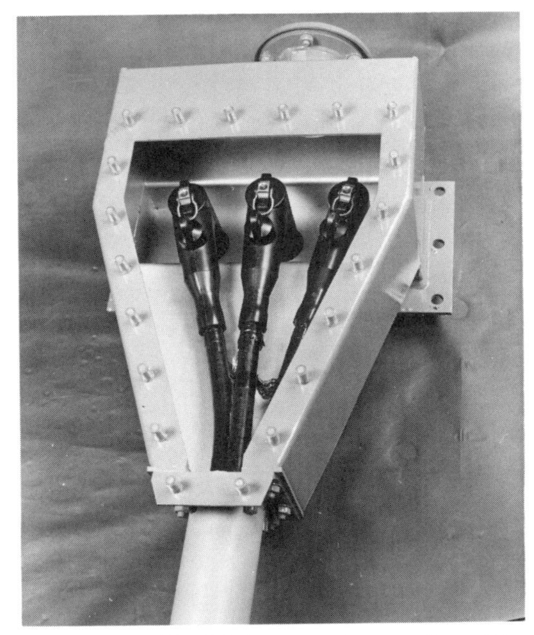

Fig. 14.3 Cables terminated with 600 A screened connectors. (*Courtesy of BICC.*)

Fig. 14.4 Accessories for screened cable termination systems. (*Courtesy of RTE Corporation, USA.*)

have now been adopted virtually as an international standard, but in recent years the French have had a need for a 400 A rating and added a third version. The original use of the current rating to denote the taper was unfortunate, although the originators of that standard could not have foreseen how popular this method of cable termination would become. Considerable confusion is likely to develop unless action is taken soon to divorce the reference to the interface from the current rating. This is because manufacturers are already using different current ratings for elbows of the same profile, depending on the material of the conductor in the associated bushing. As the connector has little influence on the temperature rise either of the cable or the bushing this development is a natural one, and it is to be hoped that a different reference system is introduced soon. Figures 14.2–14.4 show typical screened, cable-connecting components.

Many cable termination boxes on rotating machines, transformers or switchgear are fully metal-enclosed, bolted structures in which the earthed metal screen of the connectors mentioned above is superfluous. The only exception to this is where very compact plant rated at 17.5 kV or more is being cable

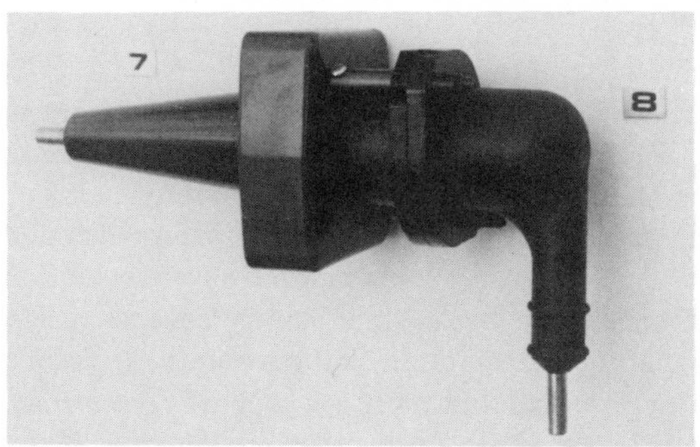

Fig. 14.5 Unscreened 200 A cable connector – DYSCON. (*Courtesy Electromoldings Ltd.*)

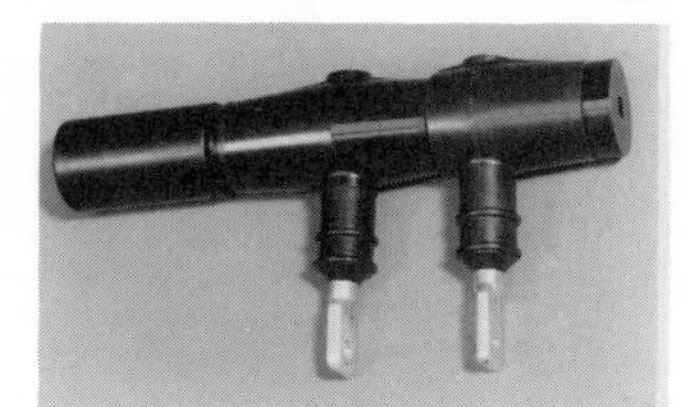

Fig. 14.6 Unscreened 400 A cable connector – DYSCON. (*Courtesy Electromoldings Ltd.*)

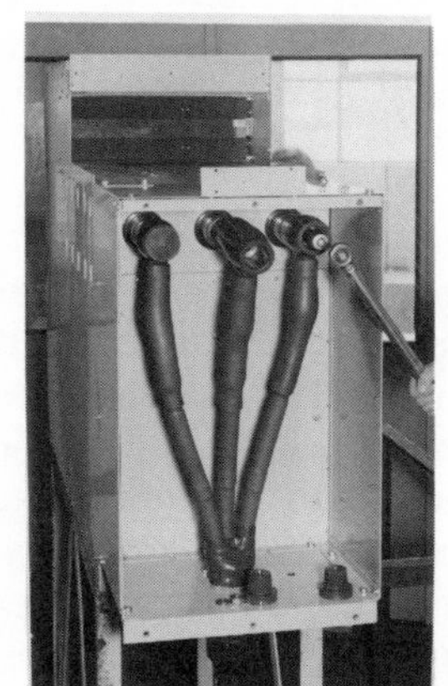

Fig. 14.7 Cable box showing unscreened cable connectors in use. (*Courtesy Yorkshire Switchgear Ltd.*)

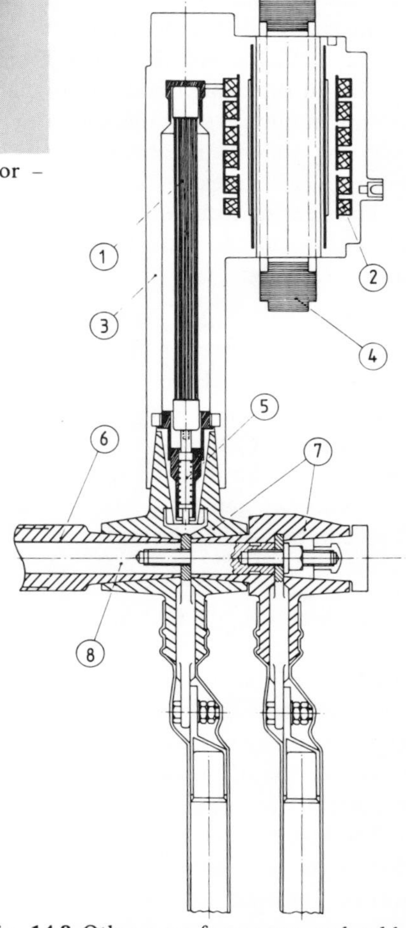

Fig. 14.8 Other uses for unscreened cable termination components. (*Courtesy Yorkshire Switchgear Ltd.*)

connected. In this case the concentration of all the electrical stress within the body of the connector allows closer spacing of the cables. In all other cases, the same facilities can be provided by using a bolted, separable, unscreened connector, as illustrated in Figs. 14.5–14.8. As the cable is not terminated within the body of these connectors, the live screen is not required either, which considerably reduces the voltage stress within the insulation of the moulding, compared with the American designed version.

INSULATION SYSTEMS

The cable is coupled to these unscreened connectors either by an in-line, crimped ferrule or by a standard cable lug. There is thus the same insulation requirement that exists for a cable termination using proprietary insulation systems, except that the difficulty of the right-angle joint has been removed. The system includes a range of components to create a comprehensive cable handling system similar to that available with the screened connectors but at significantly less cost. Since the air in the cable termination chamber forms part of the insulation system, certain minimum clearances have to observed, but these are generally in keeping with the dimensions already needed to make off the cables and for the plant bushing clearances.

The sleeving techniques for insulating cable tails have now reached a high level of sophistication, with three basic systems in use and available from several manufacturers:

1. Heat shrinkable extruded sleeves and mouldings from a variety of plastic materials. A high ratio of expanded to shrunk diameter can be obtained so that they are easy to apply.
2. 'Cold shrink' sleeves. These are elastic in nature and either rolled into place, pushed on with the assistance of internal air pressure, or kept in the expanded condition by an internal spiral that can be stripped out when in place.
3. Self-amalgamating tape systems. These are tapes of such a constitution that they can be applied by wrapping and then cure into a homogeneous material.

DISSIMILAR METALS

A complication in the termination of cables to electrical plant is in the metals employed as conductors in the bushings, connectors and cables. In the USA aluminium is employed quite extensively for the conductors in the bushings of electrical plant, while copper is still predominant in Europe. Aluminium conductors in cables, however, are in common use universally.

The bi-metal joint between copper and aluminium has to be made with care due to the insulating nature of aluminium oxide and the speed with which it forms. The significant difference in electrochemical potential between copper and aluminium makes it imperative to exclude moisture from the joint area, for the electrolytic cell that is then set up leads to accelerated corrosion. The crimping of a cable lug or ferrule onto a cable conductor successfully breaks through the oxide layer and produces a reliable joint in like materials, e.g. an aluminium ferrule on an aluminium cable. It is where the aluminium lug meets a copper conductor that care has to be taken to avoid future trouble. Satisfactory experience can be achieved by scrupulous cleaning of the surfaces, using vaseline or similar material to exclude air whilst the joint is made. Then so long as the

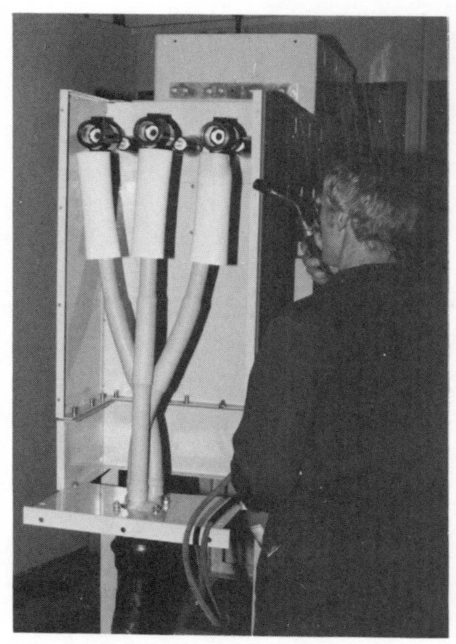

Fig. 14.9 A joint being insulated with 'heat shrink' sleeving. (*Courtesy Sigmaform Ltd.*)

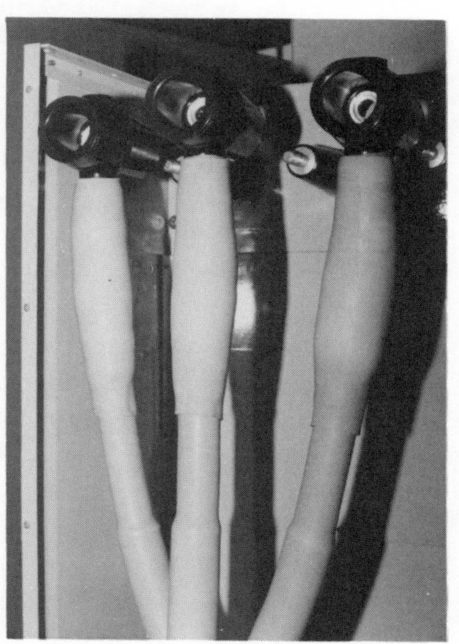

Fig. 14.10 Cable box showing use of 'heat shrink' sleeving. (*Courtesy Sigmaform Ltd.*)

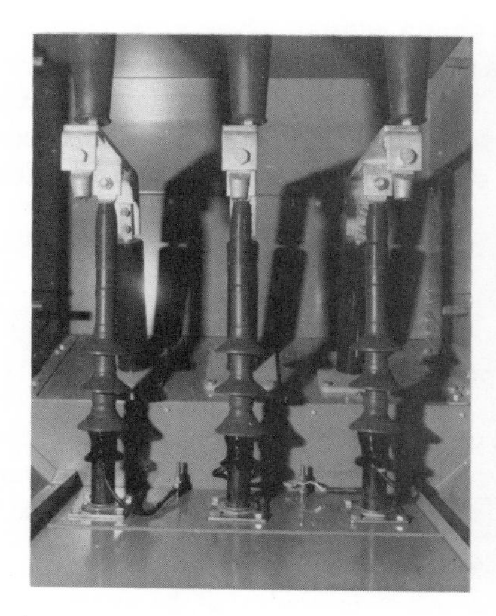

Fig. 14.11 Elastic sleeving type insulation system. (*Courtesy Pirelli General.*)

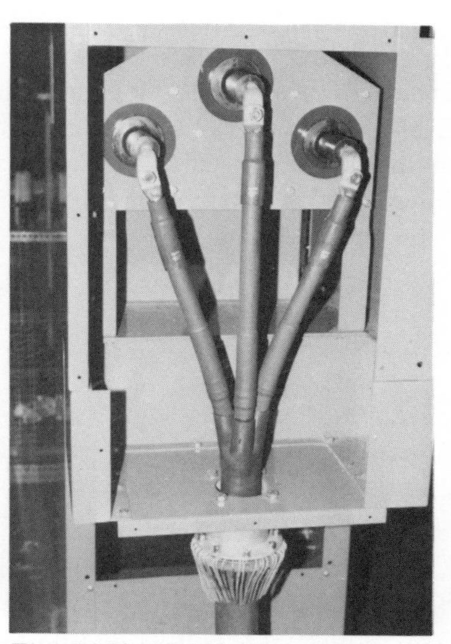

Fig. 14.12 36 kV heat shrink cable terminations. (*Courtesy Raychem Ltd.*)

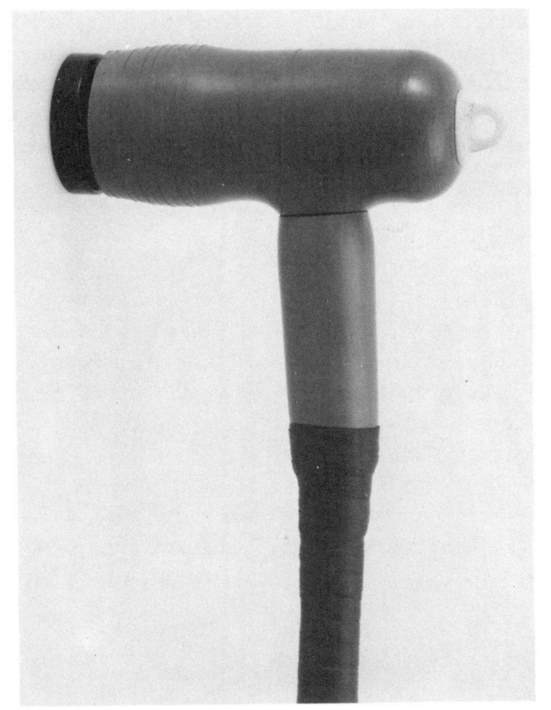

Fig. 14.13 Retractable moulded elbow and self-amalgamating tape. (*Courtesy Kabeldon Ltd, Sweden.*)

joint is effectively kept dry, no problems should arise.

To give more security it is possible to obtain jointing washers that reduce the electrochemical potential and make for a safer joint, and if in-line ferrules are in use, bi-metallic versions are available, made by friction welding, so that an aluminium tube is used on the cable and a copper tube on the plant conductor. Figures 14.9–14.13 show examples of some of the sleeving and terminating systems in use.

CABLE BOX DESIGN

The design of the cable box itself is important for the reliability of the cable terminations. The only British Standard on the subject is aimed at transformers, and similar plant, although there exists an Electricity Supply Industry Standard, which gives guidance on limiting dimensions, etc. The following points should be borne in mind by designers:

1. Phase spacings should be capable of accepting the cable lug which is used on the largest cable to be fitted into the box.
2. Clearance to the walls of the box must be sufficient for the cable jointer to obtain proper access to work on the cable.

3. The distance from the centre line of the cable lug securing bolt to the bottom of the box must be sufficient in a three-phase cable for the spacing required at the ends to be achieved without mechanically overstressing the cable insulation

4. For single-phase or three-phase cables, the dimension in (3) must be sufficient to allow the recommended length of insulation on the cable to be stripped, with the earth screen still adequately clear of the gland plate. In a switchgear box, with three-phase cables, there must be room to cross the cores, if that is needed to secure correct phase rotation. This is not normally required in a transformer box, presumably because this is connected first!

5. Where separable connectors are used, and more than one cable per phase is fitted, the gland plate should be split, and so designed that each cable can be separately pulled clear of the box, bringing its section of the gland plate with it. This is because many cables are not flexible enough for separation just with the length available within the box.

6. The mounting of the box on the switchgear should be at a height that allows the cable to be brought out of the trench without too sharp a bend. Some cable boxes allow for the gland to be attached at an angle to facilitate this. This is among the most difficult requirements for distribution switchgear, as it becomes more compact, and the possibilities for raising the cable termination high off the ground diminish.

LOW VOLTAGE TERMINATIONS

The above principles apply to any voltage rating, but become difficult to apply on low voltage switchgear. At the lower voltages, current ratings are generally higher than on medium voltage plant, so cable sizes are larger. The use of aluminium almost without exception makes this even worse. The space available is reduced and the existence of a neutral conductor as well as the three-phase conductors adds even more to the congestion. Much ingenuity has to be shown to make an LV cable box adequate for easy jointing. At these voltages, fortunately, the electrical conditions are much less onerous than at medium voltage.

A look at a large multi-panel 415 V switchboard will show just how much of the design is devoted to the termination of the cables!

CHAPTER 15

Operation and Maintenance of Switchgear

OPERATION

As has already been mentioned, switchgear spends much of its life in a quiescent state, and the designer spends all his life concerned with what might happen when it does operate!

Most operations of switchgear are concerned with controlling and protecting the network, including itself, or they are concerned with preparing to maintain some part of it. If the control activity is frequent, some form of remote facility for opening and closing the switching device is almost always provided. Certainly this is the case with primary substations. The direct safety of the operator is not therefore in question, but the consequences of his operations may endanger other people if there does not exist some form of authority and control, however elementary, to ensure that any operation does not energise a part of the system, or a piece of plant, that should not have been energised. If the operation requires direct manual attention at the location of the switching device, the consequences may be more directly obvious to the operator, but even then the switchgear is rarely close to the equipment being supplied with power.

Much of this chapter is concerned with safety, and the introduction indicates the level at which that concern should start. The first conclusion may be that no switchgear operation should take place without some system for checking on its possible consequences. The converse of this is that no-one should work on the system itself or on any plant controlled by it without direct authority from the person responsible for operating it. Once such work has started, then it should not be possible for a supply to be connected to the system or plant in question until the authority to work has been rescinded. This is why attention has been paid in earlier chapters to the need to provide interlocks and padlocking features on switching devices.

Planning for safe operating procedures must start before the switchgear is ordered. The need to be able to lock off parts of the system so that maintenance work can be done on the associated plant should be established, and arrangements provided. It is part of the design of metal-enclosed switchgear that all live conductors are behind either permanently fixed partitions or locked doors. It follows that a proper and safe control system to supervise all keys should be instituted. This must ensure that every key issued has a purpose and that issue should be recorded. The person responsible must then ensure that the relevant padlock has been restored to its proper function, locked and the key returned. In this way, given good discipline, the supervisor will know that the system is safe to operate. In an industrial environment, a system must ensure that switching devices controlling supply to any piece of plant which should not be supplied for any reason are positively locked off, and the key returned to a secure place.

Clearly there should be a written set of rules governing these operating safety principles, and it is the responsibility of the engineer in whose charge the plant and distribution network is placed to draw up such rules. Further, the content of those rules must be understood by all who work in the environment, and this requires a positive training programme. In most countries of the world there are nowadays statutory provisions in the law for safety at work, and familiarity with these provisions is essential for engineers responsible for electrical installations.

Large organisations have evolved systems of this kind over the years, electricity supply authorities being primarily concerned. Any person requiring help with creating such a system will usually find it in such a body.

Apart from the control aspects of system operation, there are additional requirements associated with earthing, testing and maintenance. Operation of the equipment in connection with the maintenance of connected plant, e.g. motors, furnaces, etc., is covered above. Now the question of using the switchgear to gain access to the system for earthing conductors, for testing cables and other equipment and for maintaining the switchgear itself is considered.

The safety rules referred to above must also cater for these actions as well. The person responsible for authorising the actions must be identified so that all those concerned know who is co-ordinating procedures. When an earth is to be applied, for example, the point of application has to be established, all possible supply routes to that point locked off and authority given for that earth to be applied.

At this point procedures differ quite widely from country to country. In the UK supply industry,and those undertakings that follow their lead, earths have to be applied to a system through a switching device with a proven fault-making capability, and in some cases this has to be preceded by a voltage check. The effect of this on the construction of the switchgear, and acceptable ways in which the required facilities can be provided, are described in the following paragraphs.

EARTHING AND TESTING FACILITIES

With fully enclosed non-withdrawable switching devices some orifices, with adequate protection, need to be provided in the construction of the cubicle so that earthing and testing connections can be made, at least to the circuit side of the switching device.

If the switching device is withdrawable, then this access can be obtained through the circuit spouts when the switching device has been withdrawn.

The philosophy of operation has an important bearing on the cubicle and switchboard construction. This philosophy ranges from giving the operator freedom to do as he wishes and to rely for his safety on thorough understanding and training, to almost total protection of the operator by preventing him from taking a wrong action through extensive interlocking and padlocking.

The latter approach has been adopted in the UK, and the full effect of this is seen on secondary switchgear, discussed in Chapter 11. America and much of Europe have tended to adopt the first approach, although there is an international trend towards safety consciousness which leads to new designs paying more attention to prevention of risk to the operator.

The operation of applying an earth illustrates some of the differences, particularly if the case of a ring main unit is taken.

Continental switches are two-position devices, so that applying an earth requires first of all taking action to disconnect the circuit to be earthed from the supply. As this is likely to be a ring cable, the far end is usually switched first, and some means of warning or locking applied there to prevent it being re-energised. Then the switch in the local cubicle is opened and the cubicle door also opened to give access to the circuit connections.

It is possible that the busbars could be live, so many continental panels of this type have facilities for inserting an insulating panel in the switch gap so that the busbars are inaccessible to the operator (Fig. 10.36). The next step is to use a voltage indicator to check that the circuit conductors are dead, and this involves using a portable voltage source to check that the voltage indicator is giving the correct indication both before and after it is applied to the circuit. Then a portable earthing cable is applied to the conductors, often manually.

The same operation is carried out quite differently on a British ring main unit, such as the oil-filled types described in Chapter 11. These switches are three-position devices, and the process of earthing a ring cable starts in the same way as described above, by making the necessary switching operations to remove the supply from the remote end. However, as the switchgear there is also likely to be a ring main unit, its facilities are used to earth the cable at that point as well as disconnecting it. Then a padlock is applied to ensure the earth cannot be disturbed. At the local end the cable switch is now opened.

There is a padlocked component preventing the operation of the switch to earth and the issue of that key is restricted. After obtaining the key, access to the mechanism can be obtained and the switch closed into the earth position. The switch has a fully proven rating for closing on to the rated fault capacity of the system, and for carrying that fault current for three seconds. This is ample time for the primary circuit-breaker supplying the ring to clear the fault, in the very unlikely event that one is created.

It will be noted that no voltage test is made, and no access to the conductors is possible at this point in the operation. When the earth has been made it is possible to unlock and open the test access cover, through which access to the conductors is now possible. Test connections are now made and the earth switch opened to enable testing to proceed. Another interlock now comes into play, to make it impossible to restore the switch into the service position without first going back into the earth position, removing the test connections and closing the test access.

On circuit-breaker switchgear, slightly different operations are performed, depending on the type of earthing facilities provided. These vary from making provision to use the circuit-breaker itself as the proven fault-making device to the use of a fault-making earth switch.

In all cases the facility to earth the circuit is barred during normal operations, usually by a padlock with a special key. It is typical also, if separate facilities for circuit earthing and busbar earthing are provided, for the relevant keys to be different. For obvious reasons the earthing of the busbars is treated even more seriously than the earthing of a circuit.

When the circuit-breaker itself is used as the earthing device, arrangements are made to ensure that the electrical tripping is disconnected while the circuit-breaker is being used for this purpose. This must be done in such a way that it is not possible to restore the circuit-breaker to the service location without

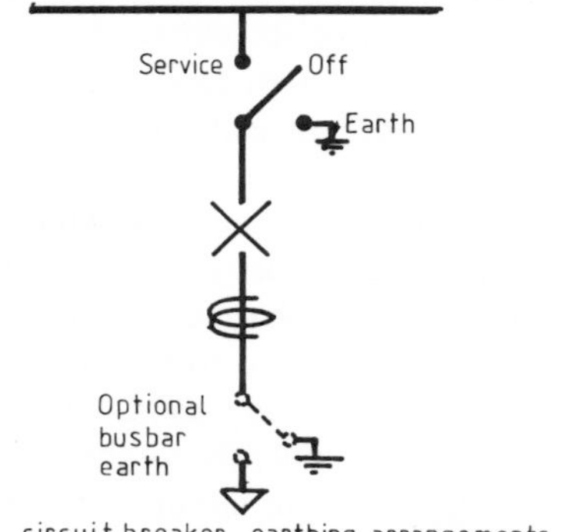

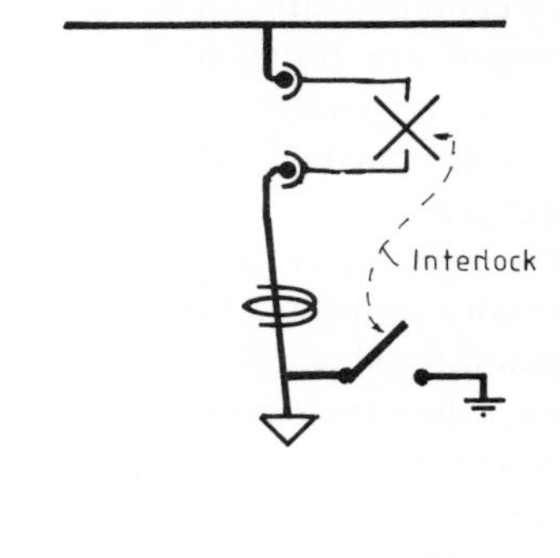

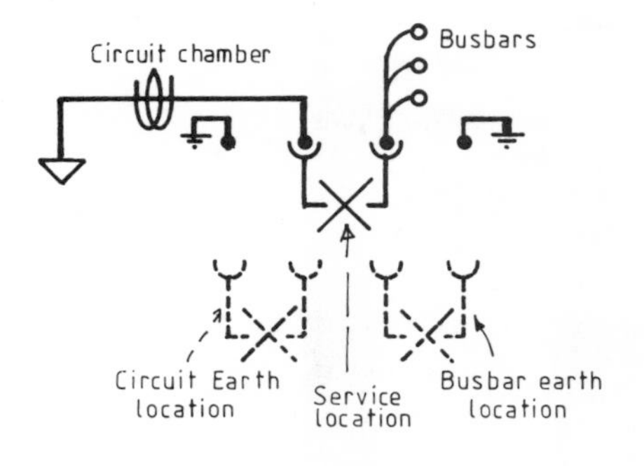

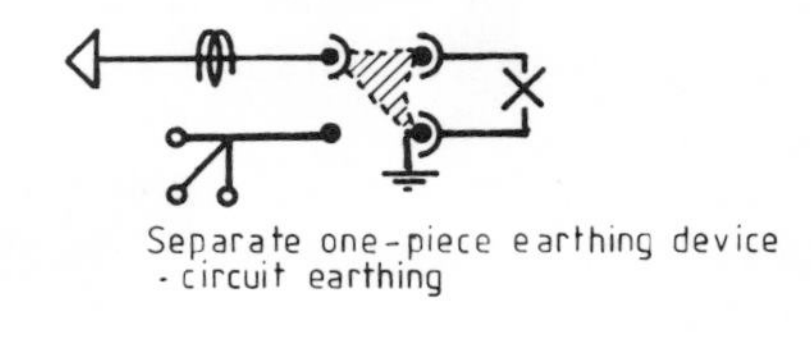

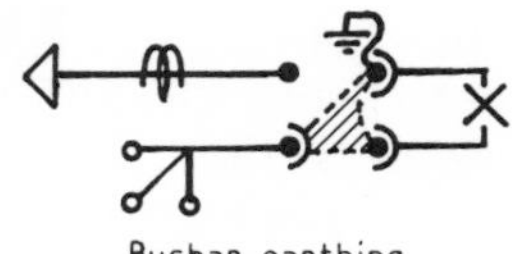

Fig. 15.1 Typical switchboard earthing arrangements.

reconnecting the trip circuit. Furthermore, the method used must not prevent the circuit-breaker from being tripped manually. All this is to ensure that, in the unlikely event of the circuit-breaker closing on to a live circuit, it cannot trip out in front of the operator. Such an arrangement obviously requires a good system of back-up protection.

This fault-making earthing facility can be provided in several ways, as shown in Fig. 15.1.

One sub-division of these is between integral and non-integral methods. The former means that the earthing operation can be performed without fetching any additional equipment to the switchboard. The non-integral methods generally mean that an earthing carriage is substituted for the circuit-breaker carriage, or that the circuit-breaker is converted into an earthing carriage by the addition of a component that extends three of its bushings to reach into the cubicle spouts, and at the same time connects the other three bushings together and to earth.

Integral earthing on withdrawable switchgear usually means that alternative locations are provided for the circuit-breaker carriage in which the circuit-breaker can connect either the circuit or the busbars to an extra set of plugs which are earthed. This is a common procedure on vertically withdrawable switchgear.

On switchgear with fixed circuit-breakers, the disconnectors which are normally connected between the circuit-breaker and the busbars have an extra position in which they connect the circuit-breaker to earth on the busbar side. Then, when the circuit-breaker is closed, an earth is applied to the circuit.

Another integral earthing technique, applicable to any design of metal-enclosed switchgear, is to fit an earthing switch permanently to the cubicle, connected to the conductors to be earthed either on the circuit side or the busbars. This earthing switch is fitted with a mechanism designed to give it the necessary fault-making capacity, and arrangements are made so that it is impossible to close it unless the circuit-breaker, or other switching device, has been disconnected from the system.

Fig. 15.2 A typical one-piece earthing device. (*Courtesy Yorkshire Switchgear Ltd.*)

The technique to be used to earth the circuit on any design of cubicle does not have to be the same as that provided for the earthing of the busbars, and indeed there are schools of thought that see a virtue in them being different. In particular, with horizontal drawout gear it is possible, without too much increase in size, to provide for the elevation of the circuit-breaker within the carriage so that integral circuit earthing can be performed, but the provision of a further location for busbar earthing as well would be both complicated and space consuming. Then there is virtue in using either a one-piece additional component, as illustrated in Fig. 15.2, to effect the earthing of the busbars, or to fit one or two fixed earthing switches on to strategically chosen panels in the switchboard for this purpose.

In double-busbar switchgear which uses circuit-breaker transfer for busbar selection it becomes virtually impossible to use the transfer feature for circuit earthing. The use of a separate earthing carriage, the conversion of the circuit-breaker into an earthing carriage by fitting a one-piece earthing extension device, or a fixed earthing switch is needed to give the required fault-making earthing capability.

Once the earth has been applied and any work carried out, it is usual to test the system before putting it back into service. If the earth has been applied by circuit-breaker transfer, it is a simple matter to remove the circuit-breaker and apply test connections through the unit spouts. If an earthing carriage has been used the same may apply, but there is a commonly used variation which allows the earth connection on the earthing carriage to be removed so that test connections can be made through the earthing carriage. Preferably the test connections should be applied before the earth connection is removed. This gets over the objection to the previously mentioned system, that if any appreciable interval of time were to elapse between removing the circuit-breaker from the cubicle and the attachment of the test leads, there is a remote possibility that the circuit might have inadvertently been re-energised.

When an integrally mounted earth switch is used for earthing, the fitting of test leads at that point and the subsequent removal of the earth connection is also a possibility. With all such methods, however, means should be provided to ensure that the earth connection is replaced before the switch can be used again for earthing.

MAINTENANCE

Maintenance covers a wide range of activities, all designed to keep the switchgear in the condition necessary to ensure that it is ready at all times to perform its protective function satisfactorily.

There are three British Standards on the subject, divided by voltage category. Those that fall within the compass of this book are BS 6423 for switchgear up to 1000 V, and BS 6626 for switchgear above 1000 V and up to and including 36 kV. The latter standard defines four separate aspects of maintenance, each stage building on the preceding one, as follows:

Inspection. A maintenance action which comprises making a careful scrutiny of an item, carried out without dismantling it and using all the senses as required to detect anything which causes the item to fail to meet an acceptable condition. It may include an operational check.
Servicing. Work carried out to ensure that the equipment is kept in an acceptable condition, which does not involve any dismantling, and is typically limited to cleaning, adjustment and lubrication.
Examination. An inspection with the addition of partial dismantling as required, supplemented by means such as measurement and non-destructive tests in order to arrive at a reliable conclusion as to the condition of the item.
Overhaul. Work done with the objective of repairing or replacing parts which are found to be below standard by examination, so as to restore the component and/or equipment to acceptable condition.

Study of these definitions shows how one builds upon another. Inspection may lead to the conclusion that servicing is desirable, or if the engineer suspects that all is not well, then an examination is called for. The result of that examination may then be that an overhaul is required. To the average individual, the maintenance function is generally seen as an overhaul. One of the principle objects of the use of new technology such as vacuum interruption, or arc extinction in SF_6 gas, is to eliminate the examination and overhaul procedures for switchgear.

The actual conditions of use in terms of environment, frequency of operation and severity of switching duty vary so widely that it is impossible for the manufacturer to give detailed guidance to the user concerning the frequency of inspection. A distinction needs to be made between the switching device with its moving parts, and the rest of the equipment which performs a supporting role of connection, containment, etc. The latter components are only likely to need periodic inspection and servicing to ensure that the equipment is free from dirt, damp and deterioration, such as corrosion of metalwork and contamination of insulation.

These British Standards draw attention to the importance of the manufacturer's handbook in establishing the methods for determining when various procedures need to be carried out. As an example, with SF_6 circuit-breakers the most likely criterion needing to be monitored occasionally is contact wear, and the timing of the opening operation. The manufacturer's handbook therefore needs to detail how to perform these operations and the acceptable levels of performance. The recommended content of the handbook is laid down as an appendix to the standard for switchgear between 1 and 36 kV.

The early clauses in the same standard contain much good advice on safe procedures to use and precautions to be taken when setting out to undertake maintenance. The manufacturer's handbook should also give any special advice specifically related to the switchgear being maintained. Typical of the latter is the procedure that should be followed to release any stored energy in closing mechanisms before beginning an examination.

In the same way as the stages of maintenance build on each other, the inspection phase should also be carried out after an examination or overhaul to ensure that the overall condition of the equipment has been restored once the work has been completed, and an operation check, or preferably a series of operation checks, carried out to establish that the correct performance has been achieved and the equipment may safely be restored to service.

With fully sealed circuit-breakers of the SF_6 and vacuum type the majority of manufacturers would not expect examination or overhaul to be necessary during the economic life of the installation under normal conditions of use. In addition, because of the need to restore the sealed-for-life condition of gas-filled equipment in particular, it is strongly recommended, in the unlikely event of it needing overhaul, that it be returned to the manufacturer for that purpose.

CHAPTER 16

Application of Switchgear

This chapter is aimed at bringing together the many aspects of switchgear design and performance characteristics that have been discussed in earlier chapters, in the task of deciding what switchgear to use in any particular application. Distribution switchgear manufacturers are not, in the main, in the business of tailoring switches and circuit-breakers to match each set of circumstances, but tend to make a limited combination of ratings, based on their experience of what their normal market requires. Thus the first steps in the economic planning of a distribution system are:

1. Be aware of what switchgear is readily available from makers' catalogues.
2. Examine alternative approaches to see whether in some cases it is better to try and alter the system characteristics to use the available standards or to ask for special designs to be made to match the system.

The American standards go further in this respect than the British or the International (IEC). A look at the rating table (2) in ANSI Standard C37.06, listing preferred rating combinations for oilless circuit-breakers up to 38 kV, shows a total of 20, while the British Standard (see Table 1.2 in Chapter 1 of this book) suggests 88!

However, it is not always possible or practical to restrict the choice too much, as there are a number of parameters to be satisfied in deciding on the type, rating and performance criteria for the switchgear in a distribution system. The standards that have been produced in the world in the switchgear field are extensive and have been based on many studies carried out by international bodies to arrive at typical characteristics that are found in the majority of installations. Each of the relevant standards will be found to draw attention to the fact that some systems may depart from the normal, and the application engineers should be wary of the areas where this can happen. In this chapter these areas are examined and attention drawn to the standard parameters and how they might vary. The effect of these variations on different types of switchgear is then discussed.

The switchgear in any installation may be used for a number of functions. The basic duty is usually to carry load current for long periods of time and be able at any time to carry a fault current for a short time or, in the case of the circuit-breaker, to interrupt it very quickly.

But there are other functions that may be more important in the life of the switching device. It may be used for frequent switching of a particular type of plant, such as a large motor or a capacitor, and that could dictate the form of switchgear used. Its principal function could be the isolation of plant for maintenance or a setting-up operation in which current interruption may not be necessary and a disconnector or switch can be used.

NORMAL SERVICE CONDITIONS

The international standard IEC 694 establishes two sets of normal conditions under which 'standard' switchgear is deemed to be installed, one for indoor gear and the other for outdoor.

For indoor switchgear the parameters are:

1. Ambient temperature not to exceed 40° C and the average over a 24 h period not to exceed 35° C.
2. Two classes for minimum temperature, one of −5° C and the other, for more severe environments, down to −25° C.
3. Maximum altitude 1000 m (3300 ft)
4. The average value of the humidity does not exceed 90% over a month or 95% over any 24 h.
5. Vibrations external to the switchgear, or earth tremors, are negligible.

The American standards do not specify any condition for humidity and make no distinction between indoor and outdoor so far as minimum temperature is concerned, which in the ANSI standard C37.04 is −30° C.

International standard service conditions for outdoor equipment are the same as for indoor switchgear so far as maximum temperatures, altitude and vibration requirements are concerned and there are again two classes for the minimum temperatures, i.e. −25° C and -40° C.

Other conditions are:

1. Three classes of permissible ice coating are recognised, 1 mm, 10 mm and 20 mm thick.
2. Wind pressure not to exceed 700 Pa (equivalent to a wind speed of 34 m/s).
3. Account should be taken of rain, condensation, rapid temperature changes, solar radiation and ambient air pollution.

ABNORMAL SERVICE CONDITIONS

OUTDOOR SITES

Air pollution in the form of ordinary dust, pollen, etc., can be expected not to cause a problem as periodic rain usually washes the insulators free of this and in any case the design allows for a moderate degree of inert dust pollution. It is when that pollution is of an industrial nature and may be corrosive or abrasive, or the site is near the sea and subject to salt spray, that such circumstances should be brought to the attention of the potential supplier of the switchgear. With outdoor switchgear regular insulator washing may be prescribed in order to avoid problems.

If the site is subject to earthquakes, then again special precautions need to be taken, and the manufacturer will need to produce evidence of seismic tests.

INDOOR INSTALLATIONS:

For any potential indoor substation site consideration needs to be given to whether any of the environmental conditions are more onerous than those envisaged in the specification. If they are, it does not necessarily follow that the switchgear available from at least some suppliers will not be suitable, but it does mean either that the attention of potential suppliers should be drawn to the condition(s), or action should be considered to make the environment within the substation comply with the 'standard' conditions.

HIGH HUMIDITY

For example, in a design of switchgear using porcelain as a main insulation material, relatively frequent condensation might not be a problem to the life of the insulation, but if another material was used, the manufacturer would probably fit electric heaters in each panel close to any vulnerable insulation to prevent condensation. In this connection it is interesting to note that although IEC 694, repeated as BS 6581, considers that with a permitted maximum level of 95% relative humidity condensation may occasionally occur, IEC 298 (BS 5227) considers that it does not normally occur. Expert opinion in this country believes this to be wrong, as any environment where temperature and humidity are high by day must inevitably be prone to condensation when the temperature falls at night, since the dew point must also be high. This is where the fitting of unit heaters, perhaps controlled by a humidistat, would often be used. As an alternative to the humidistat, switching on the heaters at the start of the 'rainy season', and then leaving them on permanently until the risk is past, is also practised.

HIGH TEMPERATURE

If both maximum temperature and humidity are more onerous than the standard, air conditioning of the substation could be considered, and may prove more economical than specifying special switchgear. Ambient temperatures higher than 40° C are common in many tropical areas of the world. It is usual to ask the manufacturer to compensate for this, which he does by derating the switchgear.

Although heat generation is proportional to the square of the current, the heat dissipation is not directly proportional to heat generation, as a study of Chapter 8 shows. Thus the derating factor used will be found to be proportional to current to the power 1.8 or something similar.

ALTITUDE

With the reduction in air density as altitude increases, the electrical strength of air gaps decreases and the effectiveness of air as a cooling medium also decreases. Therefore the voltage rating and current ratings both need to be reappraised if the altitude is significantly above sea level. The ANSI specification C37.04 has a table which indicates reduction factors for maximum rated voltage, B.I.L. and rated normal current at different altitudes. At 1500 m the voltage ratings are reduced to 95% and the current to 99%. At 3000 m the corresponding figures are 80% and 96%.

EXCESSIVE DUST, OR VERMIN

If the site is subject to heavy dust deposits which may be corrosive, or conducting, e.g. coal dust or carbon black, then special steps need to be taken to protect the switchgear. Depending on other factors, fitting effective filters to ventilation openings and arranging air lock doors may be sufficient, or air conditioning may be a wise precaution. Many tropical sites are subject to pollution by vermin, including insects, snakes and other reptiles. Attention to the degree of protection, and to operational and maintenance procedures to ensure that the protection is retained when a task is finished, may be adequate to avoid trouble from this cause. One point to be watched is that a reptile may get through a small opening when young, and then feed on small insects that get in afterwards, until it is large enough to be dangerous!

VIBRATION

On an industrial site, vibration can arise from nearby machinery and affect sensitive parts of circuit-breaker mechanisms, causing them to operate inadvertently. Trip latches are usually balanced dynamically to make them insensitive to such vibration. The worst form of vibration is that associated with earth tremors. The requirement for switchgear to be able to withstand seismic forces is increasing with the high safety standards being applied on nuclear sites such as power stations. It is common in the UK for a nuclear site contractor to present the switchgear manufacturer with the seismic profile of the substation building and for tests on a seismic test table to be performed to prove that the cubicles, etc., can withstand the resultant forces.

The frequency of the vibrations is usually low, in the order of 1–20 Hz, and these can often excite resonances in the framework, or particularly in such components as instrument panels. Ideally, it is more economic to establish minimum seismic withstand conditions for the switchgear, and then to adjust the substation seismic profile to match, thus putting seismic conditions in the same category as temperature and humidity. This matter is under discussion in several places.

SWITCHING DUTY

The rating characteristics specified in the standards are based on 'typical' substation sites and conditions. They are not average conditions, but represent values that are rarely exceeded, and therefore are generally adequate for all but the most unusual conditions.

RATED VOLTAGE

This is the highest voltage at which the switchgear is to be used, and is chosen in the standards so that it allows a tolerance above the commonly used service voltages. For example, a very common distribution voltage in the UK is 11 kV, and a standard rated voltage in BS 6581 is 12 kV. Similarly, a common service voltage in America is 13.8 kV, with a maximum rated voltage of 15 kV.

RATED NORMAL (CONTINUOUS) CURRENT

This is the highest current that the switchgear can carry without exceeding the maximum permitted temperature at any point. Since the current carried affects the temperature rise then the current rating is dependent on the maximum ambient temperature, as discussed above. Sometimes one temperature limit prevails in determining the current rating, and different countries have different limits for similar materials. As an example, the ANSI standard states that any part that can be handled by an operator in the normal course of his duties must not exceed 50° C, which could become a limitation if solar heating is taken into account. No such limits are quoted in the British or IEC standards, although commonsense also plays a part in switchgear design , but even lower temperature limits are given for handles in the BS for LV switchgear. Also the IEC standard gives 100° C as the highest permitted temperature for metallic parts in contact with oil, whereas the ANSI document limits this to 90° C.

However, these are isolated examples, and in most areas the various standards give very similar values for maximum permitted temperatures. Many of the critical limits are given in Table 8.1.

SHORT-CIRCUIT DUTY

The fault levels at the various points where short circuits may occur are calculated using techniques such as those explained in Chapter 3. Particular allowance should be made for the contribution to the short-circuit making current from any induction or synchronous motors on the system. Calculations of the X/R ratios are then made to establish whether the d.c. component of the fault current is within the figure of 14 or 15 used to establish the standard ratings in the specifications. In the event of any difference the envelope of the calculated fault current should be compared with the envelope of the standard rating next above that calculated. It may well be that the margin between the calculated symmetrical current and the nearest standard rating will be enough to accommodate the extra d.c. component. If this is not the case, then the non-standard X/R ratio should be drawn to the attention of potential suppliers, as it is quite possible that the switchgear in prospect may have margins above the standard levels.

In cases where the fault levels are high and X/R ratios are significantly higher than the typical values, the likely TRV values should also be evaluated to make sure that they are within the figures stated. Values for these are given in the IEC standards and are reproduced in Table 2.1. The ANSI specifications only quote peak values for indoor circuit-breakers below 38 kV, although there is no reason to expect American levels to be significantly different from European values.

If all these calculations give rise to fault levels at the top end of the standard ratings, it may be a valuable exercise to study the economics of introducing measures to reduce the fault levels. As the ceiling of available ratings is approached, the price of switchgear rises sharply because of the technical problems and the mechanical forces that are encountered. The addition of reactors to divide the system and reduce the short-circuit current is often cost effective. It may even be possible to arrange the operation of the system so that incoming supplies are never paralleled, thus achieving the same objective without adding any plant.

CAPACITANCE CURRENT, OR INDUCTIVE CURRENT SWITCHING

If there are circumstances when any switching device has to interrupt capacitance current, the current breaking duty is calculated and the impedance that will limit inrush currents on closing also estimated. A particular situation that is important to look for is if there is more than one bank of switched capacitors, and where one bank can be closed on to the busbars to which another bank is already connected. If the length of conductor between one bank and the other is relatively short, then it is very important to ensure that the inrush current when the second bank is connected does not exceed the rated making current of the switchgear.

When the switching of unloaded motors or transformers is to be part of the duty of a switching device, the surge impedance of the plant involved should be established to ensure that it is unlikely to give rise to any overvoltage problems. Figure 16.1 is a chart which gives approximate values of the surge impedance for motors of different powers and voltages, as well as some transformers. By taking a typical value from this graph and applying it to Fig. 16.2, an estimate of the likely surge voltage for different types of circuit-breaker can be found. This value is added to the peak phase voltage of the system to establish the total peak voltage likely to arise from a no-load switching operation. Obviously, if the

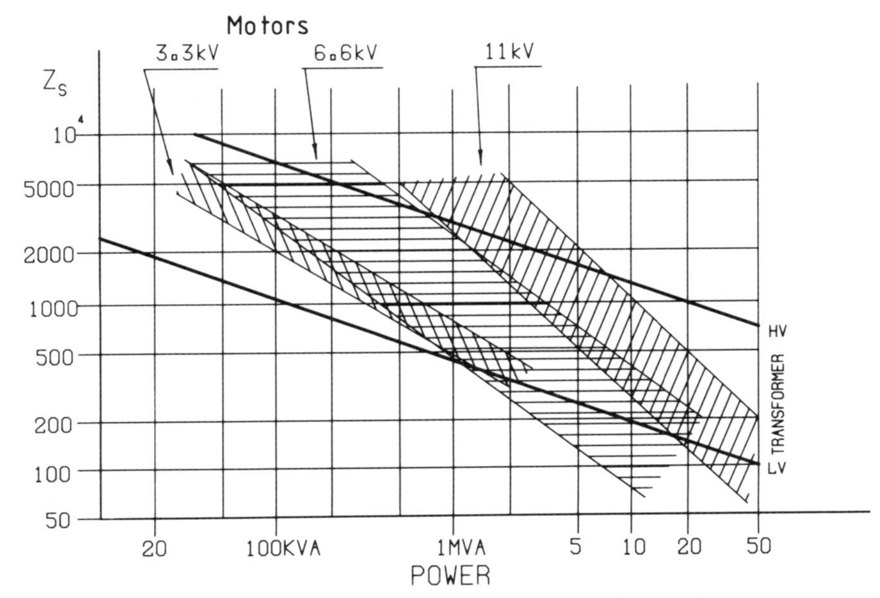

Fig. 16.1 Typical values of surge impedance for medium voltage motors and transformers.

actual figures for the plant are known, they should be used.

As an example, a 1000 kVA 11 kV motor could have an average surge impedance of 5000 Ω. From Fig. 16.2 this might produce a chopping current surge of between 6 and 9 kV if switched by an oil circuit-breaker, an SF_6 puffer breaker or a vacuum breaker with chrome-copper contacts, to be added to a peak phase voltage of 9.5 kV. This is a maximum of less than 20 kV which should not worry an 11 kV motor of that size. On the other hand, a vacuum interrupter with copper-bismuth contacts would possibly produce a surge of 50 kV plus 9.5, or nearly 60 kV, which would give cause for concern. For this reason it will be found that manufacturers using such contact materials usually fit metal oxide surge arrestors as standard practice to all feeder circuit-breakers.

OUT-OF-PHASE SWITCHING

It is unusual for out-of-phase switching operations to arise on distribution systems, and it is rare to find a specification asking for proof of this feature. It is perhaps possible on large industrial sites which have local generation and where there is connection to a local utility as well. If this is the case it is important to check whether parallel switching conditions can be prevented, or if there is a risk that phase shifts between the system voltages on the two sides of any circuit-breaker can occur, that the manufacturer be made aware of this special requirement. A circuit-breaker of higher voltage rating may well be required to meet this duty.

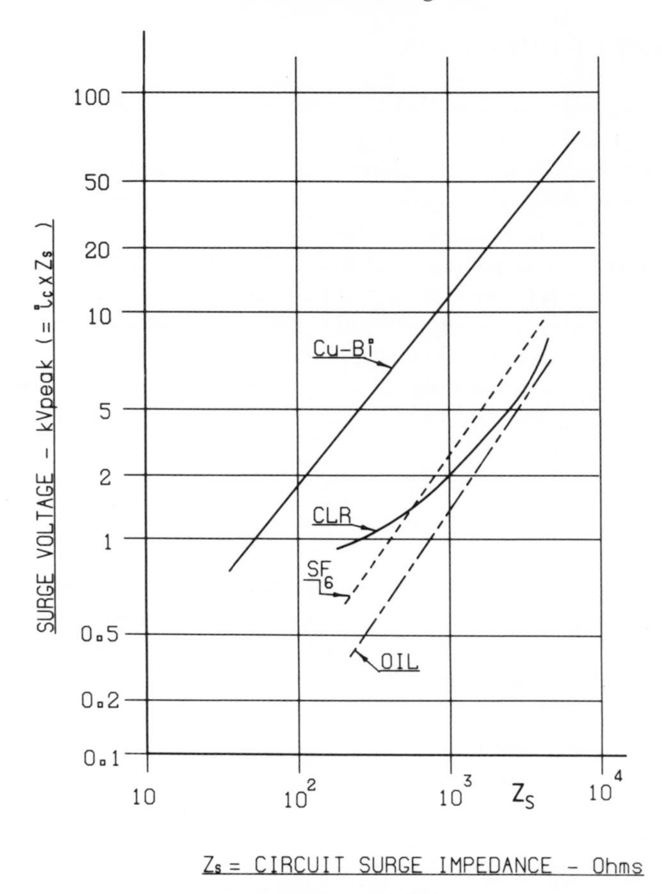

Fig. 16.2 Overvoltage for different types of circuit-breaker as a function of surge impedance.

PROTECTION REQUIREMENTS

If the system requires high speed protection then make sure that the correct rating and duty of switching device are specified. In the American standards in particular, there are different total operating times specified as alternatives. The total operating time is the time from energisation of the trip coil to the final clearance of the fault current at any level above 25% of its rated short-circuit current. Typical values are 3, 5 or 8 cycles at 60 Hz, and represent 50, 83 or 133 ms.

The effect of the faster interruption is to increase the proportion of d.c. component at contact separation and this is examined in more detail in Chapter 18, dealing with circuit-breaker testing.

FREQUENCY OF OPERATION

Switchgear is generally intended to be used for protection of distribution systems and is not expected to be operated very frequently. Normally, this is true for utility networks, but in industrial systems it happens occasionally that circuit-breakers are required to control items of plant that consume large amounts of power and for which no suitable control gear exists. Typical examples are large motors, such as those in rolling mills, and arc furnaces. These are very special conditions and careful consideration should be given to the type of circuit-breaker to use. A type that requires infrequent maintenance is very desirable, and a mechanism with low energy requirements is likely to be capable of more operations without attention than one of high energy. Oil circuit-breakers would require frequent oil changes. The choice today usually narrows to either a vacuum circuit-breaker or SF_6.

AUTORECLOSURE

If the circuit-breaker is controlling overhead lines, the need for high speed autoreclose facilities to deal with the possibility of transient faults must be considered. Again, a low maintenance type is desirable so that there is no need to limit its effectiveness by using a lock-out feature after a certain number of operations. A lightning storm on a rural network can require tens of operations over a relatively short period.

OVERLOADS

Switchgear does not have an overload rating as such, but advantage can be taken of the rating characteristics and the thermal time constant in order to deal with some special circumstances. One simple short duration condition is the starting current of large motors or energising cold loads. The time duration is usually short enough not to run any risk of exceeding the allowable maximum temperatures.

Another important case is where advantage can be taken of lower than maximum ambient temperatures. This can be particularly advantageous as low ambient temperatures often go hand-in-hand with increased loads due to increased heating requirements. So long as the maximum permitted temperatures are not exceeded there is no time limit to these overload possibilities.

However, a warning must be given here. Although switchgear is designed to carry rated current continuously while exposed to an average temperature of 35° C, the usual result of a practical application is that there is a margin between its rated current and that of the circuit in which it is applied. Also, the cyclic variation in temperature during a day will mean quite long periods at 5–10° C less than the maximum. This under-running of almost all switchgear is one important reason why it historically has a very long service life. Continuous operation at 40° C and rated current is likely to result in a life rather shorter than that normally experienced. However, this should not prevent a user taking advantage of lower than standard ambient temperatures to carry an overload current for a few days.

CIRCUIT-BREAKING TECHNOLOGY AND APPLICATION

Different circuit-breaking technologies give characteristics to the switching device that are sometimes important in the application. The feature of negligible maintenance is mentioned above as desirable for applications requiring frequent operation. Some other characteristics are listed in Table 4.2.

Another important aspect is the high arc voltage characteristic of magnetic air circuit-breakers and its effect on system parameters, particularly transient conditions such as offset current waves, high TRV values and switching small inductive currents. In power station auxiliary switchgear these aspects can be important, and weigh heavily in the choice of switchgear for this application. This subject is examined in Chapter 4, and the other features of that type of switching technology, such as a limited performance for switching capacitance currents and susceptibility to high humidity, have to be taken into account in an overall assessment.

Bearing in mind the conservative attitude of distribution engineers – for good reason – it is understandable that there has been a long-running debate on the relative merits of the two new switching technologies now very much to the forefront of circuit-breaker and switch development. Comparison between these and the 'traditional' air- and oil-break technologies is an obvious part of this.

It must be said at the outset, in view of the overall safety role played by switching devices, and circuit-breakers in particular, that most of the technical comparisons which are made derive from a study of statistically rare events, i.e. the study of the consequences of a failure or breakdown, rather than comparison of normal behaviour. Normal behaviour usually consists of the switching devices being closed and carrying a continuous current well within their rating, or being open and withstanding continuously the system voltage – something that both the systems under review can do admirably for many years at a time.

COMPARISON OF SF_6 AND VACUUM CIRCUIT-BREAKERS

The questions most often asked concerning the comparative behaviour of these two ranges of equipment are listed in Table 16.1.

Examining these questions in order, and bearing in mind the earlier point that most of these questions relate to incidents that will be extremely rare, the following comments can be made.

LEAKAGE IN VACUUM INTERRUPTERS

There is no simple way of monitoring the pressure inside a vacuum interrupter,

Table 16.1 Questions often asked concerning vacuum and SF_6 switchgear

Vacuum	*SF_6*
1. Leakage – how do we know?	1. Leakage – what are the consequences?
2. Production of overvoltages?	2. Are arc products toxic?
3. Production of X-radiation?	3. Is the circuit-breaker a pressure vessel in the statutory sense?

These issues are discussed in this chapter in relation to their application in distribution switchgear systems.

as the necessary degree of vacuum is such that laboratory equipment is needed to measure it. The only certain way of knowing whether or not a vacuum interrupter is still good is to take it out of service and apply a test voltage across its open contacts, of a value greater than the withstand level of that gap at atmospheric pressure.

In the admittedly rare event of a leak in a vacuum interrupter, the voltage strength of the gap follows the curve of Fig. 4.4, which falls to a minimum level of a few hundred volts in the glow discharge region of 0.1–10 torr, recovering to around 30 kV/cm at atmospheric pressure. It used to be presumed that a vacuum interrupter would always be full of vacuum, or quickly rise to atmospheric pressure. Experience has shown this is not always the case and it is now thought possible that atmospheric pollution drawn into microscopic leakage channels can temporarily seal the leak, and allow the glow discharge region to be a longer term situation. The earlier hypothesis predicted that a leaked vacuum interrupter would withstand the system voltage and therefore, since in most situations the two sound interrupters in the other phases would successfully switch a three-phase load, the faulty interrupter would cause no problems before a routine check discovered it.

LEAKAGE IN SF_6 CIRCUIT-BREAKERS

With the SF_6 circuit-breaker, a fall in gas pressure is accompanied by a fall in withstand voltage in a linear manner. Recognising the remote possibility of such a leak, SF_6 circuit-breaker manufacturers design the contact system in such a way that the open gap will easily withstand the service voltage even with the gas at atmospheric pressure. As SF_6 gas is five times heavier than air, the possibility of some air replacing the SF_6 even after pressure has equalised cannot be ruled out. However, such are the characteristics of SF_6 that a large proportion of air is needed to reduce significantly its breakdown strength.

So far as switching performance at reduced pressure is concerned, the full short-circuit level must be reduced, but such is the power of this remarkable gas that load-breaking capability will certainly be retained and, depending on type, some short-circuit capabilities may also still exist. In this connection, the puffer type design has the highest retained rating, for two reasons:

1. The standing gas pressure for a given rating is lower than for alternative types – so the degree of change is less.
2. The energy for arc extinction comes from the external mechanism, and this is not affected by gas loss.

In any case the gas pressure can be monitored continuously and the above information is only relevant to the action to be taken in the event of the monitoring device giving a low-pressure warning. Further, most SF_6 distribution circuit-breakers have a 'sealed-for-life' rating which places them in the same category as the vacuum interrupter.

X-RADIATION IN VACUUM INTERRUPTERS

X-rays can undoubtedly be emitted by any vacuum gap subjected to a voltage stress, but the level of emission from a vacuum interrupter at distribution voltages is normally well below the accepted tolerance levels, particularly as there is usually a metal barrier between the operator and the vacuum interrupter. Only when test voltages are applied to a vacuum interrupter with the circuit-breaker removed from its cubicle might there be any need to take precautions. The manufacturer's advice contained in the handbook should be followed when testing.

ARC PRODUCTS IN SF_6 CIRCUIT-BREAKERS

Hazards to personnel in connection with SF_6 circuit-breakers relate to the possible release of toxic arc products. It must be realised that pure SF_6 gas is totally non-toxic, non-flammable and invisible. It does not dissociate until high temperatures are reached, such as in the arcing region of a circuit-breaker, where the gas is totally dissociated into sulphur and fluorine atoms in the arc core.

Vaporising of the contact metals – usually copper and tungsten – also takes place. After arcing has ceased, recombination takes place. A number of sulphur/fluorine compounds are produced, although the re-creation of SF_6 is the most common, as this is the compound that truly satisfies the valency requirements of sulphur and fluorine. The most toxic of the possible alternatives, S_2F_{10}, is also the most unstable and, in fact, has never been found in post-arc gas analysis. The usual constituents are small quantities of sulphur tetrafluoride (SF_4), and if there has been some air present, traces of sulphonyl compounds with oxygen and fluorine. In addition there will be grey powdery residues comprising copper and tungsten fluorides. All SF_6 circuit-breakers contain active reagents which absorb the gaseous byproducts, and the small quantities of powdery residues are dispersed inside the circuit-breakers.

Such is the contact life of the typical SF_6 distribution circuit-breaker that the above information is purely academic. Most manufacturers caution the user against opening up the circuit-breaker on site, and the contact endurance means that the circuit-breaker requires no internal maintenance during its entire service life. None of the other parts associated with interruption deteriorates significantly in use, nor does the gas become either contaminated or exhausted, so the service life is a function solely of contact wear.

Even in the remote possibility of a catastrophic failure leading to fracture of the chamber, thus releasing the gases, the quantities are so small that very little hazard exists. In an enclosed space the presence of arced SF_6 can be detected by its nauseous smell at levels well below those that are harmful and vigorous ventilation soon disposes of them.

SF_6 SWITCHGEAR CONTAINERS

The small quantity of pressurised gas in an SF_6 distribution circuit-breaker usually exempts such equipment from the provisions of any statutory regulations applying to pressure vessels in those countries where such regulations exist. This is certainly the case in the UK but not quite so positively in Germany, although the German regulations make concessions to SF_6 circuit-breakers.

No conclusions are drawn from the above discussion since many non-technical factors such as source of supply, availability of after-sales service, delivery and most of all, price, will have a bearing on the final purchase decision. Technically, both 'new' technologies satisfy the objectives of removing fire risk and maintenance requirements. In normal utility service neither type of circuit-breaker requires overhaul within the normally accepted service life, i.e. 20–25 years, but the user has to decide on the importance of the foregoing issues in his particular application.

CHAPTER 17

Installation and Commissioning

The erection of switchgear described in this chapter denotes an activity remote from the manufacturing centre: the installation of the equipment in a location provided by the purchaser where the switchgear will eventually go into service. This activity may take place in a factory, a power station, a coal mine, or any other site where a substation may be situated. The switchgear often cannot be completely assembled in the factory and shipped as a complete plant, so it is necessary for it to be packed in sub-assemblies, shipped to site and then erected there. The erection includes the setting in place of all the components of the substation, their mechanical and electrical interconnection, the fitting of any extra relays or instrumentation and the completion of the control and power circuits. Some chambers of the switchgear may require filling with insulating media of various kinds.

Once the equipment has been erected, it needs to be commissioned, which is defined as testing and finally putting into service of the installed apparatus. Testing is necessary to prove that the complete installation meets the required specifications.

Although the work content for the installation of an outdoor substation is different from that for indoor switchgear, the principles that are followed are the same. For the outdoor substation, particularly a large primary substation, there is more co-ordination required between the switchgear contractor and the civil engineering contractor, as each individual piece of switchgear has its own foundation, which has to be set out reasonably accurately if the connections are to fit.

Against this is the fact that there is usually a fair degree of latitude in the fit of the main connections due to the way in which the joints are made. This subject was touched on in Chapter 10. As there are more stages to the erection of a metal-enclosed, indoor switchboard than for an outdoor substation, this chapter will concentrate on that operation, pointing out any differences in procedures as they arise.

STORAGE

As a general rule delivery of the switchgear is made to the site at a time convenient for erection to proceed almost immediately. However, it sometimes happens that the site works are delayed, perhaps by weather conditions, in which case the switchgear has to be carefully stored until it is needed.

On delivery, it is important to check that all items are present and correct in accordance with the delivery notes and then to store the components systematically, to ensure that no parts go astray. At all times, care must be taken to maintain all the items in good condition, often not easy if they are stored on a site where building work is going on. The equipment must be properly stacked so that no parts are subject to unnatural stresses, and it must be kept clean and dry.

It may be assumed that components for outdoor use are able to withstand outdoor storage, but that should be checked with the manufacturer as, for example, the omission of the filling medium in an oil circuit-breaker may render it susceptible to insulation damage, if moist air can infiltrate the interior. Indoor equipment must be stored under proper cover and, if possible, arrangements made to keep the building temperature in excess of the dew point. This will prevent condensation on the equipment and thereby protect the plant from corrosion or possible breakdown when finally energised.

When the site is ready for installation to commence, the components should be carefully inspected, with particular attention being paid to insulation parts and mechanisms, to ensure that no deterioration has occurred, as remedial work is much more difficult and leads to delays if problems are only discovered as work proceeds.

ERECTION

Before commencing erection the workforce must be in possession of all the drawings and instructions required to erect the equipment. These are normally provided by the manufacturer, and many switchgear manufacturers also provide an erection service. The use of this service would ensure a knowledgeable and practised erection team. When the user is carrying out his own erection, the manufacturer's handbook and any other specific erection instructions must be carefully studied before commencing work.

The substation in which the switchgear is to be erected must be clean and dry and all debris cleared away. During erection, particular attention must be paid to the following points:

1. Exclude dirt and debris from partially erected cubicles. All openings that are not in immediate use must be blanked off or covered by clean sheets.
2. All electrical insulation must be kept clean and dry by protective coverings and, if necessary, heated.
3. Materials that have been issued from stores and not yet used on erection should be stored carefully and tidily. If any parts are lost, ensure that exact replacements are obtained, as breakdowns have been caused by unsuitable substitutes being used unwittingly by installation staff as replacements for misplaced items.
4. When handling the cubicles and major components, care should be taken to observe the correct lifting arrangements and that slings are attached to the manufacturer's designated lifting points. This ensures that parts are not subjected to undue stresses which could result in disturbed settings or other damage.

FOUNDATIONS

The successful erection and operation of a switchboard will depend very largely on the accuracy of the foundations. The most useful form of fixing medium for indoor distribution switchgear is some form of proprietary channel embedded in the floor, containing captive adjustable nuts. These channels should be assembled truly parallel and level and project slightly above the surrounding floor. This enables the switchgear to be clamped without interference to the channels, which are known to be level. A less common arrangement nowadays is to cast holes in the floor into which suitable foundation bolts can be grouted after the switchboard has been carefully levelled.

GENERAL ASSEMBLY

It is normal to commence the assembly of a distribution switchboard from the central unit. This minimises the effect of any build-up of errors as the erection proceeds. Any interunit tie-bolts are loosely positioned as each unit is erected, all units being lined up on the centre unit. Plumb lines should be used on each unit to ensure that it is vertical. When all units have been placed in position and checked for alignment, the interunit tie-bolts may be tightened and checks made that any withdrawable portions may be entered and withdrawn smoothly. Instrument cabinets and any other components can then be added to the switchboard.

BUSBAR AND CIRCUIT CHAMBER ASSEMBLY

Long busbar assemblies are broken up into smaller sections and these must be erected in accordance with the manufacturer's drawings. Care must be taken to ensure that the sections are correctly located in their respective units. When heavy current busbars are made up of several bars in parallel, extra care must be taken to ensure that the correct number of bars are used in each unit. Contact surfaces must be carefully cleaned before assembly, using fine sandpaper (not emery cloth as the dust from this is conducting) if the surface is natural copper, or a proprietary silver polish if the surface is plated. Where the connections have screw threads these should be cleaned with a fine scratch brush. After cleaning, the joint faces must be wiped with a clean, lint-free cloth to remove any dust, and the joints assembled as soon as possible.

Depending on the design of the busbar chamber, the least accessible bar is assembled first, usually the one furthest away from the operator, and then successive bars nearer to the covers that will be put in place when assembly is complete. As each bar is placed in position, with careful reference to the drawings to make sure that the overlaps and joints are exactly as the manufacturer intended, all securing bolts are tightened. This frequently involves the use of a torque spanner, and again, care must be taken to use the recommended torque settings. These figures have been established to produce the correct joint resistances without overstressing the bolts, so careful setting of the spanner is important.

The manufacturer's instructions should be studied to see if any of the joints in the busbar chamber require to be insulated. This is a common requirement in UK metal-enclosed, medium-voltage switchboards, and should be carried out in accordance with the instructions, paying particular attention to any safety precautions. The components required are usually supplied with the switchboard.

CABLING

The preparation of cables for connection to the switchgear is a specialist function and an experienced jointer should be employed for this purpose. The cable has to be carefully laid to avoid sharp bends and the length to the cable lug accurately measured to avoid any stress on the cable or on the bushing to which it is attached, once the joint has been made. A wide range of cable termination arrangements are available nowadays and each has its own particular procedures, which must be meticulously followed.

It is increasingly the practice to terminate cables in air-filled cable compartments nowadays, and several techniques exist for this purpose. These are described in Chapter 14.

Occasionally paper cables are still terminated in compound-filled boxes and where this is the case care must be taken to ensure that the bitumen compound used is continually stirred during the melting period to avoid excessive heating and possible burning of the compound at the bottom of the boiler. The cable box must be clean and so must all the buckets and utensils used for this operation. The cable box itself and any buckets and ladles used must be adequately preheated to ensure that the compound remains fluid until the filling operation is complete.

Once the compound is at the correct pouring temperature it should be poured slowly, to avoid splashing and the inclusion of air bubbles. When the cable box has been filled to the requisite level it is allowed to cool slowly, avoiding draughts. Should the compound fall below the indicated level on cooling, it should be topped up while still warm to ensure a good bond between the main mass and the topping layer.

EARTHING

Earthing of switchgear is a particularly important operation on an outdoor site, where each individual piece of plant must be connected to the station earth bar, and the earth rods implanted in the ground. Measurements of earth impedance are made to ensure that no dangerous voltages can arise due to fault currents flowing in the earth conductors. In indoor substations the panels normally have earth bars fitted at the factory, and they are connected together on site in a similar manner to the busbars.

If the switchboard is fitted with frame-earth protection, care must be taken to follow any requirements for the insulation of individual units from one another, and the insulation of the station earth bar from the frame of the switchgear cubicles. In this case cable glands also have an insulating layer, and the effectiveness of this must be checked, and care taken to connect the cable sheaths to the earth bar.

FILLING WITH INSULATION MEDIUM

In practice this means oil filling, since distribution switchgear using any other media, such as SF_6, will be filled at the factory before despatch to site. The first step in the process of filling oil chambers is to check that the oil supplied is in accordance with the manufacturer's specification. Electrical oils must comply with BS 148 and should be tested for electrical breakdown and water content before pouring into the chambers. The chambers should be checked for cleanliness immediately before filling and the oil carefully poured up to the indicated level. It is advisable to fill voltage transformers after they have been mounted on the switchgear. Indicated levels are usually based on an ambient temperature of 15° C and if the actual temperature differs markedly from this due allowance must be made.

SMALL WIRING

It is necessary to complete small wiring connections between the panels on adjacent cubicles and to connect external multicore cables to the control cable terminating boxes which are usually mounted on the rear of the switchgear cubicles. If it is not already contained on the wiring diagram, a useful aid to ensure correct connection for multicore cables is to list the cores in order and to record the appropriate connection to be made to each core at each end of the cable, before starting work.

FINAL INSPECTION

After the switchgear erection has been completed, all the securing bolts tightened, all busbars and other connections made, all insulation procedures finished, cables (or overhead lines) connected and small wiring completed, a final inspection is made. The following is a typical, but by no means exhaustive, check-list:

1. Before fitting covers, all chambers should be checked for complete cleanliness and the absence of all foreign matter, including tools used in the erection operation.
2. Once the covers have been fitted, a check should be made to ensure that all fixing screws are in place and secure.
3. All labels, where required, are fitted and visible.

4. Any mechanical interlocks should be checked; safety shutters, where fitted, must be checked to be operative.
5. All withdrawable items shall be proved to be capable of extraction and isolation as required.
6. All fuses and links shall be inserted in the appropriate holders.
7. All exposed insulation surfaces should be examined for cleanliness and dryness.
8. Make a final check for the continuity of the earthing.
9. All tools used in the erection of the switchgear should be carefully accounted for.

ELECTRICAL TESTING AND COMMISSIONING

Routine factory tests will have been carried out during manufacture to check design criteria and the maintenance of the supplier's standards. These tests provide a valuable reference when any query is made concerning the apparatus during its commissioning tests and operational life. Commissioning tests are conducted at site after the installation of the equipment. These are to ensure that the equipment will perform its duties in service, that interconnection with other apparatus is correct, and that data is provided for future maintenance and service work.

The conducting of these tests can also provide valuable training for the purchaser's operations staff. Normally such tests are carried out once only, after erection, prior to putting the installation into service. The most common form of commissioning test is to simulate operating conditions using portable voltage and current sources. A test log should be maintained, listing all tests carried out, the objectives and the results.

Typical tests are:

1. Visual checks.
2. Impedance measurements.
3. Insulation resistance checks.
4. Current and voltage transformer checks.
5. Proving of protective schemes.
6. High voltage tests.
7. Circuit-breaker operation checks.
8. Control scheme checks.
9. Load testing.

VISUAL CHECKS

The object of these checks is firstly to ensure the mechanical integrity of the equipment, then that all the electrical connections made on site conform to the appropriate diagrams. It is also necessary to ensure that all fuses and links are in place and have been fitted with the correct cartridge.

IMPEDANCE MEASUREMENTS

There is an increasing requirement nowadays to carry out resistance measurements on switchgear conductors after erection. These measurements will pin-point any badly made joints. The resistances to be measured are in the microhm range and need specialised equipment, the 'Ducter' low resistance bridge being typical. Consistency in the figures recorded is as good a guide to proper erection as the actual values. The busbars and earth conductors are the principal conductors that need to be checked, but if the erection contract included the provision of the actual earth points, particularly on an outdoor site, then particular attention must be paid to these.

INSULATION RESISTANCE CHECKS

Insulation resistance measurements on all small wiring should be carried out at 500 V d.c. This test ensures that such wiring is in good condition. Any apparatus which may be damaged by the application of such a voltage, e.g. static protective devices, should have the appropriate terminals short circuited for this test. When measuring the insulation resistance to earth of an individual circuit, all other circuits should be normal, i.e. earth links closed and d.c. circuits normally connected. This ensures that the insulation of the particular circuit is satisfactory both to earth and to all other circuits.

It is impractical to give a definitive lower limit for the resistance measurement from this test, as such readings are dependent on the cable lengths, the size of the cores and the number of parallel paths. As a guide, any reading of less than one megohm at 500 V d.c. should be investigated.

CURRENT AND VOLTAGE TRANSFORMER CHECKS

Normally the voltage and current transformer wiring will have been completed at the manufacturer's works and only if this wiring has been disturbed will it be necessary to carry out any tests on site. Should this be the case, then one important test is to ensure that the CT secondary winding has been correctly connected in relation to the primary winding. Should this have been reversed in any phase then the operation of the protective circuits will be seriously affected.

All CTs have some means of identifying the primary and secondary terminals, common practice being to identify the primary as P1–P2 and the secondary as S1–S2 as recommended in BS 3938, with test windings numbered T1–T2. The following test is recommended to check these polarities (Fig. 17.1).

A low reading d.c. voltmeter is connected across the CT secondary winding terminals and a battery across the primary. If relative polarities are correct, on closing the circuit a positive 'flick' is observed on the voltmeter, and on opening the circuit a negative 'flick' is seen. Where CTs are mounted in transformer

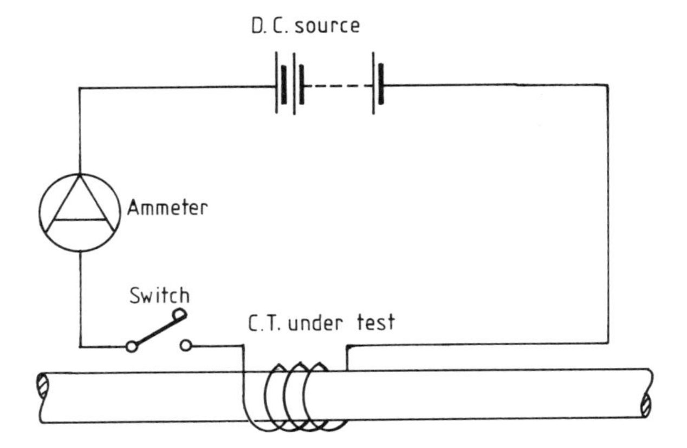

Fig. 17.1 Circuit for checking the polarity of instrument transformers.

bushings it will be necessary to short circuit the main transformer LV winding, thus reducing the overall impedance of the transformer, to give a good deflection of the instrument needle.

Voltage transformers can be checked in a similar fashion.

PROVING OF PROTECTION

The operation of the various forms of protection which can be fitted to distribution switchgear is reviewed in Chapter 13. Such operation can be simulated either by injecting heavy currents into the CT primaries or by the injection of suitably chosen currents into the CT secondaries, to represent the effect of fault currents. Sometimes the CTs are fitted with special tertiary or test windings to facilitate this current injection. Whichever method is adopted, sufficient tests should be made to ensure that the relays operate as intended on all types of fault and that any auxiliary relays also operate correctly.

To adjust the current settings it is normal to use a variable auto-transformer to provide the supply. As the waveshape of the test current can have an influence on the behaviour of the relay, as much resistance as possible should be included in the injection circuit.

HIGH VOLTAGE TESTS

If called for in the contract documents, high voltage tests can now be applied to the insulation. The test values should be those laid down in the appropriate standard for site testing, and should be applied by an a.c. high voltage test set or, particularly if the cables have been connected, by a d.c. test set. Where vacuum interrupters are in use it is usually recommended that a commissioning voltage test of about 20 kV be applied across the open contacts of each individual interrupter, as a check that the vacuum is still in good condition.

CIRCUIT-BREAKER OPERATION TESTS

Following the completion of insulation testing, the circuit-breakers should be tested for operation. These checks should include the following:

1. Trip and close operation from both local and remote positions. These tests should preferably be made at minimum and maximum supply voltages. It should be checked that correct indication of the closed and open position is given both by the mechanical and any electrical indicators that may be installed.
2. The correct operation of all the interlocks should be checked. It should be checked that all permitted operations are possible and that the interlocks prevent the carrying out of prohibited operations as intended.

Before carrying out these tests care must be taken to isolate the circuits from any source of supply.

CONTROL SCHEME CHECKS

These are particularly applicable to switchboards supplied for the control of industrial processes. These are often fitted with sophisticated sequence control systems. They require checking in association with the user and his engineers, before the switchboard can be energised.

LOAD TESTING

Once all the checks on the equipment have been carried out, the operational tests are satisfactory,the voltage tests passed, and all the functional tests on the relays, etc., completed to the customer's satisfaction, the equipment can be energised. When the circuits are on load all the instruments and indicators can be checked for correct readings.

If test blocks are available or the relays have test plug connections it may be worthwhile doing a final check to see that the CT polarity is correct and the instrumentation is giving correct readings.

CHAPTER 18
Switchgear Testing and Quality Control

Switchgear occupies a rather unusual position in the world of manufactured goods, in that practically every attribute that is of value to the purchaser is virtually impossible to test at the end of the production line. Most consumer goods for example, from toasters to television sets, can be given acceptance tests after manufacture which reproduce the job they are made to perform.

The task of the switchgear unit is to withstand system voltage, in perhaps difficult environments, for times measured in decades. In addition, after carrying varying currents up to rated full load during the preceding long period of time, they have to react instantly to a signal that tells them to open and maybe interrupt a heavy fault current.

So the switchgear manufacturer has to use indirect methods to ensure that each product he supplies has these necessary characteristics. Because the actual duty is inevitably different for every unit he sells, there have to be some standardised tests that tell him and his customers that the equipment will meet the requirements that have been discussed in detail in previous chapters. These tests are referred to as 'type tests', as they are carried out once only on a representative of a product line to prove that all products of the same type will perform in accordance with the standard.

Thus, by definition, there is a need for other tests and procedures to ensure that every example of that product will perform in the same manner as the type tested representative. The tests that are carried out on every product are called 'routine tests', and they have as their object to demonstrate that the important characteristics of each switchgear unit, be it a switch, a circuit-breaker or a cubicle for a metal-clad equipment, are like those of the prototype.

It is up to the manufacturer to assure himself that these requirements are met. The international, national and, in many countries, user standards already establish many type and routine tests, but most makers of switchgear will also add their own, particularly in the routine test area. In such a safety conscious business, each manufacturer is acutely aware of the attention that each breakdown in service is going to bring, and the time and effort it is going to take to regain acceptance of a reputation. Most of the world's switchgear manufacturers have been in the business for a very long time, and have a wealth of service experience to fall back on. It is this experience, as well as the content of standards (which, of course, are produced from a further wealth of experience), that guide the manufacturer in creating his routine testing and quality assurance pattern.

TYPE TESTS

These are listed in international recommendation IEC 694, and their applicability to individual types of switchgear is covered in the relevant IEC standards. In this chapter the IEC recommendations will be taken as a basis for discussion, since they are broadly applicable for all parts of the world except those parts influenced by American practice. Where numbers are quoted and an equivalent British standard exists, the number is quoted in brackets. In the USA, the ANSI standards are followed, and while these follow exactly the same principles as the IEC range, there are differences of detail, some of which are highlighted in Chapter 19 and some of which will be referred to in this chapter as occasion demands. The following list is based on IEC 694 (BS 6581). Some of the tests are mandatory, and must be made before a certificate of rating can be issued. Others are optional and are made if the manufacturer wishes his product to be proved for a particular duty, or if they are subject to agreement between the manufacturer and the purchaser, if the purchaser has a need to prove a particular characteristic for a particular application.

Type tests on all switchgear are:

1. Dielectric tests
   1.1 Lightning impulse tests*
   1.2 Power-frequency voltage withstand tests*
   1.3 Partial discharge tests
   1.4 Artificial pollution tests
2. Temperature rise tests*
   2.1 Measurement of the resistance of the main circuit
3. Short-time withstand and peak withstand current tests*.

To these tests are added other tests for different switching devices:

Switches:

1. Tests for breaking normal load currents*
2. Short-circuit making current tests*
3. Capacitor bank switching tests
4. Line charging current switching tests
5. Cable charging current switching tests
6. Mechanical endurance tests*.

Circuit-breakers: as for switches, plus the following:

1. Short-circuit breaking current tests*
2. Out-of-phase switching tests.

* These tests are required for certification purposes.

Before entering into a description of the content of each of these tests, some mention of the facilities required to carry them out is required. At first glance it would seem that access to a suitable distribution network would provide circumstances akin to service conditions, and therefore be ideal for the purpose. However, apart from the impossibility of continually applying fault conditions to a working system, and the inconvenience that would cause, the degree of control required to reproduce the test conditions required to meet the specification makes it essential to provide a special-purpose test plant. It will be realised that such a facility requires a large capital investment, and fortunately those in existence today were built many years ago.

Test plants first appeared in the USA, in Europe and in the UK around the year 1930. They usually derive short circuit current from specially built, low reactance generators driven by induction motors. These typically generate at medium voltages in the range up to 15 kV, and transformers are used to adjust the voltage for testing higher voltage rated switchgear. Generally speaking, the power output is sufficient to allow distribution switchgear of the types described in this book to be tested three-phase up to their full rating. These high-power test plants also require short-circuit making switches that can be controlled to fine limits in the point on the voltage wave at which the fault is initiated, back-up circuit-breakers that can clear the fault if the test equipment has a problem, a range of reactors and resistances to adjust the test values as required and comprehensive recording facilities capable of measuring all the relevant quantities in the time scales referred to in Chapters 2 and 7.

These high power test plants deal with all the switching tests; the fault currents, the load currents, inductive currents and capacitance currents. High voltage plants are also required capable of applying power-frequency and impulse test voltages to suit the range of switchgear in question. Most manufacturers have their own facilities for these tests, as the test equipment is also used for research and development work. Development work also needs to be done in the high power field, and some manufacturers have small scale plants that enable small scale testing to be carried out. All full scale testing has to be done at one of the main high power test plants.

The question of 'certification' has been mentioned, and with it go considerations of other concepts such as 'approval' and 'accreditation'. With increasing interest in the issue of quality assurance, which is closely allied to the issues of type and routine testing, there is greater demand for independent testing authorities.

Many large users of switchgear, such as nationalised electricity undertakings, oil and chemical companies, other national industries such as steel, coal and transport, produce their own specifications for switchgear and require to approve the testing of the equipment they buy. In the USA, 'self-certification' is still very much the present state of the market. This means that the manufacturer arranges for all the type tests, either carrying them out himself if he has the facilities, or taking them to another laboratory where he is not able to do the tests with his own resources. The latter usually means the high power tests. The original high power test plants were all set up by manufacturers such as GE and Westinghouse in America, AEG and Siemens in Germany, AEI, GEC, Reyrolle and English Electric in the UK, to mention only the large ones. To allow for the fact that all these plants would have to test for other, less fortunate, manufacturers from time to time, independent 'observers' were appointed to ensure that no competition entered into the testing arena.

This was followed in the UK by the creation of the organisation known as ASTA (Association of Short-circuit Testing Authorities) which controlled these independent observers, and also co-ordinated the interpretation of the test requirements so that all switchgear equipments were tested to similar standards in all the plants. This co-ordination has now become international, at least within Europe, by the setting up of a wider based association, the Short-circuit Testing Liaison (STL).

Thus, for many years now, an independent certificate of short-circuit rating has been available from the ASTA organisation which has gained worldwide recognition as a reliable reference. In Holland, there is a large nationally owned testing laboratory known as KEMA, which also has an international reputation as a certification body.

There is increasing pressure developing for independent approval of the other type tests, which until now manufacturers had undertaken within their own resources and issued their own certificates. The development of increasingly stringent product liability laws is likely to hasten this process. The process of approving a body or test laboratory as being expert and experienced enough to certify independently the compliance of switchgear equipment with its type test requirements is known as accreditation. It is undertaken by a body of experts, usually under the jurisdiction of the national standards making body, such as the British Standards Institution. It is not unknown for internationally recognised consultant organisations to function as certification witnesses and authorities.

DISCUSSION OF TYPE TESTS

Under each of the type test requirements, consideration has to be given to the setting up of the equipment for test, which usually requires a complete 'functional unit' as the specification words it – the performance of the tests, the behaviour during those tests, and the condition after the test. It is desirable to create conditions as near as possible to the service conditions, but there are matters which would complicate the test without adding to its authenticity and these can be ignored for the sake of simplicity. A case in point is the ambient temperature during the temperature rise test. This test is allowed to be performed at room temperature, as the value of the ambient temperature has very little effect on the rise.

Dielectric tests

The essential tests in this category are the impulse voltage tests and the power-frequency voltage tests. For these tests the equipment must be set up as it will be in service, so far as all the main connections, earthed enclosures and insulation components are concerned. Auxiliary components, such as instrument transformers, can be omitted unless this would increase the insulation level in a particular case.

The method of carrying out the tests is the subject of yet another international standard, IEC 60 (BS 923). Since the tests are carried out single-phase, various arrangements of the test object are needed to ensure that all the electrically stressed areas are tested. For example, the switching device has to be tested with its contacts open as well as closed, and tests are required between phases and between poles. All these permutations are worked out and presented as tables in the relevant equipment specifications.

The standard lightning impulse test voltage has a rise time to peak of 1.2 μs and should fall to half peak value in 50 μs. This is the same in both the ANSI and IEC standards. The power-frequency tests may be done at 50 or 60 Hz as the difference is not significant. Corrections need to be made to the test voltage levels to correct to standard atmospheric conditions of temperature and humidity. Correction factors are given in the standards.

Artificial pollution tests are currently specified only for outdoor open-terminal circuit-breakers and even then are subject to agreement between the manufacturer and purchaser if the circuit-breakers are to be installed in environments where salt, fog, or their equivalents, are a likely hazard.

Partial discharge tests are usually a component test, and the only class of switchgear for which the complete equipment has to be subjected to this test is the insulation-enclosed switchgear in accordance with IEC 466. Since the earth electrodes in such equipment are somewhat indeterminate, there is some logic in this requirement. There cannot be the same need for metal-enclosed equipment to be tested as a complete assembly if all the components that are individually subject to partial discharge type and routine test measurements are within the limits when tested in a representative manner. IEC 56 and IEC 265 clearly state that partial discharge tests are not required on complete circuit-breakers or switches.

Temperature rise tests

The test conditions for temperature rise tests are established logically with the object of reproducing as accurately as possible the service condition. Individually mounted equipment must be away from any objects that might interfere with normal air flow, and at a representative height above ground level. In the case of a metal-enclosed unit that is mounted in a switchboard, then either it must have its adjacent units fitted and fully loaded or, if this is not practicable (and it rarely is), then it is permissible to simulate the conditions with heaters or heat insulation. The latter is the more usual method.

The test connections must not import or export significant amounts of heat and should be proportioned so that they have a temperature drop, over a length of one metre from the equipment, of not more than 5 K. The test, at rated current, is applied until a steady temperature is reached. The temperature at salient points is measured by thermocouple, and the results judged according to the criteria discussed in Chapter 8.

Measurements of the main circuit resistances are made both before and after the temperature rise tests, and it is not expected that any significant change will occur between the two sets of results. It is usual to take readings that show the individual resistance of all the main joints, both bolted and those of the contacts. These values are useful as a basis for comparison with the routine resistance measurements eventually made on every similar equipment produced.

Tests at high-power laboratories
This groups together all those tests that involve switching operations at rated current and voltage. These involve load current breaking tests with loads of all power-factors, the interruption of short-circuit currents and the demonstration of the ability of the equipment to make and carry its rated fault current. A separate test that requires the high-power plant is the internal arcing test.

For circuit-breakers it is not usual to test for normal load breaking, as it has to be proved that the circuit-breaker will interrupt all currents up to its maximum rating at a low power-factor. Tests are still required to prove capability in respect of capacitance currents and low inductive currents, although the criteria are more concerned with overvoltages and restrikes than ability to interrupt.

With all the switching tests, tolerances in respect of all matters that can influence the capability of the switching device have to be taken into account. For gas filled equipment, for example , the gas pressure should be as low as it is ever expected to become in service. The operating energy of mechanisms must be adjusted to take account of variations in supply voltages or spring tolerances. Supply voltages for trip coils have to be set at the minimum levels expected in service, and so on. All these matters are the subject of the standards dealing with testing. For circuit-breakers this is ANSI C37.09 in the USA and the IEC recommendation is IEC 56 (BS 5311). For other plant the relevant equipment standard covers the same points, although as far as the kind of tolerances referred to here are concerned, the requirements are the same for all switching equipment.

For switches, load breaking tests are required, both to prove switching capability at the lowest power-factors likely to be met in service and to give some indication of service life between maintenance periods. The power-factor for the load breaking tests is a maximum of 0.7, and tests have to be made at 100% load current, and at 5% of the load current, except for oil switches when this lower value is 25%. Two categories of switch are recognised, one intended for standard conditions of service which does not expect frequent load switching (Category I), and the other for use where frequent load switching is expected (Category II). The former has to pass a total of 20 full load breaks and 20 reduced current breaks, while the latter has to interrupt full load a further 80 times. After performing these tests the switch may need overhaul before being fit for further service.

Circuit-breakers require a much more comprehensive cycle of testing. The variables that need to be controlled for checking the ability of a circuit-breaker to deal with all the fault conditions likely to be met with in a normal network are listed here:

1. Applied voltage
2. Short-circuit breaking current
3. Short-circuit making current
4. Power-factor
5. D.C. component
6. Transient recovery voltage
7. Power frequency recovery voltage

The required tests are divided into duties, each of which comprises a number of tests at prescribed time intervals. It is in this area that the greatest differences exist on the two sides of the Atlantic. Table 18.1 attempts to set out the corresponding duties as far as a relationship can be determined. Rather than quote formulae for the duties, a representative case has been selected for comparison as mentioned in Note (10) below the table. Both standards require tests over a range of currents from around 10% of the rating up to 100%. At the full rating, symmetrical and asymmetrical circuits are required and closing and latching at full peak making capacity has to be proved.

The issue is complicated by the major differences in rating method used by the different standards arising from the American intention to cover a wide range of voltage ratings within a given MVA rating. This is discussed in Chapter 19.

In addition to these test requirements there are additional points to watch, particularly in connection with the IEC test duty 5, the asymmetrical test duty. With modern circuit-breakers having very efficient arc interrupting means, it is difficult to obtain an asymmetrical test where the circuit-breaker interrupts at the end of a major loop. This is usually the test that creates the most difficult condition for a modern circuit-breaker and it has to be demonstrated that the circuit-breaker has the ability to interrupt after a major loop in at least one test of the three tests at this duty. It is here that the ability of the test plant to adjust both the point on wave switching of the station making switch and the trip initiation of the circuit-breaker within fine limits is put to the test.

Study of Chapter 2 will show that the current zeros in a three-phase test occur every 3.3 ms at 50 Hz (2.8 ms at 60 Hz) and with a circuit-breaker that is quite capable of interrupting if the first current zero after contact separation occurs within 2 ms, particularly if that is in a phase with a minor loop of current preceding that first current zero, it will be understood just how critical every millisecond can be.

The asymmetrical test duty is required to prove that a circuit-breaker will interrupt an offset current wave containing a significant d.c. component. This will happen if the circuit-breaker is tripped instantaneously in the event of a fault, and the amount of d.c. component at contact separation will depend on the rate of decrement of the d.c. and the opening time of the circuit-breaker. The 'standard' decrement is based on an X/R ratio of 14 which is equivalent to a power-factor of 0.07. This is explained in Chapter 2. The IEC requires the d.c. component on test to be so arranged that the d.c. component in one phase is at least equal to the value it would have in the standard case at a point in time after initiation at a voltage peak allowing for the circuit-breaker opening time plus a half-cycle relay operating time. IEC standard 56-2 contains a graph of d.c. component against time for the $X/R = 14$ decrement so that the required figures can be estimated.

The American standard recognises the same factors influencing the amount of d.c. component, and the corresponding graph in ANSI standard C37.04 (Fig. 2) is virtually identical to the one in the IEC standard. This does not appear so at first glance, since the Americans rate their asymmetrical duty in terms of total rms current, and those are the figures quoted in Table 18.1. There is a factor 's' in

Table 18.1 Comparison of short-circuit testing to American and IEC standards

Standard	*Rated voltage,* kV	*Rated breaking current,* kA	*Making current peak,* kA	*% D.C. component*	*TRV peak voltage,* kV	*TRV time to peak,* μs	*No. of phases*	*Test duty*
10% of rating								
E1	12.00	3.2–4.8	—	<20	22.08	—	3	O-t-O-t′-O
A1	15.00	2.0–3.6	—	>50	28.20	—	3	O-15 s-O
25/30% of rating								
E2	12.00	9.6–14.4	—	<20	22.08	12.8	3	O-t-O-t′-O
A2	15.00	5.6–8.4	—	<20	28.20	—	3	O-15 s-O
50/60% of rating								
E3	12.00	21.6–26.4	—	<20	22.08	25.5	3	O-t-O-t′-O
A3	15.00	11.2–16.8	—	>50	28.80	—	3	O-15 s-O
100% rated breaking current								
E4	12.00	40–44	120	<20	20.60	60.00	3	O-t-CO-t′-CO
or								
E4a	12.00	near 40	120	—	—	—	3	C-t′-C
E4b	12.00	40–44	—	<20	20.60	60.00	3	O-t-O-t′-O
A4	15.00	>28	—	<20	28.20	—	3	O-15 s-O
plus								
A6	15.00	>32.2	97.2	>50	28.20	—	3	CO-15 s-CO
and								
A5	11.50	>36	—	<20	21.60	—	3	O-15 s-O
plus								
A7B	11.50	>41.9	97.2	>50	21.60	—	3	CO-15 s-CO
Test to prove cumulative interrupting ability (400%)								
A8	11.50	<30.6	—	random	21.60	—	3	O or CO
	11.50	>41.9	—	random	21.60	—	3	1 hr – CO
100% asymmetrical rating								
E5	12.00	40	—	>36	20.60	60.00	3	O-t-O-t′-O
Single-phase test								
E s-p	6.93	40	—	<20	14.60	12.80	1	-O-
A13 s-p	8.66	32.2	—	<20	18.80	—	1	O+O
A14 s-p	8.66	37.03	—	>50	18.80	—	1	O+O
Short time rating								
E	Any	40	100	—	—	—	3	3 s
A12	Any	30	—	—	—	—	3	3 s
Critical current test								
E	12	declared critical current	—	<20	22.10	c.24	3	O-t-O-t′-O

Table 18.1 *(contd)*

Standard	*Rated voltage,* kV	*Rated breaking current,* kA	*Making current peak,* kA	*% D.C. component*	*TRV peak voltage,* kV	*TRV time to peak,* μs	*No. of phases*	*Test duty*
Delayed tripping test								
A11	11.5	36	97.2	0	21.60	—	3	C-2 s-O
Auto-reclose tests								
A9	15	30.75	—	>50	28.20	—	3	O-0 s-O
A10	11.5	30.27	—	>50	21.60	—	3	O-0 s-O

Notes

1. C = Close operation, O = Open operation.
2. t = 0.3 s for auto-reclosing circuit-breaker, otherwise 3.0 s.
3. t′ = 3.0 s.
4. s-p = Single-phase test, applied to an outer pole.
5. The ANSI standard only specifies peak voltage for the TRV parameters.
6. The rated breaking current is symmetrical rms for all the IEC om that standard.
7. E indicates tests to the IEC standard 56; part 4 and the number refers to the Test Duty from that standard.
8. A indicates tests to ANSI standard C37.09 and the number refers to the Test Duty from Table 1 of that standard.
9. ANSI Test Duty 8 requires a number of tests to be performed at any values of current less than 85% of the maximum, which, when added to the last test at full rated current, after the one hour break, total four times the maximum current, in this case 144 kA.
10. To give a recognisable comparison, a circuit-breaker rated on the IEC basis as a 12 kV rated voltage, 40 kA rated short-circuit breaking current was selected for comparison with a 15 kV maximum rated voltage, 28 kA breaking current circuit-breaker from the American standard, as these are similar in size and duty, both meeting a 750 MVA fault level. It will be seen that tests are required at two voltage levels with constant MVA, to the American standard, which complicates the testing procedure. (As noted in Chapter 19, the American standards still retain some references to circuit-breaker interrupting capability in terms of an MVA rating, obtained by multiplying the breaking current by the rated voltage and the phase factor, $\sqrt{3}$.)

the ANSI standard, which is the ratio between the total asymmetrical current and the rated symmetrical breaking current I. The figures quoted under the rated current heading of Table 18.1 are equal to (s × I) for the American test requirements where the d.c. component is recorded as >50%.

The testing to the American standard is complicated by the fact that these duties have to be carried out with a d.c. component of greater than 50%, even though the value equivalent to the relevant 's' factor may be much less. The symmetrical component of the short-circuit current then has to be reduced below the rated level, so that the total current at contact separation with the higher d.c. component is at least equal to the rated value of (s × I) (Fig. 18.1).

As also shown in Table 18.1, the IEC standard requires tests to prove that the circuit-breaker can interrupt at its 'critical current', if this exists. This is the current, within the range up to full fault capability, that creates the greatest arc length. It is valid for oil circuit-breakers, and sometimes even more so for magnetic air circuit-breakers, but puffer SF_6 and vacuum circuit-breakers do not usually have a critical current. Figure 4.9 shows an oil circuit-breaker arc characteristic with a quite marked critical current.

The American standards do not require any specific TRV rates for indoor distribution switchgear within the voltage range covered by this book, but they do prescribe the values for the voltage peak as equal to at least 1.88 times the rms test voltage. Times to peak are given for outdoor circuit-breakers. The IEC standard prescribes values for peak and time in accordance with the figures in Table 2.1. The achievement of these TRV figures as well as the voltage and short-circuit current values requires careful adjustment of the test station parameters.

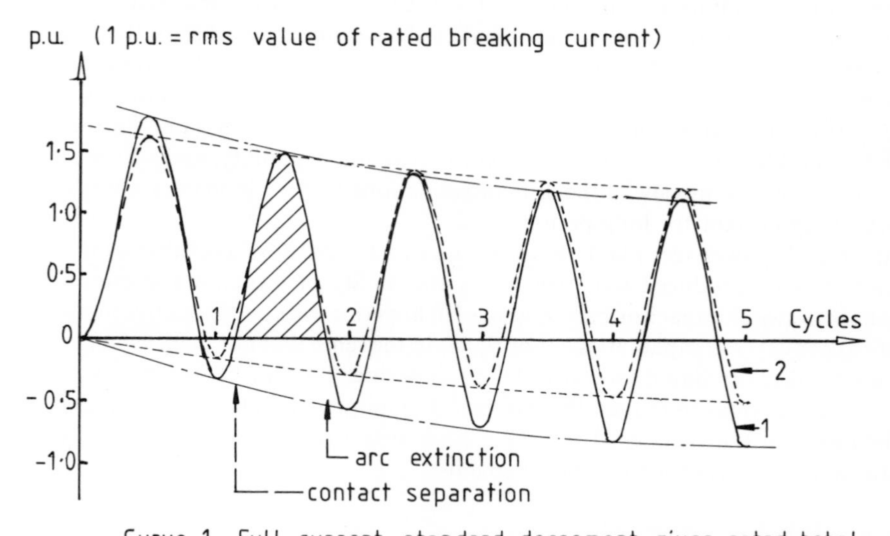

Fig. 18.1 American definition of asymmetry.

Making capacity tests and short time tests

These tests apply to switches and circuit-breakers, and are related in the sense that the service condition that is being represented is the case where a switching device closes onto a faulted system and then has to carry the short-circuit current for a time until it is interrupted. In the circuit-breaker case, the circuit-breaker that closes may be the one that also interrupts the fault.

For test purposes this duty is split into three for the circuit-breaker and two for the switch. As the test plant usually has difficulty in maintaining a continuous level of short-circuit current for up to three seconds at rated voltage, it is usual to do the short time withstand current and peak withstand current test at reduced voltage, generally a few hundred volts. (To avoid surges on the power system supplying the generator driving motors, it is common practice to disconnect them just before the test is initiated and rely on the rotational stored energy of the system for the short-circuit test. This is why a three second test would have significant decrement.) However, the making current test must be done at rated voltage, but can be interrupted by the master circuit-breaker, or by the test circuit-breaker, after a few cycles.

The circuit-breaker test schedule contains the third test, i.e. a 100% make-break test to demonstrate its capability to break the fault immediately after making on to it.

The American standard addresses itself also to the further issue of the behaviour of a circuit-breaker carrying the fault current for an appreciable time, and becoming very hot in the process, and then opening to interrupt it. This recognises the possibility that the overheating, beyond the normally accepted temperature rise limits, that must occur during the short time test may reduce the interrupting capability of the circuit-breaker. This is not referred to in the IEC standards. The ANSI standard contains maximum tripping delay limits for power circuit-breakers, and this is two seconds for circuit-breakers up to 38 kV. The normal short time rating in the ANSI standard is three seconds. The test schedule in ANSI C37.09 contains a test to prove this capability, and advice on how to perform it, in view of the difficulty mentioned earlier in maintaining the short-circuit current for long periods.

Most high power test plants have reactors and capacitors available so that circuits can be produced that will prove the ability of switches and circuit-breakers to switch capacitor banks, unloaded lines or cables and inductive loads. Tests to prove these capabilities are optional in the sense that not every switch or circuit-breaker submitted for test will need to be rated for these tasks, and it is up to the manufacturer to request these tests if the equipment under test is required to be rated for these duties. The testing standards establish the test levels and performance criteria for these requirements.

Mechanical endurance test

All types of switchgear have to undergo a mechanical endurance test, which comprises a specified number of tests to be carried out at no-load, and the rated mechanism energy, that is at rated voltage or pressure, depending on the form of the closing mechanism. This is another area where national standards differ in requirement, the IEC standard requiring 1000 trip and close operations, after which the mechanism parts shall be in good condition and not show undue wear.

The American standard is much more attuned to proving the total life of a circuit-breaker as it requires a large number of no-load operations, with permitted mechanism maintenance at specified intervals, after the completion of which the circuit-breaker opening and closing speeds shall be substantially unchanged. The service condition that switching devices with heavy normal current ratings or short-circuit ratings will not be operated as often as switchgear of lower rating is recognised in the ANSI standards. As an example, a 15 kV circuit-breaker of 1200 A rating and fault rating up to 29 kA has to be able to withstand a 10000 operation endurance test with maintenance permitted every 2000 operations. In contrast, a 2000 A circuit-breaker with a fault rating of 37 kA would only have to do half this duty – 5000 operations with maintenance every 1000.

ROUTINE TESTS

Routine tests are intended to be performed on every switchgear product after manufacture. Other tests will be made by the manufacturer in accordance with other national or international standards on many of the components as manufacture proceeds, but the following tests are required to ensure that the characteristics of the product are like those of the type tested article.

1. Power frequency voltage dry tests of the main circuit
2. Voltage tests of control and auxiliary circuits
3. Measurement of main circuit resistance
4. Mechanical operation tests
5. Operational checks on auxiliary devices and interlocks
6. Verification of the wiring.

Most manufacturers will carry out more tests than those listed and it is common practice to make travel records and timing tests which are kept as a 'finger-print' of the product so that in the event of any problem arising during its subsequent years of service, these tests can be repeated and compared with its original condition.

QUALITY ASSURANCE

Switchgear manufacturers are in the safety business, and fully aware that their product must always perform as intended, even after years of virtual idleness. This needs a close attention to detailed quality control at every stage of the product, from conception to commissioning. Whatever may be said about quality standards, quality manuals and systems, quality is an attitude of mind. Through all the years of gathering service experience, makers collect a dossier of facts concerning the things that can cause switchgear to give trouble in service.

By continuously eliminating these factors from the materials, processes and assembly techniques, and by refining the inspection routines and testing arrangements, and feeding the knowledge gained back into the design procedures, a reputation for quality can be gained.

Quality is a combination of factors, and the most succinct definition is probably fitness for purpose. The following points all contribute towards an understanding of where we look for quality. It is an amalgam of:

1. Knowing the market and the customer's needs,
2. Designing a product that meets these needs,
3. Faultless construction,
4. Using reliable suppliers of components and materials,
5. Regular sample checks on materials employed,
6. Exhaustive testing and certification,
7. Ensuring safety measures,
8. An efficient after-sales service,
9. Feedback of field experience.

CHAPTER 19

Switchgear Standards

All through the preceding chapters of this book there has been considerable reference to standards, particularly in Chapter 18 in the area of testing. The impression will have been gained that the switchgear field is very well served in the matter of national and international standards, and a look at Table 19.1 confirms that impression. In fact, the whole field of electrical engineering is probably better served in the matter of standards than most other fields of standardisation. Is this necessarily a good thing?

Ever since the first nut was fitted to the first bolt, there was a need for standards in engineering, but it was not until the turn of the century that things started to become organised on a national scale. The British Standards Institution began life in 1901 as the Engineering Standards Committee, and took on its present title in 1931, having received a royal Charter some two years previously. This was not the first in Europe, as there were enough bodies around to form the International Electrotechnical Commission in 1906. This is the body that now produces the international standards referred to as the IEC recommendations, or standards. There are now 43 countries represented by the IEC which, together, comprise 80% of the world's population, consuming 95% of the world's electrical energy. It currently has 1800 publications on issue and it only covers the electrical industry and its users. Another international body, the International Standards Organisation (ISO), looks after the rest.

Within Europe, there is another international body which harmonises those standards where there is enough similarity in application to make this a reasonable proposition, and then the member countries each publish the same standard within their own national standards organisation. This body is known as CENELEC and 17 European countries subscribe to its activities. The specific object of this organisation is to remove technical barriers to trade, in accordance with Article 100 of the Treaty of Rome, the Treaty under which the European Economic Community (the 'Common Market') was created.

AIMS OF STANDARDISATION

The objectives of the first engineering standards were no doubt just to ensure a wider degree of interchangeability between engineering components and materials, so that an interchange could take place among the earliest

Table 19.1 List of relevant IEC and British Standards for switchgear

Subject	*IEC No.*	*BS No.*
Common clauses for HV switchgear, etc.	694	6581
A.C. circuit-breakers above 1 kV	56	5311
High voltage switches	265	5463
Disconnectors and earth switches	129	5253
Metal-enclosed switchgear	298	5227
Insulation-enclosed switchgear	466	—
HV switches combined with fuses	420	—
HV fuses	282	2692
HV fuses for motor protection	644	5907
Low voltage circuit-breakers	157	4752
LV factory built assemblies	439	5486
Cartridge fuses up to 1000 V a.c./1500 V d.c.	269	88
Current transformers	185	3938
Voltage transformers	186	3941
Instruments	51	89
Relays	255	142
Switchgear maintenance – up to 650 V	—	6423
Switchgear maintenance – from 650 V to 36 kV	—	6626
Degrees of protection	529	5490
Insulation classed by thermal stability	85	2757
Insulation thermal endurance	216	5691
Insulation electrical endurance	727	—
Artificial pollution on HV insulation	507	—
Bushings	137	223
SF_6 gas	376	5207
Guide for checking SF_6 gas	480	5209
Insulating oil	296	148
HV test techniques	60	923
Partial discharge measurements	270	4828

manufacturers and some economy of scale was possible. Now the aims of standardisation are rather wider, and continue to grow, with greater emphasis on international trade, safety and product liability.

These aims are:

1. To provide a means of communication between interested parties.
2. To promote an economy of effort, materials and energy. This brings in the concepts of interchangeability and variety reduction.
3. To protect consumers by establishing quality through defining the characteristics necessary to satisfy the needs of purchasers and users.
4. To promote trade by removing barriers due to different techniques and requirements.
5 To promote safety aspects and procedures.

It is important that, before any standard is produced, a demand for it is identified. When it is in existence it is important that that demand be proved by putting the standard to use, otherwise much effort by a lot of busy and knowledgeable people will have been wasted. The timing of new standards also needs to be carefully planned and timed. There is a grave danger today that the writing of a standard be seen as an end in itself, and for it to be produced too early in the product life cycle. This leads to a risk of crystallising design parameters and stifling development.

The preparation of standards is an immense task today and involves large numbers of people. The BSI is probably typical of most national bodies and it has a number of standing committees and a comparatively small permanent staff. These committees keep a watch on the needs of specific sections of the market and initiate the preparation of new national standards when a need is established. A large part of the work load today is made up of the continuing need to revise existing standards as techniques change. Small teams of experts, drawn from manufacturers, users, government agencies and academic life, are then set up to carry out the work of drafting the specifications. The total number of these experts, who are 'donated' by their employers to spend some scores of hours a year on such work, amounts to some 28000 according to the BSI annual report for 1984. This represents a huge hidden investment in national expertise.

In fact the picture is somewhat distorted from the above, now that the IEC and CENELEC play a large part in standards making activities. Originally, national standards were prepared in this way and the first international standards would have been based on a consensus arising from a study on national standards on a particular subject, the common material being selected and then built upon. Now that a wide range of international standards exists, new standards work tends to take place at international level, national experts being nominated by the interested countries to working groups to do the drafting work. The results of their efforts are then referred to national standards committees and the resultant comments mulled over and incorporated as far as consensus allows. The final draft is then issued for voting by the national committees, for which a period of six months is allowed. This procedure is known as 'voting under the six months rule'. If less than 20% of the respondents vote against the proposals they are issued as an IEC standard.

At national level the decision will then be taken whether to issue a national standard based on the international one. The BSI tends to do this for most of the

IEC standards. The starting point for this is to issue the IEC document under a BSI reference, which is sold to anyone interested who might wish to comment. After a period of time for those comments to be received, the national committee, or a working panel if the work load justifies it, will review the comments and produce a clean draft for the national committee to approve for issue as a British standard. Frequently the result of this exercise is that the IEC standard just gets a new cover!

However, with the CENELEC organisation standing in the wings, the foregoing is an over-simplification. If that body decides that there is enough common interest in Europe for all countries to issue identical standards, then what is termed the *harmonisation procedure* is started. In simple terms this means asking each member country for any reason why this across the board issue cannot take place. Usually agreement is reached after some bargaining, and the IEC standard receives new covers in a number of different languages!

Another group of people who play a significant role on the standards scene, in many countries, are the major users of a product. In the case of switchgear, this area is dominated by large utilities, particularly in those countries with central 'nationalised' generation, transmission and distribution systems. There are also other large users, or in some cases groups of similar users such as oil companies, who have a wide enough demand to justify some standardisation. This is stronger in the distribution field than any other, because the volume of equipment purchased by such organisations is greater in that area. These standards are more application oriented than performance oriented, because the attraction of a common standard to these large users is the avoidance of writing out long technical specifications each time they order some switchgear. As such these standards are helpful to manufacturers as well, as it means that a number of similar equipments will be ordered and the manufacturer can have on file much of the application diagrams, etc., that are necessary.

The biggest danger with some of the utility type standards is the natural tendency that exists to try and write into the standard solutions to service troubles that have been experienced by a minority of users. This tends to make such standards restrictive and inhibit genuine development, arising from something that usually is a failure of quality control, not a type or design defect. It is therefore important that some negotiating machinery be set up to allow manufacturers and large users to get together and mutually agree the content of such standards.

It is also often government policy to frown on a proliferation of standards that arise in this way as diminishing the standing of the national standards. However, this is often a difficult area, as the very nature of achieving a consensus for an international standard, that then becomes the national document, is that those issues on which agreement cannot be reached will be left out. This then opens the door for large users to want to fill the gaps. As time goes on, and the use of internationally approved standards grows, manufacturers and users alike will become more accustomed to them, and that in itself will lead to a broadening of the areas of agreement.

AMERICAN STANDARDS

There are still some differences in approach to the question of switchgear standardisation between America and Europe. As the years go by the two groups of standards come closer together, but historical reasons make this a fairly slow process.

With a wider range of service voltages in that very large country with many hundreds of independent supply authorities, the ANSI (American National Standards Institution) standards still adhere to the principle of rating distribution switchgear in terms of MVA breaking capacity, and working on the basis that this MVA level remains constant over a range of voltage. This is not specifically stated today, but it can be seen, for example in a 15 kV circuit-breaker according to ANSI standard C37.06, that a 750 MVA rating corresponds to 28 kA at 15 kV and 36 kA at 11.5 kV. The ratio between the two voltages is the K factor, in this case 1.3. This and other differences in respect of ratings have a lot of practical logic behind them, all dependent on the type of distribution system used in the USA. For interest, another difference is the ratio of peak making current to the symmetrical breaking current which is 2.7 in the ANSI standard. In the IEC standard this is 2.5; it used to be 2.55 in the old British standard that the IEC document superseded. This figure of 2.55 is the theoretical value for a fully offset sine wave with a decrement derived from the X/R ratio of 14.

The ANSI X/R ratio is 15, but that is not a big enough difference to produce the 2.7 figure, and one explanation is that the higher value is to make some contribution for the extra input to fault levels at the onset of fault conditions from connected rotating plant.

Another difference is the American use of rms currents where the IEC standard uses instantaneous values. Reference has already been made, in the chapter on short-circuit testing, to the specification of asymmetrical ratings by the use of total asymmetrical rms values rather than a combination of symmetrical current and a d.c. component. Another rating in which the same principle is used is the making capacity, which is also quoted as an rms value. This practice is a throwback to the principle of some years ago, when the whole rating structure of the American standards was based on total current. Now, in fact, the making capacity is given as both a peak current and an rms total current, evidence that thinking is changing. Also, at the higher voltages, the breaking capacity is coming into line with IEC principles by choosing values from the R10 series.

In normal current ratings for voltages in the range up to 38 kV the older 600 and 1200 A ratings are retained. On the other hand, as has been referred to elsewhere, the list of preferred ratings in ANSI C37.06 is much shorter than the corresponding table in BS 5311, so they are paying more attention to the variety reduction objective of standardisation.

REGULATIONS AND SAFETY

A regulation is a binding document which has legislative or legal authority.

Many countries in the world now have regulations which cover a number of areas in which standards also have an interest. There is a growing relationship between regulations and standards in some areas, of which safety is one of major interest. The existence of the Health and Safety at Work Act, 1974, in the UK, and corresponding legislation in other countries, brings interest to the role that standards can play in enhancing the safety of electrical products. Too close a relationship between standards and legislation can be very restrictive, as changes to standards of this kind become much more difficult. An approach that is being fostered in many areas is the 'deemed to satisfy' approach. The object of this is that if a court case arises from a mishap with electrical plant, its compliance with a suitably drafted national standard would be taken as satisfying the safety requirements, and be a defence in law.

This is being taken quite seriously, and a natural result is that the corresponding government officials responsible for safety legislation are playing an important role in the standards making process.

CHAPTER 20

Future Trends

The switchgear industry is subject to very slow evolutionary change, the 'last resort' function of most switchgear products and the safety and security objectives that are high on the list of the user's priorities making them reluctant to accept change. Long periods of trial service are normal for any product that is significantly different from the traditional.

This can be illustrated with respect to circuit-breaker switching media, and the availability of vacuum and SF_6 within the last two decades. The two major advantages of these media in virtually eliminating the cost of maintenance and fire precautions from the operating expenses of users might have been expected to generate a strong enthusiasm for their acceptance, in spite of some initial price premium. Figure 20.1 shows what actually happened, which was as predicted by many manufacturers at the time of introduction. There was a guarded response by a few customers willing to try the new types of switchgear. This was followed by a pause as these initial installations were evaluated, and then a gradual spread of interest as the claims of the originators of these new technologies were seen to be justified. The curves in Fig. 20.1 show the change in mix of interrupting technology in the UK and in the European Economic Community over the years from 1979 to 1984. The UK figures are also incorporated in the EEC graph, of course. The starting point for these curves is the beginning of the 1970s, so it took six years or more for the position shown at the beginning of these graphs to be reached.

In the UK, SF_6 is a comparatively recent introduction at distribution voltages, and this is seen as adding to the rate at which the application of the oil circuit-breaker has diminished from 1981. It has not apparently slowed the gradual growth of the vacuum switchgear in the UK. In the EEC, SF_6 was introduced in France as the first new arc interrupting medium at about the same time as vacuum in the UK. The combination of its application in Italy, Switzerland and France sets the beginning of the EEC graph. The vacuum part of the EEC graph owes its content mostly to the UK, supplemented somewhat later by Germany. The conclusion, which underlines the opening comments of this chapter, is that it has taken almost 15 years for the advantages of the new media to be recognised to the point that oil circuit-breakers have just about fallen to the level of representing 50% of new business. Obviously, on a basis of the proportion of circuit-breakers actually in service at distribution voltages up to 24 kV, this must still be quite small.

So any trend for the future can only be a gradual evolution from present techniques, and the factors which govern the rate of change are:

1. Compatibility with plant already installed
2. Experience gained with the new features put forward
3. Price – if cheaper, acceptance is accelerated.

The opportunities for change are reviewed in the order of the preceding chapters.

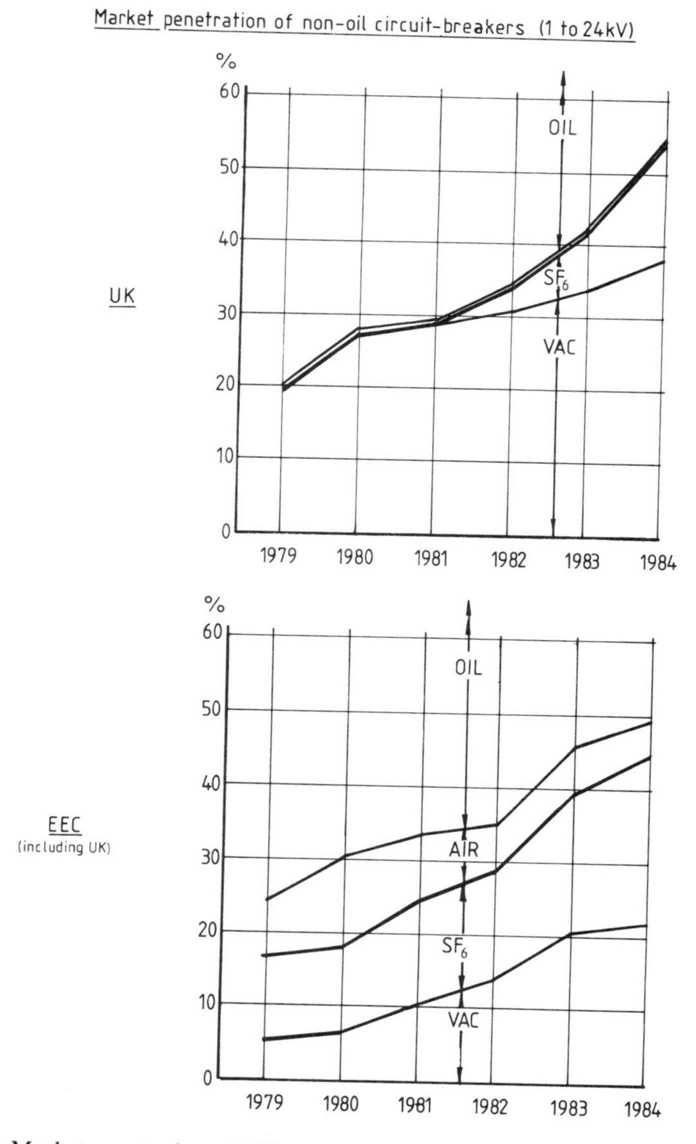

Fig. 20.1 Market penetration of different interruption technologies, 1979 to 1984.

CIRCUIT-BREAKING

No revolutionary new circuit-breaking techniques are foreseen in the near future, and this is seen as a period of consolidation with vacuum and SF_6. Refinement of the ways in which the technologies are applied will continue, and the advances from this development work are more likely to be seen in the SF_6 field, where at present there are quite a lot of exciting prospects being evaluated. With vacuum interrupters, the technology has been exhaustively researched and the advances will have to come from production engineering to enable low cost volume manufacture to bring the price down. This is a technology that can match any foreseeable requirement for breaking current at modest voltages. It is unlikely to attract much interest above 33 kV, due to the difficulty of achieving the longer contact travel required through a metal bellows. The use of interrupters in series leads to unacceptable costs compared with SF_6 circuit-breakers, where a single break can handle very high voltages.

Low voltage switchgear will probably continue with air break devices, as they are economic and the high arc voltage has value in modifying system electrical constants to give advantages such as current limitation and overvoltage limitation.

MATERIALS

The area for development lies mostly in the insulation field, and new plastics appear fairly frequently, although the pace has now slowed compared with ten years ago. The present preoccupation with the environmental resistance of some plastics materials and casting resins will have to be resolved, again largely by modifying application techniques where dimensional reduction has gone a little too far.

PRIMARY SUBSTATIONS

Not much change is envisaged here except for the continuing increase in the introduction of vacuum and SF_6 circuit-breakers. The impact of digital electronics in the control and monitoring systems is another area where evolution can be expected. The technology to make substations virtually automatic already exists, so it is the question of proving the reliability of this technology to keep running for years at a time without failure that has to be answered to the user's satisfaction.

SECONDARY SUBSTATIONS

There is more scope here for changes in operation philosophy and application in the secondary switchgear field than in the primary substations. In the UK, and probably in much of the rest of Europe and in America, the question of replacement of ageing switchgear will become a growing need before long. Switchgear has proved itself as having a very good record for long trouble-free service, and the only reasons for replacement are likely to be:

1. *Obsolescence*. Only where the features do not match modern needs or safety principles.
2. *Deterioration of insulation*. Depending on the environment, the risk of insulation failure as the materials age must increase.
3. *Wear of mechanical parts*. If operation has been frequent, mechanical wear might be reaching a point where operating tolerances are exceeded.

In many instances, when the decision to replace part of a secondary distribution system is taken, the new system could often be different from the old. New ways of operating the network, and experience in the reliability of modern plant, give rise to changes in the planning of the system and the application of the switchgear.

An example in ring main systems in the UK is that it has been shown that approximately 80% of the operations of the switches of the ring main units have been for the purpose of maintaining other ring main units. One conclusion from this is that elimination of some ring main units would lead to a reduction in maintenance cost and first installation costs, without any great penalty in operational terms. Similarly, the provision of fuse-switches on the ring side of the distribution transformers has led to more trouble with those fuses and the fuse-switches than there is with the transformers. This raises the question of how reliable the system will be without local transformer protection. This has been discussed in Chapter 11 to some degree. The problem of detecting transformer faults and LV faults on the LV busbars from the primary substation, has already been mentioned also, and research on this problem is continuing. In the meantime a proposal to detect faults locally with small current transformers round the transformer bushings, which then trigger off a fault-throwing switch on one phase of the ring system at the transformer, would perhaps provide an economic way of signalling a fault condition that the primary circuit-breaker

Fig. 20.2 A fault-thrower and cable termination assembly. (*Courtesy Electromoldings Ltd.*)

would recognise. Also, it being an earth fault, the primary circuit-breaker protection would operate quickly. Used in association with something like Visitrace, which would quickly lead the operational staff to the point where the fault was, remedial action to restore supply to the majority of the consumers would soon be accomplished. Figure 20.2 shows an assembly of components which include a simple fault-thrower released by a chemical actuator. Three current transformers detect the fault and time limit fuses give some delay for small faults, but if serious, the energy is directed to the explosive actuator, which releases a spring-loaded contact system. This connects to earth the phase on which it is mounted, and the primary circuit-breaker trips immediately. The fault-thrower is filled with SF_6 to reduce pre-arcing and to keep dimensions small.

System trials with this equipment and a similar version using screened components and a vacuum fault making device are taking place. The cables are looped into the cable box using bolted separable connectors as discussed in Chapter 14. This facilitates the disconnection of the cables from the faulty transformer, and accessories exist so that the ring can be completed without the transformer. The application of many of the new ideas in secondary switchgear application depends on the degree of LV interconnection. If this is at a reasonably high level the removal of one distribution transformer has no serious effect on the consumers.

Where there is not a degree of LV interconnection there is still a case for individual transformer protection, and the fuse is useful for this purpose, if two problems can be overcome. Most fuses employed at present for transformer protection cannot deal with faults in the overload range, which is why they are fitted in fuse-switches which can interrupt these comparatively low currents. When mounted in a fuse-switch it is easy to arrange for a 'striker-pin' type of fuse blowing indicator to trip the associated switch, and thus avoid single-phasing. There are now available fuses with a 'full range' characteristic which overcome the first of these problems, but still leave the single-phasing issue. There are parts of the world which are not so bothered about this issue, but the worry in the UK, due to the amount of cable connected to the system and its capacitance, is that there is a risk of producing serious overvoltages due to ferroresonance under single-phase conditions.

Elimination of the ring switches may have attractions from the point of view of reducing maintenance, but the new switches now available, mostly using SF_6, are maintenance-free and thus remove this problem. So various proposals have been made for combinations of single switches and separable cable connectors which permit the splitting of the ring in the event of cable faults.

In addition to modifications to the sand-filled fuse to make it more flexible in application, there are developments in 'fuse technology' based on the new arc interruption techniques. If a vacuum or SF_6 circuit-breaker has its contact system removed and a fuse-wire substituted, then a single-shot protective device is produced. The time delay in the event of a fault is similar to that of a fuse, as it depends on the time it takes to melt the wire. The current limitation that occurs in the sand-filled current-limiting fuse will not take place, but the arcing time will

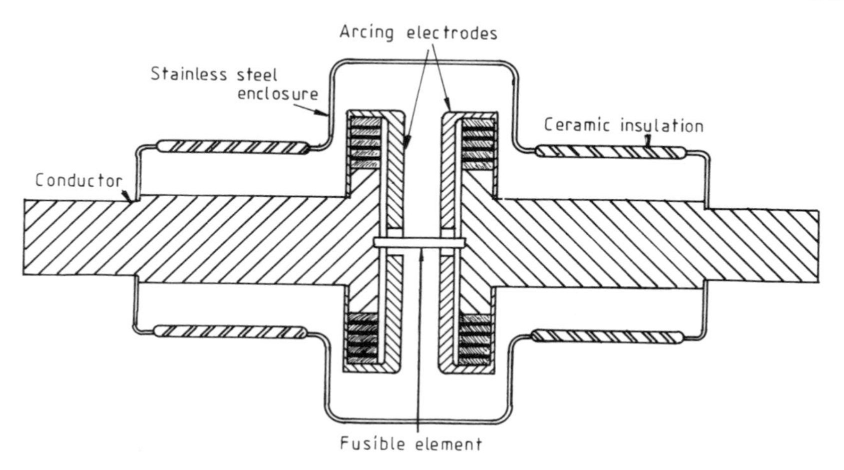

Fig. 20.3 A vacuum fuse.

be a maximum of one, probably asymmetric, loop. The technical press contains descriptions of vacuum devices of this type, such as the one shown in diagrammatic form in Fig. 20.3. This has plain contacts and magnetic coils fixed behind them to create an axial magnetic field between the arcing rings. The coils are normally shorted out by the arcing wire, so the impedance can be kept low. When the wire melts, the coils are switched in and the current is extinguished at the first zero. It is not important that the arcing rings be made of a material that will prevent current chopping as the fuse cannot operate on low inductive currents. However, the inherent expense of a vacuum interrupter means that a throw-away device like this cannot have much attraction as a protective device for transformers. A solution to the problem of tripping the healthy phases is also difficult to see with such a device.

It is conceivable that a more economical 'fuse' could be made using SF_6, perhaps by using rotating arc technology in a simple form. Such a device fitted into a cable box and used in conjunction with separable cable connectors would give transformer protection and the means of isolating the transformer in the rare event of a fault, particularly if the possibility of single-phasing could be overcome.

Over the next few years, developments to reduce switchgear utilisation will occur in secondary distribution networks.

CABLE TERMINATION SYSTEMS

In this field the use of sleeving systems to insulate cable tails so that air-filled cable boxes can be used is sure to increase. Paper cables at distribution voltages are diminishing, even in the UK, and this will accelerate the trend. Separable connectors offer advantages that will make them a growing part of the cable termination scene.

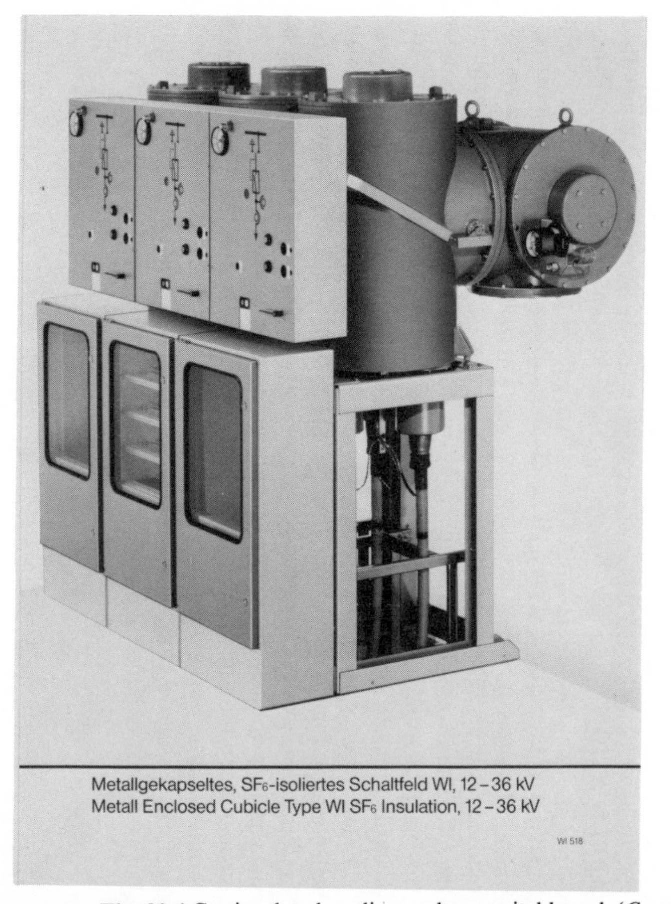

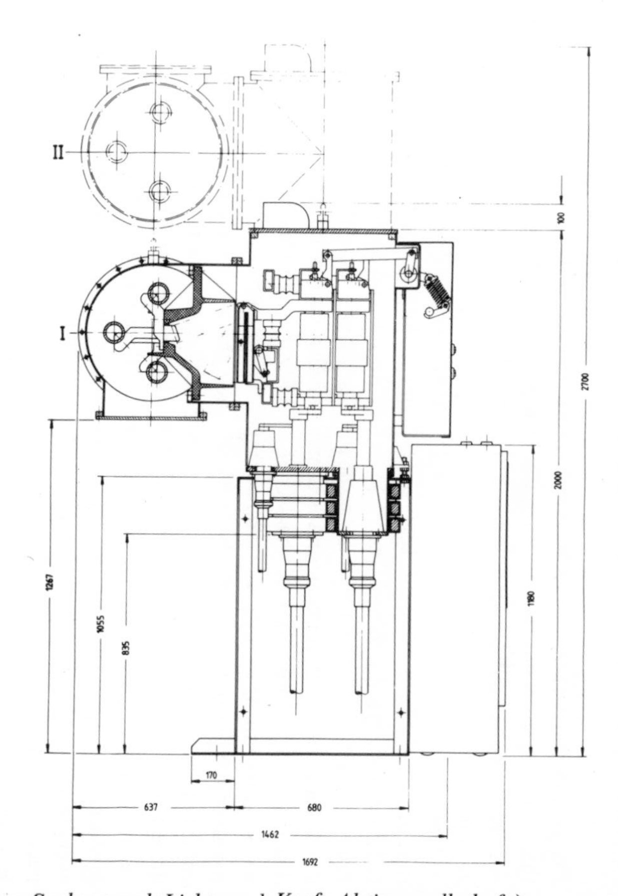

Fig. 20.4 Gas insulated medium voltage switchboard. (*Courtesy Sachsenwerk Licht- und Kraft-Aktiengesellschaft.*)

INSTRUMENT TRANSFORMERS AND PROTECTION

Protection must continue to be converted to static methods, particularly now that the price premium has virtually disappeared. The problem of reliable secondary power supplies is small in countries where the availability of electricity is universal.

The size of indoor metal-enclosed, circuit-breaker switchgear could be reduced if the requirement for current and voltage transformers could be reduced, or virtually eliminated. The actual power and ratio requirements for modern digital relays is tiny but the need to isolate the delicate electronics from the voltage surges that occasionally appear on the main conductors requires the discarding of a large part of the signal. If a solution to this can be found, and work done in the area of instruments and meters, so that their power requirements could be similarly reduced, then the circuit chamber of a metal-enclosed unit could be substantially reduced. Small, multi-turn CTs of small dimensions that could be incorporated in circuit-breaker bushings and allow ratio changes to be made in the software, and capacitance taps, also on the bushings, could have a substantial effect on size and consequently, cost. Full acceptance of the sealed-for-life concept for switching devices will remove the need for withdrawal of the switching device. This is already seen in the latest European metal-enclosed (gas-insulated) equipment (Fig. 20.4).

APPLICATION

Safety will continue to play a large part in establishing and modifying operational practices. For instance, low voltage fuse cabinets and pillars are tending towards fully protected designs in Europe and versions are available in the UK. The advent of increasing legislation in the field of the use of electricity in the place of work, and in the field of product liability, will accelerate trends towards protected installations and remote operation.

Finally, there can be no doubt that the development of distribution switchgear will continue for years to come. The present trend for under-developed countries in the 'third world' to introduce switchgear assembly plants, followed by switchgear manufacture, will grow and the present pattern of a limited number of manufacturers in a few highly industrialised countries exporting these products to the world will change. The latter countries will then have to fall back on further new developments in order to continue an export activity, so the business must continue to advance.

Index